# Learn OpenAI Whisper

Transform your understanding of GenAI through robust and accurate speech processing solutions

**Josué R. Batista**

# Learn OpenAI Whisper

**Group Product Manager**: Niranjan Naikwadi
**Publishing Product Manager**: Tejashwini R
**Book Project Manager:** Neil D'mello
**Senior Editor**: Mark D'Souza
**Technical Editor**: Reenish Kulshrestha
**Copy Editor**: Safis Editing
**Proofreader**: Mark D'Souza
**Indexer**: Manju Arasan
**Production Designer**: Ponraj Dhandapani
**DevRel Marketing Coordinator**: Vinishka Kalra

First published: May 2024

Production reference: 1150524

Published by Packt Publishing Ltd.
Grosvenor House
11 St Paul's Square
Birmingham
B3 1RB, UK.

ISBN 978-1-83508-592-9

www.packtpub.com

*To the memory of my "Abuelita," Doña Luisa, my mother, Luisa, and my father, Roberto, for their sacrifices and for exemplifying the power of determination, grit, love, faith, and hope. To my siblings, Moisés, Priscila, and Luisa, who are always in my mind and heart. To Chris and Sharon, I love you guys! And most especially, to my wife, Hollis Renee, for being my loving and supportive partner throughout our unbelievable joint-life journey.*

*– Josué R. Batista*

# Foreword

Six years ago, as an invited speaker on quantum processor hardware in Riyadh, and later in Al Khobar for Vision 2030, I met a machine learning co-presenter, who had a unique combination of technology enthusiasm and human care: Josué R. Batista (author of *Learn OpenAI Whisper* and series producer of *What and Why First*).

Now, fast-forward back to Pittsburgh. Josué has been involved in BraneCell AI Chip technology meetings, is a senior generative AI specialist, and remains passionate about describing artificial intelligence to whoever will listen.

**Automatic speech recognition** (**ASR**) and Transformer technologies are humanity's codification of a commonplace activity. Whisper's multilingual capabilities, integration with OpenAI technologies, and the ability to glean insights on data do what we must all do to make sense of what feels like a plethora of babble. Josué often mentions word-activated software coding technologies – reminding me of the 4,000-year-old Book of Job, saying: "Thou shalt also decree a thing, and it shall be established." The implications are far-reaching; we have arrived at the time when we whisper what we want on topics ranging from writing code to rendering images and it is established.

In addition to topics related to computer hardware and software, Josué and I talk about other subjects, such as family and classical music. He likes life's beauty, so he is an appropriate person to evaluate and announce the elegance of Whisper. Genies should be kept bottled up, but that is not always the case in the world. The aforementioned power of the tongue ought to be decreed from goodness and truth. As we all should do, Josué approaches his work with sincerity and endeavors to write with technical accuracy.

In writing this book, Josué basically got himself another master's degree, this time on OpenAI Whisper. He put in the requisite work to provide readers with a valuable educational experience, true to his sincerity: to give all he can, so that you, the reader, may receive the piece you need.

The content of *Learn OpenAI Whisper* provides a comprehensive navigation for the reader through the innovative world of OpenAI's Whisper technology. Josué has divided the book into three parts, each focusing on a different aspect of Whisper. *Part 1* introduces readers to the basic features and functionalities of Whisper, including its setup and usage.

*Part 2* explores the underlying architecture of Whisper, the transformer model, and techniques for fine-tuning the model for domain and language specificity.

*Part 3* addresses real-world applications and Whisper use cases. Readers will learn how to apply Whisper in various contexts, such as transcription services, voice assistants, and accessibility features.

The book also covers advanced topics such as quantization, real-time speech recognition, speaker diarization using WhisperX and NVIDIA's NeMo, and harnessing Whisper for personalized voice synthesis.

The final chapter provides a forward-looking perspective on the evolving field of ASR and Whisper's role in shaping the future of voice technologies.

The book exudes the author's enthusiasm for Whisper and artificial intelligence. Even after writing this book, Josué will continue to blaze this trail. He will continue to expand on ASR-related themes via his *What and Why First* blog and be your ongoing companion on this exciting journey.

These days, so many things are happening in the world and technology, yet much comes down to the same basic point: more than any other time, now is the time to be vigilant about whether you send forth sweet or bitter water from your whisper.

*Christopher Papile, Ph.D.*

*Chris founded BraneCell (*`https://branecell.com/`*) and is an author of several dozen patents and scientific publications on topics like quantum neural network hardware, decarbonized chemicals, hydrogen, catalyst nanomaterials, and artificial intelligence mechanisms.*

`https://www.linkedin.com/company/branecell` | `https://twitter.com/BraneCell`

# Contributors

## About the author

**Josué R. Batista**, a senior AI specialist and solution consultant at ServiceNow, drives customer-centric adoption of generative AI solutions, empowering organizations to reimagine processes and create impactful value using AI. Before this, he was a digital transformation leader at Harvard Business School, supporting the industrialization of generative AI and LLMs. Josué also served as a technical programmatic leader for Meta's Metaverse initiative, integrating computer vision, deep learning, and telepresence systems. At PPG Industries, he led AI/ML transformation, driving impact through big data, MLOps, and deep reinforcement learning. Passionate about leveraging AI for innovation, Josué continues to push boundaries in the AI field.

*I want to thank my friends Naval Katosh and Chris Papile, who willingly responded to my call for assistance on this project. To my editor, Mark D'Souza, thanks for your guidance in refining my writing. To my early-years mentor, Eduardo Silva, and friends and colleagues, Bob Fawcett, Gonzalo Manchego, Eric Dickerson, Rob Kost, Sandesh Sukumaran, Andreas Wedel, Brian Coughlin, Steve Sian, and Kirk Wilcox, thank you for engaging in thought-provoking discussions and expanding my understanding of what is possible.*

# About the reviewers

**Naval Katoch** is an artificial intelligence professional with a master's degree in information systems management from Carnegie Mellon University. With more than 10 years of experience, he began his career in AI at IBM as a data scientist, later becoming a **machine learning operations** (**MLOps**) lead and solution architect at PPG Industries, where he currently works. Naval specializes in AI for manufacturing and supply chain domains, and he enjoys deploying machine learning projects at scale. Beyond work, he's an avid guitarist and Brazilian jiu-jitsu practitioner and enjoys reading books on science, technology, geopolitics, and philosophy.

*I would like to thank Josué Batista for the opportunity to review his book, which deepened my understanding of ASR. Thank you to my family – Preeti, Kamlesh, Nihaar, and Taffy – for their never-ending support.*

**Marty Bradley** is the visionary behind Evergreen AI, where he champions the use of OpenAI's Whisper and cutting-edge generative AI technologies. His mission? To revolutionize how mid-sized to Fortune 100 companies leverage their colossal data reserves. With a career that has spanned crafting machine-level code for IBM mainframes to pioneering big data architectures across sprawling systems, Marty's expertise is as vast as it is deep. At the heart of his work is a passion for steering organizations through transformative journeys, shifting from project-based to product-focused mindsets. Marty believes fervently in AI's power not just as a technological marvel but also as a catalyst for crafting or enhancing business capabilities. Outside the realm of AI wizardry, Marty's world is anchored by his family. He is a proud father to two remarkable sons – one a cybersecurity maestro and the other a retired military cryptolinguist. He shares an unbreakable bond with his wife, Sandy, with whom he navigates the complexities of life with strength and humor.

# Table of Contents

# Part 2: Underlying Architecture

3

## Diving into the Whisper Architecture          65

# 4

## Fine-Tuning Whisper for Domain and Language Specificity        109

# Part 3: Real-world Applications and Use Cases

# 5

## Applying Whisper in Various Contexts        139

# 6

# Expanding Applications with Whisper                                                    177

# 7

# Exploring Advanced Voice Capabilities                                                   209

# 8

## Diarizing Speech with WhisperX and NVIDIA's NeMo 237

# 9

## Harnessing Whisper for Personalized Voice Synthesis 269

# 10

## Shaping the Future with Whisper 307

# Preface

Welcome to the world of **automatic speech recognition** (**ASR**) and OpenAI's groundbreaking Whisper technology! In this book, *Learn OpenAI Whisper*, we will embark on a comprehensive journey to explore and master one of the most advanced ASR systems available today.

OpenAI's Whisper represents a significant leap forward in speech recognition, offering unparalleled accuracy, versatility, and ease of use. Whether you are a developer, researcher, or enthusiast, this book will equip you with the knowledge and skills needed to harness the power of Whisper and unlock its full potential.

Throughout the chapters, we will dive deep into Whisper's core concepts, underlying architecture, and practical applications. Starting with an introduction to the basics of ASR and Whisper's critical features in *Part 1*, we will lay a solid foundation for understanding this cutting-edge technology.

In *Part 2*, we will explore the intricate details of Whisper's architecture, including the transformer model, multitasking capabilities, and training techniques. You will gain hands-on experience in fine-tuning Whisper for domain and language specificity, enabling you to tailor the model to your needs.

*Part 3* is where the real excitement begins as we delve into Whisper's vast array of real-world applications and use cases. From transcription services and voice assistants to accessibility features and advanced techniques such as speaker diarization and personalized voice synthesis, you will learn how to leverage Whisper's capabilities across various domains.

As you progress through the chapters, you will acquire technical skills and gain insights into the ethical considerations and future trends shaping the landscape of ASR and voice technologies. By the end of this book, you will be well equipped to tackle the challenges and opportunities that lie ahead in this rapidly evolving field.

Whether you want to enhance existing applications, develop innovative solutions, or expand your knowledge in ASR, *Learn OpenAI Whisper* is your comprehensive guide. This book leaves no stone unturned, ensuring you thoroughly understand Whisper and its applications. Get ready to embark on an exciting discovery, mastery, and innovation journey with OpenAI's Whisper!

# Who this book is for

*Learn OpenAI Whisper* is designed for developers, data scientists, researchers, and business professionals who want to gain practical insights into leveraging OpenAI's Whisper for ASR tasks.

The three primary personas who are the target audience of this book are as follows:

- **ASR enthusiasts**: Individuals who are passionate about exploring the potential of advanced speech recognition technologies and want to stay abreast of the latest developments in the field

- **Developers and data scientists**: Professionals who want to integrate Whisper into their projects, enhance existing applications with speech recognition capabilities, or build new solutions from scratch

- **Researchers and academics**: Individuals in academia or research institutions interested in studying Whisper's inner workings, conducting experiments, and pushing the boundaries of ASR technology

Throughout the book, readers will learn how to set up Whisper, fine-tune it for specific domains and languages, and apply it to real-world scenarios. They will gain a comprehensive understanding of Whisper's architecture, features, and best practices for effective implementation.

# What this book covers

*Chapter 1, Unveiling Whisper – Introducing OpenAI's Whisper*, outlines Whisper's key features and capabilities, helping readers grasp its core functionalities. You'll also get hands-on with initial setup and basic usage examples.

*Chapter 2, Understanding the Core Mechanisms of Whisper*, delves into the nuts and bolts of Whisper's ASR system. It explains the system's critical components and functions, shedding light on how the technology interprets and processes human speech.

*Chapter 3, Diving into the Architecture,* comprehensively explains the transformer model, the backbone of OpenAI's Whisper. You will explore Whisper's architectural intricacies, including the encoder-decoder mechanics, and learn how the transformer model drives effective speech recognition.

*Chapter 4, Fine-tuning Whisper for Domain and Language Specificity,* takes readers on a hands-on journey to fine-tune OpenAI's Whisper model for specific domain and language needs. They will learn to set up a robust Python environment, integrate diverse datasets, and tailor Whisper's predictions to align with target applications while ensuring equitable performance across demographics.

*Chapter 5, Applying Whisper in Various Contexts,* explores OpenAI's Whisper's remarkable capabilities in transforming spoken language into written text across various applications, including transcription services, voice assistants, chatbots, and accessibility features.

*Chapter 6, Expanding Applications with Whisper,* explores expanding OpenAI's Whisper's applications to tasks such as precise multilingual transcription, indexing content for enhanced discoverability, and utilizing transcription for SEO and content marketing.

*Chapter 7, Exploring Advanced Voice Capabilities,* dives into advanced techniques that enhance OpenAI Whisper's performance, such as quantization, and explores its potential for real-time speech recognition.

*Chapter 8, Diarizing Speech with WhisperX and NVIDIA's NeMo,* focuses on speaker diarization using WhisperX and NVIDIA's NeMo framework. You will learn how to integrate these tools to accurately identify and attribute speech segments to different speakers within an audio recording.

*Chapter 9, Harnessing Whisper for Personalized Voice Synthesis,* explores how to harness OpenAI's Whisper for voice synthesis, allowing readers to create personalized voice models that capture the unique characteristics of a target voice.

*Chapter 10, Shaping the Future with Whisper,* provides a forward-looking perspective on the evolving field of ASR and Whisper's role. The chapter delves into upcoming trends, anticipated features, and the general direction that voice technologies are taking. Ethical considerations are also discussed, providing a well-rounded view.

The following section will discuss the technical requirements and setup needed to get the most out of this book. It covers the software, hardware, and operating system prerequisites and the recommended environment for running the code examples. Additionally, it guides you in accessing the example code files and other resources available on the book's GitHub repository. By following these instructions, you will be well prepared to dive into the world of OpenAI's Whisper and make the most of the practical examples and exercises in the book.

## To get the most out of this book

For most of the book, you only need a Google account and internet access to run the Whisper AI code in Google Colaboratory (Colab). No paid subscription is required to use the free version of Colab and GPU. Those familiar with Python can run this code example in their local environment instead of using Colab.

| Software/hardware covered in the book | Operating system requirements |
|---|---|
| Google Colaboratory (Colab) | |
| Google Drive | |
| YouTube | |
| RSS | |
| GitHub | |
| Python | |
| Hugging Face | |
| Gradio | Web browser on Windows, macOS, or Linux |
| Foundational models:<br><br>Google's gTTS<br><br>StableLM Zephyr 3B – GGUF<br><br>LlaVA | |
| Intel's OpenVINO | |
| NVIDIA's NeMo | |
| Microphone and speakers | |

Whisper's small model requires at least 12 gigabytes of GPU memory. Thus, let's try to secure a decent GPU for our Colab! Unfortunately, accessing a good GPU with the free version of Google Colab (i.e., Tesla T4 16 GB) is becoming much harder. However, with Google Colab Pro, we should have no issues in being allocated a V100 or P100 GPU.

**If you are using the digital version of this book, we advise you to type the code yourself or access it from the book's GitHub repository (a link is available in the next section). Doing so will help you avoid any potential errors related to copying and pasting code.**

Fine-tuning Whisper in *Chapter 4* will take at least one hour. Thus, you must monitor your running notebook in Colab regularly. Some notebooks implement a Gradio app with voice recording and audio playback. A microphone and speakers connected to your computer might help you experience the interactive voice features. Another option is to open the URL link Gradio provides at runtime on your mobile phone; from there, you might be able to use the phone's microphone to record your voice.

By meeting these technical requirements, you will be prepared to explore Whisper in different contexts while enjoying the streamlined experience of Google Colab and the comprehensive resources available on GitHub.

# Download the example code files

You can download the example code files for this book from GitHub at https://github.com/PacktPublishing/Learn-OpenAI-Whisper/. If the code is updated, it will be updated in the GitHub repository.

We also have other code bundles from our rich catalog of books and videos available at https://github.com/PacktPublishing/. Check them out!

# Code in Action

This book's Code in Action videos can be viewed at https://packt.link/gGv9a.

# Conventions used

Several text conventions are used throughout this book.

Code in text: Indicates code words in text, database table names, folder names, filenames, file extensions, pathnames, dummy URLs, user input, and Twitter handles. An example is "Users can even provide audiovisual formats such as .mp4 as inputs, as Whisper will extract just the audio stream to process."

A block of code is set as follows:

```
from datasets import load_dataset, DatasetDict
common_voice = DatasetDict()
common_voice["train"] = load_dataset("mozilla-foundation/common_
voice_11_0", "hi", split="train+validation", use_auth_token=True)
common_voice["test"] = load_dataset("mozilla-foundation/common_
voice_11_0", "hi", split="test", use_auth_token=True)
print(common_voice)
```

When we wish to draw your attention to a particular part of a code block, the relevant lines or items are set in bold:

```
[default]
exten => s,1,Dial(Zap/1|30)
exten => s,2,Voicemail(u100)
exten => s,102,Voicemail(b100)
exten => i,1,Voicemail(s0)
```

Any command-line input or output is written as follows:

```
!pip install --upgrade pip
!pip install --upgrade datasets transformers accelerate soundfile
librosa evaluate jiwer tensorboard gradio
```

**Bold**: Indicates a new term, an important word, or words that you see onscreen. For instance, words in menus or dialog boxes appear in **bold**. Here is an example: "To get a GPU, within Google Colab's main menu, click **Runtime | Change runtime type**, then change the **Hardware accelerator** from **None** to **GPU**."

> **Tips or important notes**
> Appear like this.

# Get in touch

Feedback from our readers is always welcome.

**General feedback**: If you have questions about any aspect of this book, email us at customercare@packtpub.com and mention the book title in the subject of your message.

**Errata**: Although we have taken every care to ensure the accuracy of our content, mistakes do happen. If you have found a mistake in this book, we would be grateful if you would report this to us. Please visit www.packtpub.com/support/errata and fill out the form.

**Piracy**: If you come across any illegal copies of our works in any form on the internet, we would be grateful if you would provide us with the location address or website name. Please get in touch with us at copyright@packt.com with a link to the material.

**If you are interested in becoming an author**: If there is a topic that you have expertise in and you are interested in either writing or contributing to a book, please visit authors.packtpub.com.

## Share Your Thoughts

Once you've read *Learn OpenAI Whisper*, we'd love to hear your thoughts! Scan the QR code below to go straight to the Amazon review page for this book and share your feedback.

```
https://packt.link/r/1-835-08592-X
```

Your review is important to us and the tech community and will help us make sure we're delivering excellent quality content.

# Download a free PDF copy of this book

Thanks for purchasing this book!

Do you like to read on the go but are unable to carry your print books everywhere?

Is your eBook purchase not compatible with the device of your choice?

Don't worry, now with every Packt book you get a DRM-free PDF version of that book at no cost.

Read anywhere, any place, on any device. Search, copy, and paste code from your favorite technical books directly into your application.

The perks don't stop there, you can get exclusive access to discounts, newsletters, and great free content in your inbox daily

Follow these simple steps to get the benefits:

1.  Scan the QR code or visit the link below

https://packt.link/free-ebook/9781835085929

2.  Submit your proof of purchase
3.  That's it! We'll send your free PDF and other benefits to your email directly

# Part 1:
# Introducing OpenAI's Whisper

This part introduces you to OpenAI's **Whisper**, a cutting-edge automatic speech recognition (ASR) technology. You will gain an understanding of Whisper's basic features and functionalities, including its key capabilities and setup process. This foundational knowledge will set the stage for a deeper exploration of the technology and its applications in real-world scenarios.

This part includes the following chapters:

- *Chapter 1, Unveiling Whisper – Introducing OpenAI's Whisper*
- *Chapter 2, Understanding the Core Mechanisms of Whisper*

# 1

# Unveiling Whisper – Introducing OpenAI's Whisper

**Automatic speech recognition** (ASR) is an area of **artificial intelligence** (AI) that focuses on the interaction between computers and humans through speech. Over the years, ASR has made remarkable progress in speech processing, and **Whisper** is one such revolutionary ASR system that has gained popularity recently.

Whisper is an advanced AI **speech recognition** model developed by OpenAI, trained on a massive multilingual dataset. With its ability to accurately transcribe speech, Whisper has become a go-to tool for voice applications such as assistants, transcription services, and more.

In this chapter, we will explore the basics of Whisper and its capabilities. We will start with an introduction to Whisper and its significance in the ASR landscape. Then, we will uncover Whisper's key features and strengths that set it apart from other speech models. We will then cover fundamental guidelines for implementing Whisper, including initial system configuration and basic usage walkthroughs to get up and running.

In this chapter, we will cover the following topics:

- Deconstructing OpenAI's Whisper

- Exploring key features and capabilities of Whisper

- Setting up Whisper

By the end of this chapter, you will have first-hand experience with Whisper and understand how to leverage its core functionalities for your speech-processing needs.

## Technical requirements

As presented in this chapter, you only need a Google account and internet access to run the Whisper AI code in **Google Colaboratory**. No paid subscription is required to use the free Colab and the GPU version. Those familiar with Python can run this code example in their environment instead of using Colab.

We are using Colab in this chapter as it allows for quick setup and running of the code without installing Python or Whisper locally. The code in this chapter uses the small Whisper model, which works well for testing purposes. In later chapters, we will complete the Whisper installation to utilize more advanced ASR models and techniques.

The code examples from this chapter can be found on GitHub at `https://github.com/PacktPublishing/Learn-OpenAI-Whisper/tree/main/Chapter01`.

# Deconstructing OpenAI's Whisper

In this section, we embark on a journey through the intricate world of voice and speech, unveiling the marvels of human vocalization. Voice and speech are more than sounds; they are the symphony of human communication orchestrated through a harmonious interplay of physiological processes. This section aims to provide a foundational understanding of these processes and their significance in speech recognition technology, particularly on Whisper. You will learn how Whisper, an advanced speech recognition system, emulates human auditory acuity to interpret and transcribe speech accurately. This understanding is crucial, as it lays the groundwork for comprehending the complexities and capabilities of Whisper.

The lessons in this section are valuable for multiple reasons. First, they offer a deep appreciation of voice and speech's biological and cognitive intricacies, which are fundamental to understanding speech recognition technology. Second, they provide a clear perspective on the challenges and limitations inherent in these technologies, using Whisper as a prime example. This knowledge is not just academic; it's directly applicable to various real-world scenarios where speech recognition can play a transformative role, from enhancing accessibility to breaking down language barriers.

As we proceed, remember that the journey through voice and speech is a blend of art and science – a combination of understanding the natural and mastering the technological. This section is your first step into the vast and exciting world of speech recognition, with Whisper as your guide.

## The marvel of human vocalization – Understanding voice and speech

In the vast expanse of human capabilities, the ability to produce voice and speech is a testament to our biological makeup's intricate complexity. It's a phenomenon that transcends mere sound production, intertwining biology, emotion, and cognition to create a medium through which we express our innermost thoughts and feelings. This section invites you to explore the fascinating world of voice and speech production, not through the lens of an anatomist but with the curiosity of a technologist marveling at one of nature's most sophisticated instruments. As we delve into this subject, consider the immense challenges technologies such as OpenAI's Whisper face in interpreting and understanding these uniquely human attributes.

Have you ever pondered the complexity of the systems at play when you casually conversed? The effortless nature of speaking belies the elaborate physiological processes that enable it. Similarly, when interacting with a speech recognition system such as Whisper, do you consider the intricate coding and algorithmic precision that allows it to understand and process your words?

The genesis of voice and speech is rooted in the act of breathing. The diaphragm and rib cage play pivotal roles in air inhalation and exhalation, providing the necessary airflow for voice production. This process begins with the strategic opening and closing of the vocal folds within the larynx, the epicenter of vocalization. As air from the lungs flows through the vocal folds, it causes them to vibrate, generating sound.

Speech, on the other hand, materializes through the meticulous coordination of various anatomical structures, including the velum, tongue, jaw, and lips. These structures sculpt the raw sounds produced by the vocal folds into recognizable linguistic patterns, enabling the expression of thoughts and emotions. Mastering the delicate balance of muscular control necessary for intelligible communication is a protracted journey that requires extensive practice.

Understanding the complexities of human voice and speech production is paramount in the context of OpenAI's Whisper. As an advanced speech recognition system, Whisper is engineered to emulate the auditory acuity of the human ear by accurately interpreting and transcribing human speech. The challenges faced by Whisper mirror the intricacies of speech development in humans, underscoring the complexity of the task at hand.

## Understanding the intricacies of speech recognition

The human brain's capacity for language comprehension is a marvel of cognitive processing, which has intrigued scientists and linguists for decades. The average 20-year-old is estimated to know between 27,000 and 52,000 words, typically increasing to 35,000 and 56,000 by age 60. Each of these words, when spoken, exists for a fleeting moment – often less than a second. Yet, the brain is adept at making rapid decisions, correctly identifying the intended word approximately *98%* of the time. How does the brain accomplish this feat with such precision and speed?

### The brain as a parallel neural processor

The brain's function as a **parallel processor** is at the core of our speech comprehension abilities. Parallel processing means it can handle multiple tasks simultaneously. Unlike sequential processors that handle one operation at a time, the brain's parallel processing allows for the simultaneous activation of numerous potential word matches. But what does this look like in the context of neural activity?

The general thinking is that each word in our vocabulary is represented by a distinct processing unit within the brain. These units are not physical entities but neuronal firing patterns within the cerebral cortex, **neural representations** of words. When we hear the beginning of a word, thousands of these units spring into action, each assessing the likelihood that the incoming auditory signal matches their corresponding word. As the word progresses, many units deactivate upon realizing a mismatch, narrowing down the possibilities. This process continues until a single pattern of firing activity remains – this is the **recognition point**. The active units suppress the activity of others, a mechanism that saves precious milliseconds, allowing us to comprehend speech at a rate of up to eight syllables per second.

### Accessing meaning and context

The goal of speech recognition extends beyond mere word identification; it involves accessing the word's meaning. Remarkably, the brain begins considering multiple meanings before a word is fully articulated. For instance, upon hearing the fragment "cap," the brain simultaneously entertains various possibilities such as "captain" or "capital." This explosion of potential meanings is refined to a single interpretation by the recognition point.

Context plays a pivotal role in guiding our understanding. It allows for quicker recognition and helps disambiguate words with multiple meanings or homophones. For bilingual or multilingual individuals, the language context is an additional cue that filters out words from other languages.

### The nighttime integration process

How does the brain incorporate new vocabulary without disrupting the lexicon? The answer lies in the **hippocampus**, a brain region where new words are initially stored, separate from the cortex's central word repository. Through a process believed to occur during sleep, these new words are gradually woven into the cortical network, ensuring the stability of the existing vocabulary.

While our conscious minds rest at night, the brain actively integrates new words into our linguistic framework. This nocturnal activity is crucial for maintaining the dynamism of our language capabilities, preparing us for the ever-evolving landscape of human communication.

## OpenAI's Whisper – A technological parallel

In AI, OpenAI's Whisper presents a technological parallel to the human brain's speech recognition capabilities. Whisper is a state-of-the-art speech recognition system that leverages deep learning to transcribe and understand spoken language with remarkable accuracy. Like the brain processes speech through parallel processing, Whisper utilizes **neural networks** to analyze and interpret audio signals.

Whisper's neural networks are trained on vast datasets, allowing the system to recognize various words and phrases across different languages and accents. The system's architecture mirrors the brain's recognition point by narrowing down potential transcriptions until the most probable one is selected.

Whisper also exhibits the brain's ability to integrate context into comprehension. The system can discern context from surrounding speech, improving its accuracy in real-time transcription. Moreover, Whisper is designed to learn and adapt continuously, just as the human brain integrates new words into its lexicon.

Whisper's algorithms must navigate a myriad of variables, from accents and intonations to background noise and speech irregularities, to convert speech to text accurately. By dissecting the nuances of voice and speech recognition, we gain insights into the challenges and intricacies that Whisper must navigate to process and understand human language effectively.

As we look to the future, the potential for speech recognition technologies such as Whisper is boundless. It holds the promise of breaking down language barriers, enhancing accessibility, and creating more natural human-computer interactions. The parallels between Whisper and the human brain's speech recognition processes underscore the sophistication of our cognitive abilities and highlight the remarkable achievements in AI.

## The evolution of speech recognition and the emergence of OpenAI's Whisper

The quest to endow machines with the ability to recognize and interpret human speech has been a formidable challenge that has engaged the brightest minds in technology for over a century. From the rudimentary dictation machines of the late 19th century to the sophisticated algorithms of today, the journey of speech recognition technology is a testament to human ingenuity and perseverance.

### The genesis of speech recognition

The earliest endeavors in speech recognition concentrated on creating vowel sounds, laying the groundwork for systems that could potentially decipher phonemes – the fundamental units of speech. The iconic Thomas Edison pioneered in this field with his invention of dictation machines that could record speech, a technology that found favor among professionals inundated with documentation tasks.

> **What are phonemes?**
>
> Phonemes refer to the smallest sound units in a language that hold meaning. Changing a phoneme can change the entire meaning of a word. Some examples of phonemes are the following:
>
> - The word "cat" has three phonemes: /c/, /a/, and /t/.
>
> - The word "bat" also has three phonemes: /b/, /a/, and /t/. The /b/ phoneme changes the meaning from "cat."
>
> - The word "sit" has three phonemes: /s/, /i/, and /t/. Both the /s/ and /i/ phonemes distinguish it from "cat."

It was in the 1950s that the field took a significant leap forward. In 1952, Bell Labs created the first viable speech recognition system, Audrey, which recognized digits 0–9 spoken by a single voice with 90% accuracy. IBM followed in 1962 with Shoebox, which recognized 16 English words. In the 1960s, Japanese researchers made advances in phoneme and vowel recognition. However, this accuracy was contingent on the speaker, highlighting the inherent challenges of speech recognition: the variability of voice, accent, and articulation among individuals.

## The advent of machine understanding

A significant breakthrough came in the 1970s from the **Defense Advanced Research Projects Agency (DARPA) Speech Understanding Research (SUR)** program. At Carnegie Mellon University, Alexander Waibel developed the Harpy system, which could understand over 1,000 words, a vocabulary on par with a young child. Harpy was notable for using **finite state networks** to reduce the search space and **beam search** to pursue only the most promising interpretations.

---

### Finite state networks

Finite state networks are computational models comprising states connected by transitions. They can recognize patterns in input while staying within the defined states. Their job is to reduce the search space for speech recognition by limiting valid transitions between speech components. This simplifies decoding possible interpretations.

Examples include the following:

- Phoneme networks that restrict transition between valid adjacent sounds.

- Word networks that connect permissible words in a grammar.

- Speech recognition uses nested finite state networks spanning different linguistic tiers.

---

### Beam search

Beam search is an optimization algorithm that pursues only the most promising solutions meeting some criteria, pruning away unlikely candidates. It focuses computations on interpretations likely to maximize objective metrics. This is more efficient than exhaustively evaluating all options.

Examples include the following:

- Speech recognition beam search, which pursues probable transcriptions while filtering out unlikely word sequences.

- Machine translation beam search, which ensures translations adhere to target language rules.

- Video captioning beam search, which favors captions that fit the expected syntax and semantics.

---

Waibel was motivated to develop Harpy and subsequent systems such as Hearsay-II to enable speech translation, converting speech directly to text in another language rather than using dictionaries. Speech translation requires tackling the complexity of natural language by leveraging linguistic knowledge.

Other key developments in the 1970s included Bell Labs building the first multivoice system. The 1980s saw the introduction of **hidden Markov models (HMMs)** and statistical language modeling. IBM's Tangora could recognize 20,000 words by the mid-1980s, enabling early commercial adoption. Conceived initially as a voice-operated typewriter for office use, Tangora allows users to speak text aloud that would then be transcribed. This functionality drastically boosted productivity among office staff. The technology marked meaningful progress toward the voice dictation systems we know today.

### The era of continuous speech recognition

Until the 1990s, speech recognition systems relied heavily on template matching, which required precise and slow speech in noise-free environments. This approach had obvious limitations, as it needed more flexibility to accommodate the natural variations in human speech.

Accuracy and speed increased rapidly in the 1990s with neural networks and increased computing power. IBM's Tangora, leveraging HMMs, marked a significant advancement. This technology allowed for a degree of prediction in phoneme sequences, enhancing the system's adaptability to individual speech patterns. Despite requiring extensive training data, Tangora could recognize an impressive lexicon of English words. Commercial adoption began.

In 1997, Dragon's NaturallySpeaking software, the world's first continuous speech recognizer, arrived as a watershed moment. This innovation eliminated pauses between words, facilitating a more natural interaction with machines. As computing power increased, neural networks improved accuracy. Systems such as Dragon NaturallySpeaking could process 100 words per minute with 97% accuracy.

Google's foray into speech recognition, with its Voice Search app for iPhone, harnessed machine learning and cloud computing to achieve unprecedented accuracy levels. Google further refined speech recognition with the introduction of Google Assistant, which now resides in many smartphones worldwide. By 2001, consumer adoption increased through systems such as BellSouth's voice-activated portal.

However, the most significant impact came after widespread smart device adoption in 2007, with accurate voice assistants using cloud-based deep learning. In 2010, Apple's Siri captured the public's imagination by infusing a semblance of humanity into voice recognition. Microsoft's Cortana and Amazon's Alexa, introduced in 2014, ignited a competitive landscape among tech giants in the speech recognition domain.

### The connection to OpenAI's Whisper

In this innovation continuum, OpenAI's Whisper emerges as a pivotal development. Whisper is a deep learning-based speech recognition system that builds upon the aforementioned historical advancements and challenges. It leverages vast datasets and sophisticated models to accurately interpret speech across multiple languages and dialects. Whisper embodies the culmination of efforts to create a system that is not only highly adaptable to individual speech patterns but also capable of contextual understanding, a critical aspect that has long eluded previous technologies.

The evolution of speech recognition technology, from Edison's dictation machines to OpenAI's Whisper, represents a relentless pursuit of a more intuitive and seamless interface between humans and machines. As we reflect on this journey, it might be timely for us to ask: What new frontiers will the next generation of speech recognition technologies explore? The potential for further advancements is vast, promising a future where the barriers between human communication and machine interpretation are virtually indistinguishable. The progress we have witnessed thus far is merely the prologue to an era where voice recognition technology will be an integral, ubiquitous part of our daily lives.

In the next section, you will learn about Whisper's key features and capabilities that enable its precise speech recognition prowess. You'll discover Whisper's robust capabilities that set it apart in various applications. From its exceptional **speech-to-text** (**STT**) conversion to its adeptness in handling diverse languages and accents, Whisper exemplifies state-of-the-art performance in ASR. We'll delve into the mechanics of how Whisper converts speech to text using advanced techniques, including the encoder-decoder transformer model and its training on a vast and varied dataset.

# Exploring key features and capabilities of Whisper

In this section, we dive into the heart of OpenAI's Whisper, uncovering the core elements that make it a standout in ASR. This exploration is not merely a listing of features; it is an insightful journey into understanding how Whisper transcends traditional boundaries of STT conversion, offering an unparalleled blend of accuracy, versatility, and ease of use.

The capabilities of Whisper extend beyond mere transcription. You will learn about its prowess in real-time translation, support for a wide array of file formats, and ease of integration into various applications. These features collectively make Whisper not just a tool for transcription but a comprehensive solution for global communication and accessibility.

This section is crucial for those seeking to understand the practical implications of Whisper's features. Whether you're a developer looking to integrate Whisper into your projects, a researcher exploring the frontiers of ASR technology, or simply an enthusiast keen on understanding the latest advancements in AI, the lessons here are invaluable. They provide a concrete foundation for appreciating the technological marvel that is Whisper and its potential to transform how we interact with and process spoken language.

As you engage with this section, remember that the journey through Whisper's capabilities is more than an academic exercise. It's a practical guide to harnessing the power of one of the most advanced speech recognition technologies available today, poised to fuel innovation across diverse fields and applications.

## Speech-to-text conversion

The cornerstone feature of Whisper is its capability to transcribe spoken language into text. Imagine a journalist recording interviews in the field, where they could swiftly convert every word spoken into an editable, searchable, and shareable text format. This feature isn't just convenient; it's a game-changer in environments where quick dissemination of spoken information is crucial.

The latest iteration of Whisper, called `large-v3` (Whisper-v3), was released on November 6, 2023. Its architecture uses an encoder-decoder transformer model trained on 1 million hours of weakly labeled audio and 4 million hours of pseudo-labeled audio collected from real-world speech data from the web, making it adept at handling diverse recording conditions. Here's how Whisper converts speech to text:

1. The input audio is split into 30-second chunks and converted into log-Mel spectrograms.
2. The encoder receives the spectrograms, creating audio representations.

3.  Training of the decoder follows to predict the corresponding text transcript from the encoder representations, including unique tokens for tasks such as language identification and timestamps.

---

**Log-Mel spectrograms**

Log-Mel spectrograms are obtained by taking the logarithm of the values in the **Mel spectrogram**. This compresses the spectrogram's dynamic range and makes it more suitable for input to machine learning models.

Mel spectrograms represent the power spectrum of an audio signal in the frequency domain. They are obtained by applying a **Mel filter bank** to the signal's power spectrum, which groups the frequencies into a set of **Mel frequency bins**.

Mel frequency bins represent sound information in a way that mimics low-level auditory perception. They capture the energy at each frequency band and approximate the spectrum shape.

Whisper-v3 has the same architecture as the previous large models, except that the input uses 128 Mel frequency bins instead of 80. The increase in the number of Mel frequency bins from 80 to 128 in Whisper-v3 is significant for several reasons:

- **Improves frequency resolution**: Whisper-v3 can capture finer details in the audio spectrum using more Mel frequency bins. This higher resolution allows the model to distinguish between closely spaced frequencies, potentially improving its ability to recognize subtle acoustic differences between phonemes or words.

- **Enhances speech representation**: The increased number of Mel frequency bins provides a more detailed representation of the speech signal. This richer representation can help the model learn more discriminative features, leading to better speech recognition performance.

- **Increases compatibility with human auditory perception**: The Mel scale is designed to mimic the non-linear human perception of sound frequencies. Using 128 Mel frequency bins, Whisper-v3 can more closely approximate the human auditory system's sensitivity to different frequency ranges. This alignment with human perception may contribute to improved speech recognition accuracy.

- **Allows the learning of complex patterns**: The higher-dimensional input provided by the 128 Mel frequency bins gives Whisper-v3 more data. This increased input dimensionality may enable the model to learn more complex and nuanced patterns in the speech signal, potentially improving its ability to handle challenging acoustic conditions or speaking styles.

While increasing the number of Mel frequency bins can provide these benefits, it also comes with a computational cost. Processing higher-dimensional input requires more memory and computation, which may impact the model's training and inference speed. However, the improved speech recognition performance offered by the increased frequency resolution can outweigh these computational considerations in many applications.

This end-to-end approach allows Whisper to convert speech to text directly without any intermediate steps. The large and diverse training dataset enables Whisper to handle accents, background noise, and technical language much better than previous speech recognition systems. Some critical capabilities regarding STT conversion are as follows:

- Whisper can transcribe speech to text in nearly 100 languages, including English, Mandarin, Spanish, Arabic, Hindi, and Swahili. Whisper-v3 has a new language token for Cantonese. This multilingual transcription makes it useful for international communications.

- The model is robust with accents, background noise, and technical terminology, making it adept at handling diverse recording conditions.

- Whisper achieves state-of-the-art performance on many speech recognition benchmarks without any fine-tuning. This zero-shot learning capability enables the transcription of new languages not seen during training.

- The transcription includes punctuation and capitalization, providing properly formatted text output. Timestamps are an option if the goal is to align transcribed text with the original audio.

- A streaming API enables real-time transcription with low latency, which is essential for live captioning and other applications requiring fast turnaround.

- The open source release facilitates research into improving speech recognition and building customized solutions.

Overall, Whisper provides highly robust and accurate STT across many languages and use cases. The transcription quality exceeds many commercial offerings without requiring any customization.

## Translation capabilities

In addition to transcription, Whisper can translate speech from one language into another. Key aspects of its translation abilities are as follows:

- Whisper supports STT translation from nearly 100 input languages into English text. This feature allows transcription and translation of non-English audio in one step.

- The model auto-detects the input language, so users don't need to specify the language manually during translation.

- Translated output aims to convey the whole meaning of the original audio, not just word-for-word substitution. This feature helps capture nuances and context.

- Multitask training on aligned speech and text data allows the development of a single model for transcription and translation instead of separate systems.

- The translation quality approach uses dedicated machine translation models tailored to specific language pairs. However, Whisper covers far more languages with a single model.

In summary, Whisper pushes the boundaries of speech translation by enabling direct STT translation for many languages within one multitask model without compromising accuracy. Whisper makes content globally accessible to English speakers and aids international communication.

## Support for diverse file formats

Whisper's versatility extends to its support for various audio file formats, including MP3, MP4, MPEG, MPGA, M4A, WAV, and WebM. This flexibility is essential in today's digital landscape, where audio content comes in many forms. For content creators working with diverse media files, this means no extra file conversion step, ensuring a smoother workflow.

Specifically, Whisper leverages FFmpeg under the hood to load audio files. As FFmpeg supports reading many file containers and codecs, Whisper inherits that versatility for inputs. Users can even provide audiovisual formats such as .mp4 as inputs, as Whisper will extract just the audio stream to process.

Recent additions to the officially supported formats include the open source OGG/OGA and FLAC codecs. Their inclusion underscores Whisper's commitment to supporting community-driven and freely licensed media formats alongside more proprietary options.

The current file size limit for uploading files to Whisper's API service is 25 MB. Whisper handles larger local files by splitting them into segments under 25 MB each. The wide range of formats – from standard compressed formats to CD-quality lossless ones – combined with the generous file size allowance caters to virtually any audio content needs when using Whisper.

In summary, Whisper sets itself apart by the breadth of audio formats it accepts while maintaining leading-edge speech recognition capability. Whisper empowers users to feed their content directly without tedious conversion or conditioning steps. Whether producing podcasts, audiobooks, lectures, or other speech-centric media, Whisper has users covered on the file support side.

## Ease of use

OpenAI's release of Whisper represents a significant step in integrating ASR capabilities into applications. The Python code snippets available at OpenAI and other sites demonstrate the seamless ease with which developers can incorporate Whisper's functionalities. This simplicity enables innovators to leverage ASR technology to create novel tools and services with relative simplicity.

Specifically, the straightforward process of calling Whisper's API and passing audio inputs showcases the accessibility of the technology. Within minutes, developers can integrate a production-grade speech recognition system. Multiple model sizes allow the fitting of speech-processing capacity for the infrastructure. Whisper scales to the use case from lightweight mobile device apps to heavy-duty backends in the cloud.

Beyond sheer technical integration, Whisper simplifies the process of leveraging speech data. The immense corpus of training data produces remarkable off-the-shelf accuracy without user fine-tuning, and built-in multilingualism removes the need for language specialization. Together, these attributes lower the barrier to productive employment of industrial-strength ASR.

In summary, by delivering state-of-the-art speech recognition primed for easy assimilation into new systems, Whisper stands poised to fuel a Cambrian explosion of voice-enabled applications across domains. Its potential to unlock innovation is matched only by the ease with which anyone can tap it. The combination of power and accessibility that Whisper provides heralds a new era where speech processing becomes a readily available ingredient for inventive problem solvers. OpenAI has opened the floodgates wide to innovation.

## Multilingual capabilities

One of Whisper's most impressive features is its proficiency in numerous languages. As of November 2023, it supports 100 languages, from Afrikaans to Welsh. This multilingual capability makes Whisper an invaluable global communication, education, and media tool.

For example, educators can use Whisper to transcribe lectures in multiple languages, aiding students in language learning and comprehension. Interview journalists can transcribe and translate conversations, removing language barriers. Customer service agents can communicate with customers in their native tongues using Whisper's speech translation.

Whisper achieves its multilingual prowess through training on a diverse dataset of 680,000 hours of audio in 100 languages collected from the internet. This exposure allows the model to handle varied accents, audio quality, and technical vocabulary when transcribing and translating.

While Whisper's accuracy varies across languages, it demonstrates competitive performance even for low-resource languages such as Swahili. Whisper leverages its knowledge of other languages for languages with limited training data to make inferences. However, there are still challenges in achieving equal proficiency across all languages. Performance is weakest for tonal languages such as Mandarin Chinese. Expanding the diversity of Whisper's training data could further enhance its multilingual capabilities.

Whisper's support for nearly 100 languages in a single model is remarkable. As Whisper's multilingual performance continues improving, it could help bring us closer to seamless global communication.

## Large input handling

Whisper's ability to handle audio files of up to 25 MB directly addresses the needs of those dealing with lengthy recordings, such as podcasters or oral historians. Whisper can process segmented audio for longer files, ensuring no context or content quality loss.

### Flexible file size limits

The default 25 MB file size limit covers many standard audio lengths while optimizing for fast processing. For files larger than 25 MB, Whisper provides options to split the audio into segments under 25 MB each. This chunking approach enables Whisper to handle files of any length. Segmenting longer files is recommended over compression to avoid degrading audio quality and recognition accuracy. When segmenting, it's best to split on pauses or between speakers to minimize loss of context. Libraries such as `pydub` simplify audio segmentation.

### Maintaining quality across segments

Whisper uses internal algorithms to reconstruct context across audio segments, delivering high-quality transcriptions for large files. The OpenAI team continues to improve Whisper's ability to provide coherent transcriptions across segments with minimal discrepancies.

### Expanding access to long-form content

Whisper's robustness with large files unlocks transcription capabilities for long-form content such as lectures, interviews, and audiobooks. Longer files allow creators, researchers, and more to leverage audio content efficiently for various downstream applications at any scale. As Whisper's segmentation capabilities improve, users can accurately transcribe even extremely lengthy recordings such as multiday conferences.

In summary, Whisper provides a flexible transcription solution for short- and long-form audio through its segmented processing capabilities. Careful segmentation preserves quality while enabling Whisper to handle audio files of any length.

## Prompts for specialized vocabularies

Whisper's ability to utilize prompts for enhanced transcription accuracy makes it extremely useful for specialized fields such as medicine, law, or technology. The model can better recognize niche vocabulary and technical jargon during transcription by providing a prompt containing relevant terminology.

For example, a radiologist could supply Whisper with a prompt full of medical terms, anatomical structures, and imaging modalities. The prompt would prime Whisper to transcribe radiology reports and interpretive findings accurately. Similarly, an attorney could include legal terminology and case citations to improve deposition or courtroom proceeding transcriptions.

Here's an example of a prompt that a radiologist could supply to Whisper to transcribe radiology reports and interpretive findings accurately:

```
"Patient is a 45-year-old male with a history of hypertension and
hyperlipidemia. The patient presented with chest pain and shortness of
breath. A CT scan of the chest was performed with contrast. The scan
revealed a 2.5 cm mass in the right upper lobe of the lung. The mass
is well-circumscribed and has spiculated margins. There is no evidence
of mediastinal lymphadenopathy. The patient will undergo a biopsy of
the mass for further evaluation."
```

This prompt contains medical terms such as "hypertension," "hyperlipidemia," "CT scan," "contrast," "mass," "right upper lobe," "spiculated margins," "mediastinal lymphadenopathy," and "biopsy." It also contains anatomical structures such as "lung" and "mediastinum." Finally, it includes imaging modalities such as "CT scan" and "contrast."

By providing such a prompt, the radiologist can train Whisper to recognize and transcribe these terms accurately. This can help improve the accuracy and speed of transcribing radiology reports and interpretive findings, ultimately saving time and improving radiologists' workflow.

Prompts do not need to be actual transcripts – even fictitious prompts with relevant vocabulary can steer Whisper's outputs. Some techniques for effective prompting include the following:

- Using GPT-3 to generate mock transcripts containing target terminology for Whisper to emulate. This trains Whisper on the vocabulary.

- Providing a *spelling guide* with proper spellings of industry-specific names, products, procedures, uncommon words, acronyms, etc. This helps Whisper learn specialized orthography.

- Submitting long, detailed prompts. More context helps Whisper adapt to the desired style and lexicon.

- Editing prompts iteratively based on Whisper's outputs, including missing terms or correct errors, further refine the model.

Prompting is not a panacea but can improve accuracy for niche transcription tasks. With the technical vocabulary provided upfront, Whisper can produce highly accurate transcripts, even for specialized audio content. Its flexibility with prompting is a crucial advantage of Whisper over traditional ASR systems.

## Integration with GPT models

Whisper's integration with large language models such as GPT-4 significantly enhances its capabilities by enabling refined transcriptions. GPT-4 can correct misspellings, add appropriate punctuation, and improve the overall quality of Whisper's initial transcriptions. This combination of cutting-edge speech recognition and advanced language processing creates a robust automated transcription and document creation system.

By leveraging GPT-4's contextual understanding and language generation strengths to refine Whisper's STT output, the solution can produce highly accurate written documents from audio in a scalable manner. The postprocessing technique using GPT-4 is particularly more scalable than that depending solely on Whisper's prompt parameter, which has a token limit.

This integration paves the way for automated documentation of meetings, interviews, podcasts, and other verbal content. The resulting transcripts can feed into different systems, such as search engines, for enhanced discoverability. They also enable detailed analysis of oral communications using **natural language processing** (**NLP**) techniques.

Overall, combining Whisper and GPT-4 forms an end-to-end solution that unlocks the richness of audio data and makes it accessible for a wide range of applications, from personal productivity to enterprise knowledge management. This combination showcases the immense potential of composing multiple AI systems to create emergent capabilities.

## Fine-tunability

Fine-tuning is a great way to customize Whisper to improve accuracy, support new languages, and adapt the model to specific use cases. At a high level, fine-tuning takes a pre-trained model, such as Whisper, and trains it further on a downstream task using additional data. To perform the tuning, we need an ASR pipeline consisting of three components:

- A feature extractor for preprocessing the raw audio inputs
- The model, which performs sequence-to-sequence mapping
- A tokenizer for postprocessing the model outputs to text format

Fortunately, the Whisper model has an associated feature extractor and tokenizer called *WhisperFeatureExtractor* and *WhisperTokenizer*. We will cover this topic in more depth in *Chapter 4, Fine-Tuning Whisper for Domain and Language Specificity*.

Tuning allows the model to specialize and adapt to a particular use case. The main reasons to fine-tune Whisper are the following:

- Improve accuracy for a specific domain or use case such as meetings, call center data, and so on
- Support new languages not in the original training data
- Customize the model for an application's specific vocabulary, audio conditions, and so on
- Leverage transfer learning to perform better with less data than training from scratch

Fine-tuning is well suited for Whisper because it is trained on diverse data and can benefit from specializing further in a particular task or dataset. Tuning can happen over the entire Whisper model or at the higher layers closest to the output.

By leveraging transfer learning instead of training from scratch, fine-tuning allows the development of high-quality speech recognition with less data and computing resources. The active open source community provides ample resources for fine-tuning Whisper using Hugging Face Transformers.

## Voice synthesis

Whisper plays a vital role in the one-shot voice synthesis workflow by transcribing small voice samples to text for model training. Combined with Ozen and Tortoise TTS, it enables high-quality voice synthesis with minimal data.

> **One-shot voice synthesis**
>
> One-shot voice synthesis is a technique for creating a **text-to-speech** (**TTS**) system that can synthesize speech in a target voice using only a single recording of that speaker's voice. The process involves training an ML model on a corpus of speech from the target speaker and then using that model to generate new speech based on text input. One-shot voice synthesis is an active area of research, and there are many different approaches to implementing it.

The Ozen toolkit leverages Whisper to preprocess audio data by extracting speech segments, transcribing them with Whisper, and saving them in the LJSpeech format. Tortoise TTS uses the preprocessed data to fine-tune a personalized voice synthesis model.

> **LJSpeech format**
>
> This format comes from the one used in the *LJSpeech Dataset*, a public-domain speech dataset comprising 13,100 short audio clips of a single speaker reading passages from 7 non-fiction books. A transcription is provided for each clip. These clips are between 1 and 10 seconds in length, with a total length of approximately 24 hours (`https://keithito.com/LJ-Speech-Dataset`).

Tortoise TTS is a neural TTS model that enables high-quality voice synthesis with minimal data, even a single audio sample of the target voice. After preprocessing data with Ozen and Whisper, Tortoise TTS can be fine-tuned on the new voice and used to synthesize speech mimicking that voice.

The combination of Whisper, Ozen, and Tortoise TTS enables the building of personalized voice synthetic inferences from just a few seconds of audio data without extensive data collection or cleaning. Whisper's robust ASR handles transcription, Ozen preprocesses the data, and Tortoise TTS regulates voice synthesis.

## Speech diarization

Whisper provides robust speech recognition, while external libraries such as `pyannote.audio` can be used on top of Whisper for speaker diarization by utilizing word-level timestamps from Whisper.

Diarization is partitioning an audio recording into homogeneous segments according to the speaker's identity. It answers the question "Who spoke when?" in an audio recording. The goal is to separate speech segments belonging to different speakers without knowing who the speakers are.

Out of the box, Whisper does not support speaker diarization. It generates transcriptions without speaker labels. However, Whisper outputs timestamps at the word level in transcriptions. These timestamps, along with external speaker diarization libraries such as `pyannote.audio`, match the transcriptions with the speaker segments and thus enable speaker labeling.

In conclusion, OpenAI's Whisper is a testament to the incredible advancements in speech recognition technology. Its capabilities, from multilingual transcription to integration with advanced language models, offer a glimpse into a future where the spoken word seamlessly integrates with the digital world. As we continue exploring and expanding its applications, Whisper promises to revolutionize our process of understanding and utilizing human speech.

In the next section, we take a practical turn, guiding you through the first steps of deploying OpenAI's Whisper. This section is pivotal for anyone eager to harness the power of Whisper for audio transcription, as it lays out the straightforward procedures to get started. Here, you will learn how to set up and use Whisper through a user-friendly web interface and a more hands-on approach using Google Colab.

# Setting up Whisper

The journey begins with exploring how to access Whisper via Hugging Face's web interface, designed for simplicity and convenience. This method is perfect for those who prefer to avoid the intricacies of coding and software installation. You will learn to easily upload audio files and receive transcriptions directly through a web browser, making Whisper accessible to a broader audience.

Next, we will show you how to install and run Whisper in a cloud environment such as Google Colab. This approach is tailored for those who seek a more involved experience and wish to understand Whisper's workings from a closer perspective. We will walk through the Whisper and FFmpeg installation for audio and video support, demonstrating how to transcribe files and view the results within the Colab environment.

Importantly, this section concerns the *how* and the *why*. The ease of setting up Whisper underscores its potential for widespread application, from academic research to real-world business solutions. By the end of this section, you will have gained the technical know-how to start using Whisper and an appreciation of its accessibility and versatility. As you progress, remember that these initial steps are crucial in unlocking Whisper's full potential, paving the way for more advanced exploration and innovation in speech recognition.

## Using Whisper via Hugging Face's web interface

To use Whisper for audio transcription, you don't need to create an OpenAI account or obtain API keys. Whisper is an open source project available on GitHub, so you can use it independently of the OpenAI API. You can install and run Whisper on your local machine or in a cloud environment such as Google Colab without any OpenAI account or API keys. This accessibility is part of what makes Whisper a convenient tool for STT transcription.

To provide a more straightforward and user-friendly experience with Whisper, we will start by accessing it through web interfaces, which don't require dealing with repositories or Python libraries.

Here's a simplified guide:

1.  **Access Whisper**: Visit the Hugging Face Whisper space at `https://huggingface.co/spaces/openai/whisper`.

2.  **Upload audio**: Upload or record your audio file directly on the website. There is an audio file available at `https://github.com/PacktPublishing/Learn-OpenAI-Whisper/blob/main/Chapter01/Learn_OAI_Whisper_Sample_Audio01.m4a`.

3.  **Transcribe**: Whisper will automatically transcribe the audio into text.

4.  **Review and download**: If needed, you can review and download the transcription.

You can see an overview of the Hugging Face Whisper space here:

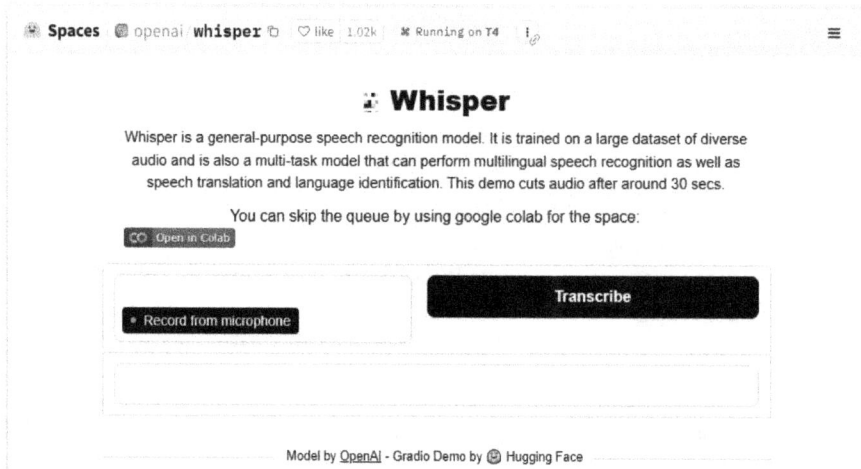

Figure 1.1 – Whisper: A Hugging Face space by OpenAI

This method provides an easy way to access Whisper's capabilities without the technical requirements of setting up the software locally. It is perfect for those who want to transcribe audio quickly without installation hassles.

## Using Whisper via Google Colaboratory

This following step-by-step guide will help you effectively use Whisper AI in Google Colab for transcribing speech to text. Here's a step-by-step guide on using OpenAI's Whisper AI in Google Colab, based on your provided text and formatted with markdown for clarity:

1.  Installing Google Colab:

    I.    Visit Google Drive and set up your Google account if you don't already have one.

    II.   In the top left-hand corner, click **New** | **More** | **Connect more apps**.

III.   Search for Google Colaboratory.

IV.   Select the first option, **Colaboratory**, and click **Install**.

V.   After installation, click **Done** and close the **Connect more apps** window.

2.   Configuring Google Colab:

I.   Open Google Drive.

II.   Click **New | More | Colaboratory** to open Colab.

III.   Rename the file by selecting Untitled.ipynb and giving it a more descriptive name.

IV.   Click the **Runtime** menu, select **Change runtime type**, and set the **Hardware accelerator** option to **T4 GPU**. (If you are using the free version of Google Colab, then a T4 GPU should be an option.)

3.   Installing Whisper AI on Google Colab:

I.   Open your Colab notebook.

II.   Paste the following code to install Whisper and FFmpeg (for audio and video file support):

```
!pip install git+https://github.com/openai/whisper.git
!sudo apt update && sudo apt install ffmpeg
```

III.   Run the code by selecting the **Run** icon.

4.   Running Whisper AI:

I.   In Colab, click the **Files** icon in the left-hand navigation menu.

II.   Drag and drop the audio or video file you want to transcribe.

III.   Click **OK** to acknowledge that uploaded files will be deleted when the runtime is recycled.

IV.   Your file should now appear under the **Files** section. You might need to press the **Refresh** icon to make the file appear.

V.   Paste the following code to transcribe the file with Whisper:

```
!whisper your-audio-file-here --model small.en
```

VI.   Right-click on the filename listed under the **Folder** menu and select **Copy path**.

VII.   Replace your-audio-file-here with the name of your filename, including path, no quotes.

VIII.   For testing and based on your memory, processing, and GPU availability, use the small.en Whisper model. However, there are other model sizes: tiny, base, small, medium, and large.

IX.   Run the code by clicking the **Run** icon.

5.  Viewing and downloading the transcript:

    After running the code, the transcription results will be displayed. You'll find `your-audio-file-here.txt` (displays the transcription text), `your-audio-file-here.vtt` (displays timed text tracks using the WEBVTT format), `your-audio-file-here.tsv` (displays text tracks using the tab-separated format), `your-audio-file-here.json` (displays the transcription text using the JSON format), and `your-audio-file-here.srt` (displays the transcription text using the SubRip format) in the **Files** section of Colab. If you do not see them, then you might need to press the **Refresh** icon in Colab. To download any of these files, hover over the file, select the ellipsis menu, and click **Download**.

---

**Whisper's output formats**

In addition to plain text (TXT), Whisper supports various output formats, including JSON, WEBVTT, SRT, and TSV. Each format serves a different purpose and is suitable for other use cases:

- **JSON (JavaScript Object Notation)**: This is a versatile and widely used data interchange format. In the context of Whisper, the JSON output includes detailed information about the transcription, such as the task, language, duration, segments, and other metadata. Each segment contains the start and end times, the transcribed text, and other details such as average log probability, compression ratio, and no-speech probability.

- **WEBVTT (Web Video Text Tracks)**: This is a popular format for displaying captions or subtitles for HTML5 videos. It's designed to be easy to read and write, making it a good choice for web developers. Whisper's output in this format can be directly used as video captions.

- **SRT (SubRip Text)**: This is another widely used format for subtitles and captions. Most video players and video editing software support it. Each entry in an SRT file includes a sequence number, start and end times, and the corresponding text. Whisper can generate SRT files that can be used to add subtitles to videos.

- **TSV (Tab-Separated Values)**: This is a simple text format for storing data in a tabular structure, similar to CSV, but with tabs as separators. It's not as commonly used as the other formats in the context of Whisper, but it can be helpful in specific applications where a simple, tabular format is needed.

Each of these formats has its strengths and is suited to different applications. JSON is great for applications needing detailed transcription metadata, while WEBVTT and SRT are ideal for video captioning or subtitling applications. TSV, on the other hand, provides a simple, tabular representation of the data.

---

Now that you have mastered the basics of using OpenAI's Whisper AI in Google Colab, it's time to explore its more advanced capabilities. The following section will introduce you to additional parameters and options you can run in Google Colab. These enhancements enable you to customize the transcription process more precisely, cater to specific language requirements, and handle various audio conditions. Let's dive deeper and unlock the full potential of Whisper's advanced features.

## Expanding on the basic usage of Whisper

You can leverage more advanced parameters with the `!whisper` command in Google Colab to customize the transcription process. Here are some additional options you can utilize:

- **Specifying language**: You can specify the language of the audio file for more accurate transcription. Replace `–model small.en` with the language code. For instance, for Spanish, use `--model small --language Spanish`.

- **Enabling automatic language detection**: Whisper can automatically detect the audio's language. If unsure, just run the model without specifying the language. In that scenario, you should see an initial output message: `Detecting language using up to the first 30 seconds...`. Try it by running the following, for example:

  ```
  !whisper "YOUR_FILE_NAME.mp3" --model small
  ```

- **Controlling verbose output**: Use the `--verbose` flag to suppress some of the output, including confidence scores and other metadata:

  ```
  !whisper "YOUR_FILE_NAME.mp3" --model small --verbose False
  ```

- **Sending output to a specific file**: Instead of saving the transcription output in the same directory location as the file being processed, you can direct the output to a specific directory using the `--output_dir` flag:

  ```
  !whisper "YOUR_FILE_NAME.mp3" --model small --output_dir "/
  whisper_output"
  ```

- **Modeling specific tasks**: Whisper can handle different tasks such as transcription and translation. Specify the task using the `--task` flag. Use `-- task translate` for translation from foreign audio to English transcription. Whisper will not translate to any other target language than English. Whisper will always transcribe from whatever source spoken language to the same language:

  ```
  !whisper "YOUR_NON_ENGLISH_FILE_NAME.mp3" --model small --task
  translate
  ```

- **clip_timestamps**: This allows for comma-separated list start, end, start, end,... timestamps (in seconds) of clips to process from the audio file. For example, use `--clip_timestamps` to process the first 5 seconds of the audio clip:

  ```
  !whisper "YOUR_FILE_NAME.mp3" --model small --clip_timestamps
  0,5
  ```

- **Controlling the number of best transcription candidates**: Whisper's `--best-of` parameter controls how many candidate transcriptions Whisper returns during decoding. The default value is 1, which returns just the top predicted transcription. Increasing to 3–5 provides some alternative options:

  ```
  !whisper "YOUR_FILE_NAME.mp3" --model medium --best-of 3
  ```

- **Adjusting temperature**: The `temperature` parameter controls the randomness in generation tasks such as translation. Lower values produce more predictable results:

  ```
  !whisper "YOUR_FILE_NAME.mp3" --model medium --temperature 0.5
  ```

- **Adjusting the beam size for decoding**: Whisper's `--beam-size` flag controls the beam search size during decoding. Beam size affects the accuracy and speed of transcription. A larger beam size might improve accuracy but will slow down processing:

  ```
  !whisper "YOUR_FILE_NAME.mp3" -model medium --temperature 0
  --beam-size 2
  ```

These advanced parameters allow you to fine-tune the Whisper AI transcription to your specific needs, improving accuracy and tailoring the output to your requirements. Experiment with these options to see which combination works best for your audio files.

## Summary

As we conclude *Chapter 1*, we have traversed a comprehensive path that laid the foundation for understanding and utilizing this advanced speech recognition system. Here are the milestones of our journey together.

We began our journey with a deep dive into the marvel of human vocalization, exploring the complex interplay of biology, emotion, and cognition in voice and speech production. This exploration was about understanding the physiological processes and appreciating the immense challenges technologies such as OpenAI's Whisper face in interpreting these uniquely human attributes. This understanding is vital for enjoying Whisper's capabilities and the sophistication required to transcribe human speech accurately.

Next, we delved into Whisper's key features and capabilities, which set it apart as a significant leap in the realm of ASR. Whisper demonstrates its robustness and versatility, from its exceptional ability to convert speech to text across nearly 100 languages and handle accents and background noise to its capacity for real-time transcription and support for a wide range of audio file formats. This section illuminated Whisper's transformative power in various applications, from journalism to international communications, showcasing its state-of-the-art performance and ease of integration into diverse projects.

Lastly, we explored the practical aspects of setting up and using Whisper through Hugging Face's web interface for a straightforward experience and via Google Colab for a more hands-on approach. This section provided a step-by-step guide to effectively use Whisper for transcribing speech to text, highlighting its accessibility and convenience for users.

Having reached the end of this chapter, you should have gained a comprehensive understanding of Whisper's functionalities and acquired the skills to apply this technology in various contexts. The knowledge and insights gleaned here are invaluable for anyone looking to harness the power of advanced speech recognition.

As we look forward to *Chapter 2, Understanding the Core Mechanisms of Whisper*, we prepare to delve into the nuts and bolts of Whisper's ASR system. This chapter will shed light on Whisper's critical components and functions, enhancing our ability to optimize its performance and implement best practices. Whether for voice assistants, transcription services, or other innovative applications, this foundational knowledge is essential for efficiently harnessing Whisper's capabilities. Prepare to deepen your understanding of how Whisper functions at a high level, dissect its components, and discover practical techniques for optimizing its performance.

# 2
# Understanding the Core Mechanisms of Whisper

Welcome to *Chapter 2* of our journey to mastering Whisper's groundbreaking speech recognition capabilities. In the previous chapter, we explored the value propositions of production-grade speech recognition and why Whisper marks a pivotal advancement in conversational AI.

Now, it's time to demystify the technology under the hood. This chapter offers a comprehensive yet accessible overview of Whisper's technical architecture and functions. Consider it your guidebook for navigating the ASR landscape as we dismantle Whisper piece by piece.

Our goals for this chapter are threefold:

- **Develop literacy** in the critical components of modern ASR systems, including Whisper's unique approach. We'll survey the techniques and data flows fueling today's speech recognition.

- **Cultivate intuition** around the systemic interactions that enable translating speech into text and downstream natural language understanding. We'll map the associations and data flows between crucial components, such as **acoustic models**, **language models**, and **decoders**, to reveal their intricate interdependence in the speech recognition pipeline. By tracing audio input through incremental processing steps and demonstrating how later stages rely on earlier ones, you'll organically grasp the cumulative effect of these interconnected systems working in symphony.

- **Enable optimization** by illuminating Whisper's internal mechanisms. Grasping Whisper's strengths, limitations, and trade-offs allows for the precise tuning of system configurations to achieve ideal accuracy, speed, and cost targets.

We won't dive into the complex math fueling innovations such as **recurrent neural networks** (RNNs) and **transformers**. Instead, we'll focus on digestible conceptual frameworks so you can hit the ground running applying Whisper. With technology demystification comes informed strategy and impact.

In this chapter, we will cover the following topics:

- Delving deeper into ASR systems
- Exploring the Whisper ASR system
- Understanding Whisper's components and functions
- Applying best practices for performance optimization

By the end of this chapter, you will understand the critical elements of Whisper's ASR system, dissect its components and functions, and learn practical techniques for optimizing its performance.

## Technical requirements

To harness the capabilities of OpenAI's Whisper for advanced applications, this chapter leverages Python and Google Colab for ease of use and accessibility. The Python environment setup includes the Whisper library for transcription tasks.

**Key requirements**:

- **Google Colab notebooks**: The notebooks are set to run our Python code with the minimum required memory and capacity. If the **T4 GPU** runtime type is available, select it for better performance.

- **Python environment**: Each notebook contains directives to load the required Python libraries, including Whisper.

- **GitHub repository access**: All Python code, including examples, is available in the chapter's GitHub repository (`https://github.com/PacktPublishing/Learn-OpenAI-Whisper/tree/main/Chapter02`). These Colab notebooks are ready to run, providing a practical and hands-on approach to learning.

By meeting these technical requirements, you will be prepared to explore Whisper in different contexts while enjoying the streamlined experience of Google Colab and the comprehensive resources available on GitHub.

## Delving deeper into ASR systems

What exactly happens behind the scenes when we talk to Siri or Alexa? How does a computer transform the ambiguous sounds of natural language into correctly identified words and phrases? Well, that is where ASR systems come in.

ASR is playing an increasingly vital role in our daily lives. ASR powers many interactions with technology, from smart speakers to voice assistants on our phones. It facilitates hands-free control, enables voice search, and supports other voice-driven functionalities. The rise of conversational AI, including chatbots and virtual assistants, depends heavily on accurate and efficient speech recognition.

ASR systems can identify and process human speech and convert it into machine-readable text. In other words, they transcribe spoken audio into written words. This technology enables voice interfaces and verbal communication with computer systems.

## Definition and purpose of ASR systems

On a basic level, ASR systems bridge the gap between human speech and machine understanding. Their role is to analyze an acoustic audio signal, identify the linguistic content, and output a textual translation that computers can process.

More specifically, here are the key objectives ASR solutions aim to achieve:

- **Convert audio to text**: The core purpose is transcribing spoken words into equivalent written text that software applications can intake. This text can then undergo further NLP.

- **Understand natural language**: Humans sometimes speak differently. We slur words, stutter, or talk over each other. ASR must handle these complexities and discern meaning from ambiguous audio.

- **Enable voice interfaces**: ASR powers the voice **user interfaces** (**UIs**) that allow us to interact with technology through speech. These UIs include voice assistants, smart speakers, and dialogue systems.

- **Improve accessibility**: For those with disabilities such as blindness or impaired motor function, ASR enables alternative input methods beyond keyboards or touchscreens. Voice control significantly expands accessibility.

- **Drive efficiency**: Automating speech transcription relieves humans of tedious audio/video captioning and documentation. ASR saves massive time and effort.

ASR delivers speech analytics that fuel **voice UIs** (**VUIs**), quantified self-applications, and other voice-driven interactions. As the demand for ubiquitous voice interfaces booms, improving ASR accuracy and capabilities remains imperative.

ASR allows us to communicate with machines as we do with other humans when implemented harmoniously with complementary technologies such as **natural language understanding** (**NLU**), dialogue management, and **text-to-speech** (**TTS**). This natural interaction paradigm is essential to the vision of an intelligent assistant in every home.

## ASR in the real world

Automated speech recognition already enables many common hands-free interfaces through a paradigm known as VUIs.

> **Voice user interfaces**
>
> VUIs allow people to interact with devices through conversational speech instead of touch, typing, or clicking. They comprise the speech recognition and NLU stacks, enabling systems such as Alexa and Siri to intake raw voice queries before responding intelligently. Effective VUIs combine ASR transcriptions with downstream dialogue systems to handle natural commands, questions, and instructions using only spoken utterances. This hands-free control paradigm powered by voice makes interacting with technology faster, easier, and more accessible.

While mostly invisible to users, ASR already enables many common scenarios through VUIs:

- **Smart speakers** such as Amazon Echo and Google Home rely on ASR to understand and respond to voice commands, allowing hands-free music playback, household control via the **Internet of Things (IoT)**, information queries, and more.

- **Virtual assistants** such as Siri, Alexa, Cortana, and Google Assistant use ASR to transcribe user queries. After speech recognition, they execute commands, answer questions, or make recommendations using downstream natural language and dialogue processing.

- **Captioning and documentation** tools employ ASR to rapidly transcribe videos, podcasts, lectures, medical reports, legal proceedings, and more.

- **Hands-free control** of smartphones, tablets, laptops, TVs, and in-vehicle infotainment happens through ASR **application programming interfaces (APIs)** that can navigate apps, input text, place calls, adjust volume, and so on via voice instructions rather than touchscreens.

- **Speech analytics** solutions extract insights from customer call transcripts generated via ASR to understand sentiment, trends, compliance, agent performance, and other metrics.

ASR thus plays a profound role in human-computer interaction. Its accuracy and robustness directly impact the user experience of many popular intelligent products and services. Under the hood, ASR feeds the voice-driven revolution through speech transcription capabilities that enable verbal system control and analytics.

After seeing the profound real-world impacts of modern ASR systems such as virtual assistants and smart speakers, one may wonder how we got here. What seminal breakthroughs in algorithms, data, and compute architectures catalyzed today's flexible and accurate speech recognition solutions? The following section will chart the rapid progression of core methodologies through pivotal eras that brought us to the cutting-edge innovations in the neural networks powering Whisper. Understanding this development arc provides context around ongoing challenges and remaining headroom as the field continues pushing further boundaries. Equipped with historical perspectives, we can better anticipate future directions amidst this technological Cambrian explosion.

# Brief history and evolution of ASR technology

The concepts behind automated speech recognition date back to the 1930s, when Bell Laboratories built machines to recognize digits spoken over the telephone. However, widespread commercial adoption of the technology we know today only occurred in the 1990s and 2000s.

After nearly a century of innovation, speech recognition capabilities have advanced enormously thanks to transformative approaches in machine learning and the availability of big data. The accuracy and versatility of ASR continue to progress at a remarkable pace.

## The early days – Pattern recognition approaches

The first significant wave of innovation in ASR came during the 1950s at Bell Laboratories. Researchers focused on isolated word recognition using heuristic techniques to match acoustic patterns by examining audio waveforms and identifying distinguishable speech components.

Bell Labs built specialized machines to interpret spoken digit sequences over the telephone. For example, callers could verbally provide a bank account number to route their request. These primitive Audrey systems represented early examples of pattern matching without modern machine learning.

> **Audrey systems**
>
> The Audrey systems developed by Bell Laboratories in the 1950s were pioneering speech recognition devices aimed at deciphering digits spoken over the telephone. They used analog circuits to match acoustic patterns to individual numbers, enabling the routing of calls based on verbally provided account or contact numbers. While limited in scope, these specialized machines represented some of the first attempts at ASR through template matching. Audrey marked an early milestone, though substantial innovation was still needed for more flexible systems that could handle continuous speech.

Over the following decades, researchers developed rule-based approaches using signal processing and acoustic fingerprinting. However, these methods struggled to fully accommodate the dynamic complexities of natural language and the variability of speech patterns across diverse speakers. They also were heavily burdened with extensive expert feature engineering, which was challenging to scale across languages.

This early progress was promising but needed more sophistication to handle continuous speech, diverse accents, environments, and vocabulary beyond a few words. More advanced techniques would be necessary to deliver today's flexible ASR.

## Statistical approaches emerge – Hidden Markov models and *n*-gram models

A paradigm shift occurred in the 1970s and 80s with the introduction of probabilistic modeling using **hidden Markov models (HMMs)** and *n*-**gram** language models.

> ### Hidden Markov models
>
> HMMs are statistical models that analyze sequences by modeling underlying states *hidden* from the observer. In ASR, they model generating speech sounds as transitions between hidden states over time, tracking the probability of particular phonemes or words occurring given previous acoustic cues. Rather than definitive rules, HMMs provide the computational framework for statistically handling speech ambiguities.

> ### *N*-gram language models
>
> *N*-gram language models calculate conditional word probabilities by examining historical sequences of 1–3 previous words. For example, a 3-gram model estimates the likelihood of each possible next word following every unique consecutive word pair. Language models can use these probability distributions to predict and refine interim ASR transcriptions into more probable phrases. However, *n*-grams fail to model longer-range contexts.

Rather than solely pattern matching, researchers adopted principles from *Bayesian statistics* to compute likelihood scores and make predictions under uncertainty. That approach enabled more graceful handling of the ambiguities and variances inherent to speech signals.

Using HMMs, researchers modeled speech components such as phonemes and words as Markov processes, allowing tracking of transitional probabilities from one sound to another. *N*-gram language models then predicted the following words based on previous word sequences. Combined with acoustic models, these key innovations could process continuous speech recognition for small vocabularies.

In the commercial realm, Dragon Systems launched its Dragon NaturallySpeaking dictation software in 1990 using HMM. That launch represented a significant milestone as one of the first large-vocabulary continuous speech recognition systems for **personal computers (PCs)**.

However, successful adoption faced challenges such as limited accuracy, lack of environment robustness, extensive training requirements, and little language context. Significant improvements would come in the following decades with neural networks and increased computational power.

## The deep learning breakthrough

While HMM/*n*-gram systems represented significant progress, they relied heavily on manual feature engineering and limited model capacities. In contrast, deep learning approaches could automatically discover intricate representations and patterns from raw data at scale.

The late 2000s saw the introduction of **deep neural networks (DNNs)** to speech recognition, delivering exceptional boosts in accuracy. Deep feedforward and recurrent networks overcame previous limitations using multilayered artificial neurons.

Then, in 2016, Microsoft achieved an industry milestone, reaching human parity in conversational speech recognition through extensive neural network training using their proprietary **Computational Network Toolkit (CNTK)** framework, now known as the Microsoft Cognitive Toolkit. The researchers reported a **word error rate (WER)** of 5.9 percent, equal to that of people who were asked to transcribe the same conversation. The key to Microsoft's success was using **long short-term memory (LSTM)** acoustic models combined with a novel spatial smoothing method and **lattice-free maximum mutual information (LF-MMI)** acoustic training. They also employed multiple RNN language models and large amounts of data, including Bing voice search logs, to train their DNNs. This data-driven approach allowed the system to learn from the variations and nuances in speech, improving its ability to recognize and transcribe spoken words accurately. Microsoft's breakthrough demonstrated capabilities on par with human listeners, unlocking new potential for commercial voice assistants and dialogue agents to reach new versatility and utility.

### Word error rate

WER is a standard metric used to measure the performance of a speech recognition or machine translation system. It is calculated as the ratio of errors in a transcript to the total words spoken. The errors are categorized into three types: substitutions (when a word is replaced with another), insertions (when a word that wasn't originally spoken is added), and deletions (when a word is omitted). The formula for WER is as follows:

*Word Error Rate = (Substitutions + Insertions + Deletions)/Number of Words Spoken*

For example, if there are 10 substitutions, 5 insertions, and 5 deletions in a transcript of 100 words, the WER would be 20%. A lower WER indicates better accuracy in recognizing speech. It's important to note that while WER is a widely used metric, it is not the only measure of the effectiveness of a speech recognition system.

### Long short-term memory

LSTM is a type of RNN designed to remember information for extended periods. Unlike traditional RNNs, which struggle to maintain long-term dependencies due to the vanishing gradient problem, LSTMs can learn these dependencies, making them well-suited for tasks involving sequential data with long-term temporal dependencies. This includes language translation, speech recognition, and time series forecasting applications. An LSTM network includes memory blocks, which are recurrently connected and contain one or more memory cells along with three multiplicative units, allowing the network to regulate the flow of information.

> **Lattice-free maximum mutual information**
>
> LF-MMI is a method used for sequence-level training of speech recognition acoustic models. The MMI objective function aims to maximize the mutual information between the observed acoustic features and the corresponding word sequences in the training data. The *lattice-free* aspect refers to the fact that this method does not require the generation of lattices (a type of graph used in traditional speech recognition systems) during training, which makes it more efficient and suitable for GPU-based training. LF-MMI has been shown to achieve state-of-the-art results on many speech recognition tasks, and it is particularly effective for training DNNs used in ASR systems.

State-of-the-art systems today use different neural architectures:

- Convolutional layers learn translation-invariant features directly from spectrograms.

- Recurrent layers include LSTMs, model speech sequences, and long-range context.

- Transformers capture global dependencies through self-attention, removing recurrence constraints.

In addition to excellent statistical foundations, these neural advances catalyzed the commercial success of virtual assistants and widespread voice interfaces.

## Ongoing innovations

ASR remains a highly active research area as we find new ways to improve flexibility, reduce latency, and enhance accuracy. Exciting innovations continue to emerge, such as the following:

- **End-to-end modeling**: Separate acoustic and language models have been replaced with single integrated networks, simplifying training and optimizing the entire pipeline.

- **Multimodal learning**: Audio, visual, and textual data can now be combined to improve robustness. Lip movements and other visual cues provide additional signals.

- **Personalization**: Models can be adapted to individual speakers' voices and accents for tailored performance. Unique vocal profiles enhance recognition.

- **Low-resource languages**: Progress in ASR for languages with limited training data is facilitated through *cross-lingual transfer learning*. That means high-resource languages (e.g., English) help bootstrap those languages with limited training data.

- **On-device deployment**: Thanks to model compression and acceleration hardware, real-time ASR is now available on mobile phones and edge devices.

Thanks to advancements in machine learning, ASR technology now delivers incredible utility after decades of iteration. With enhanced robustness to diverse speech and environments, broad language support, and scalable deployment, ASR promises to enable even more voice-driven experiences in the years to come. Whisper sits at the cutting edge, pushing boundaries ever further.

Most relevantly, Whisper represents a massive leap in applying state-of-the-art self-supervised learning to achieve human-level capabilities using only open datasets. This unprecedented accuracy and language coverage sets a new standard for production speech recognition systems.

## Exploring the Whisper ASR system

Now that we've surveyed the landscape and capabilities of automated speech recognition, it's time to demystify Whisper's technical inner workings. This section offers an accessible yet comprehensive overview of the algorithms, data pipelines, and innovations unlocking Whisper's unprecedented transcription abilities.

We'll highlight approaches in acoustic modeling, self-supervised pre-training strategies, model architectures, and performance optimizations that set Whisper apart. Collectively, these techniques enable robust real-world speech recognition across languages, environments, and hardware.

While we won't dig into granular mathematical equations, you'll develop an intuitive grasp of Whisper's competitive advantages, such as the following:

- Handling fuzzy sound-to-symbol mapping with **connectionist temporal classification (CTC)** acoustic models

- Incorporating global language context via transformers

- Streamlining low-latency deployments across devices

Understanding Whisper's mechanics empowers practical tuning for your targets around accuracy, speed, and cost. Architectural literacy breeds informed strategy.

> **Connectionist temporal classification**
> CTC acoustic models handle fuzzy sound-to-symbol mapping in speech recognition systems. These models are designed to handle the inherent uncertainty in aligning input sequences (such as audio frames) with output sequences (such as phonemes or words). This is particularly challenging in speech recognition because the boundaries between spoken words are unclear, and the same word can be pronounced differently depending on the context.

Let's dive in! We'll explore Whisper's fusion with CTC and transformers, as well as the integration of linguistic knowledge, to understand how it blends end-to-end and hybrid techniques. These are not separate discussions but rather interconnected aspects of Whisper's architecture.

## Understanding the trade-offs – End-to-end versus hybrid models

When architecting an ASR system, architects face a pivotal decision: end-to-end versus hybrid modeling strategies. This initial fork impacts everything from accuracy and speed to adaptability. Before implementing Whisper or any production-grade speech platform, developers should consider the critical trade-offs between the two paradigms below the surface.

> **Is Whisper an end-to-end or hybrid ASR system?**
>
> OpenAI's Whisper is an ASR system that uses an end-to-end approach. The Whisper architecture is implemented as an encoder-decoder transformer, where input audio is processed in a two-step process. First, it generates a mathematical representation of the audio and then decodes this representation into a sequence of text tokens. This process is characteristic of an end-to-end approach, where the entire task – from raw audio input to text output – is handled by a single, integrated model. Therefore, Whisper can be classified as an end-to-end ASR system.
>
> Some examples of hybrid ASR systems include the following:
>
> - *Kaldi-based ASR systems*: Kaldi is an open source toolkit for speech recognition research. It supports both DNN-HMM hybrid models and end-to-end deep learning models. It often uses LF-MMI training for its hybrid models.
>
> - The *Vicomtech Speech Transcription Systems*: These systems were used for the *Albayzín-RTVE 2020 Speech to Text Transcription Challenge* and likely included hybrid ASR components.
>
> - *Wav2Vec 2.0*: This was developed by Facebook AI and is a self-supervised learning model for speech recognition used in end-to-end and hybrid ASR systems. In the context of hybrid ASR models, Wav2Vec 2.0 can generate high-quality acoustic representations that can be used as input to a traditional ASR system, such as an HMM or a DNN.
>
> These hybrid systems are designed to effectively deal with the temporal variability in speech signals by leveraging neural networks' discriminative training capabilities and deep feature extraction while utilizing HMMs' robust sequence modeling.

In theory, end-to-end modeling seems ideal. A unified model optimizes the complete mapping of acoustic signals to transcriptions. This elegantly sidesteps glue code and intermediate representations. However, as we explore in this section, hybrid architectures still dominate industry systems due to other constraints such as latency, customization, and robustness.

*Figure 2.1* compares traditional hybrid-based and more recent end-to-end ASR systems. Both systems take air traffic controller (ATCO) voice communication as input and produce transcripts as output.

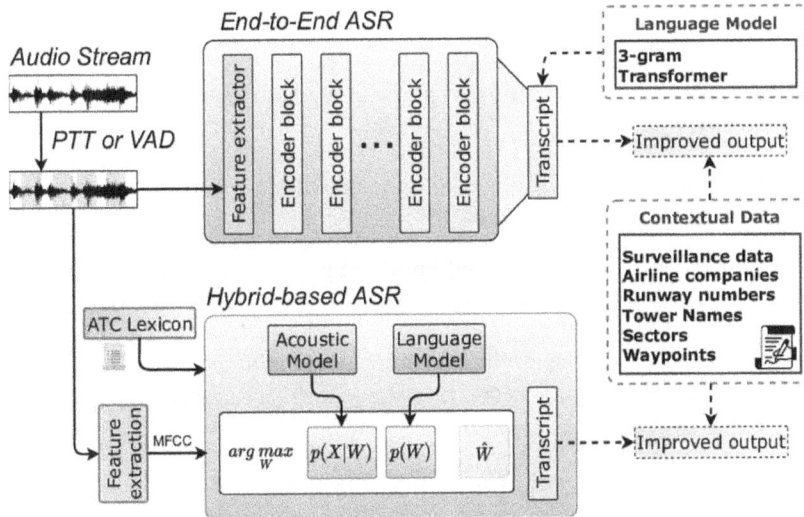

Figure 2.1 – Traditional hybrid-based and more recent end-to-end ASR systems (A Virtual Simulation-Pilot Agent for Training of Air Traffic Controllers - Scientific Figure on ResearchGate. Available from: https://www.researchgate.net/figure/Traditional-hybrid-based-and-more-recent-end-to-end-automatic-speech-recognition-systems_fig1_370961598)

In the traditional hybrid-based ASR system, the process begins with feature extraction from the input voice communication. This is followed by acoustic modeling, which maps the acoustic features to phonetic units. The next step is pronunciation modeling, which maps the phonetic units to words. Finally, language modeling is used to predict the probability of a sequence of words, producing the final transcript.

On the other hand, the end-to-end ASR system directly maps the input voice communication to the output transcript, bypassing the intermediate steps of acoustic, pronunciation, and language modeling. This simplifies the system and can potentially improve performance.

The figure also shows optional modules (represented by dotted blocks) that can be added to increase the system's overall performance. These include surveillance data or other data types such as sectors or waypoints. These additional data sources can provide context that helps the ASR system better understand and transcribe the ATCO voice communication.

In summary, *Figure 2.1* illustrates the differences between traditional hybrid-based and end-to-end ASR systems and shows how additional data sources can enhance their performance in the context of air traffic control.

There are no universally superior options for end-to-end versus hybrid-based ASR systems. The approach taken impacts adaptability, deployment requirements, and scalability. Let's analyze the critical considerations when determining the right strategic direction.

### Accuracy and output quality

For many applications, recognition precision is paramount. Architectural decisions significantly influence output quality:

- **End-to-end strengths**: Jointly trained components directly target the final objective: no suboptimal pipelines or disjoint errors.

- **Hybrid advantage**: Mix and match best-of-breed components (acoustic, linguistic). Customize the balance of errors.

In practice, hybrid models achieve state-of-the-art accuracy by combining an optimal acoustic model such as *Wav2Vec 2.0* with advanced transformer language models. Specialization beats generalization, but end-to-end models are quickly catching up.

Output quality also depends on factors such as the volume of training data, model size, and personalization techniques. But hybrid systems edge out end-to-end at scale, partially thanks to their customizability.

### Latency and throughput

Real-time recognition necessitates optimizing for low latency and streaming ASR calls for quick incremental outputs, not just batch offline decoding. End-to-end networks tend to be more extensive and computationally intensive without modular components. This introduces latency challenges for real-time systems:

- **End-to-end difficulty**: The joint model applies full sequence context. Outputs lag audio inputs.

- **Hybrid advantage**: Separate acoustic and language models enable streaming low-latency recognition.

That said, innovations such as convolutional neural architectures and model distillation continue improving end-to-end latency profiles. Cloud acceleration hardware mitigates computational constraints.

Throughput is less concerning since modern systems handle high volumes of audio streams in parallel. Overall, hybrid approaches currently achieve faster incremental outputs.

### Customization and control

Customizability allows tailoring ASR capabilities to specific domains such as medicine, law, or customer support centers. This requires interfacing separate speech and language components. End-to-end systems offer less flexibility to substitute specialized modules or inject domain knowledge:

- **End-to-end difficulty**: Entangled components limit modularity and custom inputs.

- **Hybrid advantage**: Mix and match acoustic, pronunciation, and language models.

The ability to swap modules makes it simpler to bias models toward specific vocabularies or formats in hybrid systems. This greater control and specialization increase applicability for niche use cases.

## Deployment requirements

Hardware constraints around memory, compute, and power consumption determine feasible deployment environments for ASR systems:

- **End-to-end difficulty**: Large, resource-intensive models strain edge devices.
- **Hybrid advantage**: Distribute pipelines across systems and specialized devices.

For example, researchers execute acoustic models on low-powered end devices while offloading language models to the cloud, effectively allowing split processing. Compression techniques such as *quantization* can shrink models, but hybrid systems provide more deployment flexibility.

In many ways, end-to-end modeling represents a philosophically purer approach. But hybrid systems make practical trade-offs, unlocking modular, customizable architectures. This positions them to deliver greater accuracy, lower latency, and more flexible deployments across diverse speech recognition applications.

Whisper charts an exciting path forward, unlocking many benefits of integrated end-to-end modeling while retaining a hybrid acoustic/language split. As research in conversational AI continues rapidly advancing, we may one day see end-to-end networks rivaling and even overtaking hybrid approaches thanks to sufficient data and computing. But for now, hybrid architectures rule production speech recognition thanks to their balance of quality, speed, and control.

As we just explored, Whisper strikes an optimal balance by blending end-to-end and modular components. Let's delve into the specific neural network architectures that Whisper unites, including CTC models and transformers, to create such a high-performing hybrid speech recognition pipeline.

# Combining connectionist temporal classification and transformer models in Whisper

Behind the scenes, Whisper fuses two powerful neural network architectures to unlock robust speech recognition:

- CTC, which excels at labeling unsegmented sequential data such as audio streams
- Transformers, which encode global dependencies in sequences via self-attention, capturing long-range linguistic context

This hybrid CTC/transformer scheme builds on decades of research into recurrent networks, computer vision, and NLP. The resulting system handles both aligned and unaligned speech with greater context. Let's explore Whisper's technical foundations.

## Connectionist temporal classification

The CTC algorithm, introduced in 2006, provides an elegant framework for transcribing unsegmented sequences. It can identify labels from raw streams without knowing alignment boundaries.

Speech recognition must handle continuous inputs with fuzzy transitions between words and sounds. Unlike text, audio signals don't come pre-split into semantic units. CTC learns to detect phonemes and transcripts directly from feature sequences.

CTC frames the labeling task to predict a probability distribution over all possible label sequences. An initial output layer emits label candidates for each timestep. Post-processing then consolidates these into the final, most probable transcript using dynamic programming.

For example, CTC's initial network layer might output a messier sequence such as "*th—e c—a—t s—a—t.*" Repeated labels and blanks collapsed into the final prediction: "*the cat sat.*"

CTC's algorithmic approach essentially *warps* predictions onto the correct ground truth sequence. Powerful acoustic models such as *Wav2Vec 2.0* now leverage CTC to frame speech recognition as a sequence transduction problem solved by deep learning.

## Transformers and self-attention

Transformers were introduced in 2017 for machine translation. They deliver breakthrough results by modeling sequences in radically new ways. Rather than recurrence and convolutions, transformers process inputs using multiheaded self-attention.

This mechanism calculates representations of sequence positions by relating them to every other element. Models learn which contextual relationships matter most to focused tasks such as translation.

For example, a transformer translates a sentence by heavily weighting the essential self-attentions for generating the next output word based on the inputs. This gives it a global purview of long-range dependencies unavailable to RNNs and **convolutional neural networks** (**CNNs**).

Transformers now underpin state-of-the-art NLP across machine translation, question-answering, dialogue systems, and more. Models such as *GPT-4* reveal their excellent linguistic abilities.

## Uniting CTC, transformers, and speech

Modern ASR systems blend these advanced neural architectures. CTC handles unlabeled audio streams with fuzzy sound alignments, and transformers encode robust language representations and output text corrections.

Specifically, Whisper infuses transformers after the CTC acoustic model during decoding. This two-step pipeline maximizes both auditory and linguistic learning:

**Two-step pipeline in Whisper ASR system**

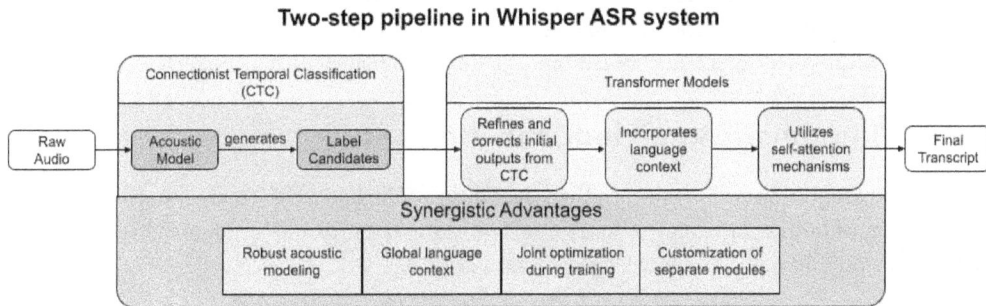

Figure 2.2 – Two-step pipeline in the Whisper ASR system

The CTC model first generates label candidates from raw audio. This handles the fuzzy sound-to-symbol transduction challenges. Downstream transformers then refine and correct the initial CTC outputs by better incorporating language context. Human speech often deviates from formal textual language, so additional language-specific conditioning is needed to improve transcript accuracy.

Jointly optimized during training, this architecture fits language structure onto imperfect acoustic outputs. Whisper bridges the auditory and linguistic domains to handle real-world speech recognition.

In summary, fusing CTC, transformers, and speech unlocks synergistic advantages:

- Robust acoustic modeling from CTC specialization
- Global language context from transformer self-attentions
- Joint optimization between all components
- Customization of separate modules

Together, CTC and attention mechanisms provide the best of both worlds. Whisper banked on their complementary superpowers to drive state-of-the-art capabilities. This technical combo meal fuels Whisper's reliability and accuracy in recognizing speech in the wild. The all-neural design also simplifies training by co-optimizing the entire pipeline end-to-end.

Expect transformer architectures to dominate as foundations for advancing conversational AI. Combined with complementary specialization techniques, as shown in Whisper, their flexible modeling capacities unlock ever-improving language-aware speech recognition systems.

Now that we have explored how Whisper combines the strengths of CTC and transformer models to handle the acoustic challenges of transcribing speech signals, we are ready to examine the other half of the equation – integrating linguistic knowledge for translating signals into coherent language. After all, accurate speech recognition requires more than precise acoustic signal decoding – outputs must conform to the constructs and conventions of natural languages such as English to be usable.

## The role of linguistic knowledge in Whisper

Robust speech recognition requires more than decoding audio signals. Systems must account for the complexities of human language to handle real-world variability and ambiguity. This is where integrating meaningful linguistic knowledge becomes critical.

State-of-the-art solutions such as Whisper enhance accuracy by combining acoustic predictions with **language-specific conditioning**. Language models provide the statistical probabilities of potential word or phoneme sequences based on the language's constructs. This allows for selecting the most likely text transcript fitting the acoustic signal among multiple guess candidates.

Language-specific conditioning then adapts the models to the characteristics and conventions of the target language, including the following:

- Vocabulary – the valid words and lexical constructs
- Grammar – how words fit together into phrase structures
- Pronunciation modeling – the plausible speech sounds and patterns
- Non-native pronunciation challenges – adapting to accents of second-language speakers
- Dialects – handling different global dialects

By tailoring to these linguistic attributes, Whisper develops an enriched understanding that facilitates recognizing language elements robustly, from core sounds to semantics. The customization allows for the graceful handling of real-world speech complexity.

---

**Language-specific conditioning**

Language-specific conditioning makes transcriptions more precise by resolving acoustic uncertainties, such as reducing confusion between homophones such as "they're," "their," and "there." The ASR system models probability distributions from potential words and the context provided by language-specific conditioning. For example, **phonotactic constraints**, which describe the allowable combinations of phonemes in a particular language, can guide the ASR system, especially when acoustic cues are missing or distorted. **Semantic analysis** can also bias the speech recognizer toward sentences appropriate to a particular task or domain and away from meaningless sequences of words.

---

Let's explore the various facets of language-specific conditioning that empower Whisper's supervised and semi-supervised training.

## Vocabulary encoding

Recognizing speech requires mapping audio to semantic symbols such as words or subword units. Whisper encodes vocabulary from diverse textual datasets spanning web pages, books, code, and more.

At the time of writing, Whisper's latest iteration, called `large-v3`, was released on November 6, 2022. The model was trained on 1 million hours of weakly labeled audio (weakly supervised pre-training) and 4 million hours of pseudo-labeled audio collected using `large-v2`. Whisper's weakly supervised pre-training process engraves **parseable language constructs** into the model. These constructs are essentially patterns and structures in the language that the model learns to recognize and reproduce. This learning process is not as precise as fully supervised learning, but it provides enough information for the model to learn effectively. When the model is later fine-tuned on downstream speech recognition tasks, these learned language constructs transfer to the new tasks. This means the model can apply the patterns and structures discovered during pre-training to recognize and transcribe speech in the downstream tasks. In essence, the weakly supervised pre-training process allows Whisper to learn a broad understanding of language vocabulary from a large and diverse dataset and then apply this understanding of vocabulary to specific tasks during fine-tuning.

Whisper's language-specific conditioning and vocabulary encoding are key to engraving nuanced statistical representations around lexical and phrasal *shapes* in the language model. By having encoded permissible linguistic forms and structures, Whisper can segment continuous speech signals and selectively surface plausible word candidates matching learned vocabulary patterns. This linguistic familiarity helps resolve uncertainty during acoustic decoding by restricting outputs to plausible lexical selections that align with the encoded vocabulary. In other words, by thoroughly modeling the shapes and contours of a language's lexical norms, Whisper can smoothly map noisy speech signals to valid textual candidates that agree with its vocabulary knowledge.

## Grammars and structures

Beyond individual words, processing natural language requires encoding permissible grammatical patterns. Structures such as part-of-speech sequences (e.g., noun→verb→adverb) and multiword phrases (e.g., "on the other hand") constitute allowable syntax.

Whisper's pre-training exposures ingest common English language text structures across genres such as news articles, literature, emails, code, etc. The diversity captures constructions that are both simple and complex.

Resulting language models steer outputs toward valid utterances. For example, grammatical knowledge informs the proper expansion of abbreviations and acronyms based on context (e.g., knowing NASA refers to the National Aeronautics and Space Administration). This goes beyond basic vocabulary familiarity.

Structured language representations reduce false positive transcripts that fail to conform to accepted grammar and phrasing conventions.

## Pronunciation modeling

Humans pronounce words differently across regions and contexts. Modeling diverse accents, speech impediments, coarticulation, and other variabilities improves recognition.

Whisper demonstrates remarkable adaptation to unique pronunciation styles. Its self-supervised pre-training leverages audio narration data containing diverse voices.

Exposure to different speakers teaches the correlations between raw acoustic signals and their associated words, regardless of minor variations. Patterns still emerge around customary pronunciation deviations.

By ingesting many voices, Whisper builds acoustic-linguistic connections resilient to reasonable deviation. This gives decoding flexibility without excessive brittleness.

## Non-native pronunciation challenges

However, modeling fluent non-native speech poses added challenges. Second-language speakers learn pronunciation patterns that can deviate more significantly than others from native norms.

For example, Mandarin Chinese speakers consistently substitute /l/ for /r/ sounds when speaking English. Other syllabic mismatches trip up language learners in relatively systematic ways.

Handling these non-native patterns requires even more diversity during training to capture a long tail of accents. Thankfully, Whisper's self-supervised pre-training leverages English narration data from international sources.

The model encodes correlations between common second-language speech deviations and correct standard transcripts. This exposure teaches associations despite incorrectly pronounced words or misordered *visemes*.

The result is a more globally relevant system forgiving non-native, accent-influenced speech. Whisper demonstrates marked gains in recognizing learners compared to previous benchmarks lacking sufficient dialectal range during training.

## Dialectal fluency

Beyond pronunciation, more considerable dialectal differences characterize groups speaking the same root language, whether regional British English or Singaporean English; supporting diverse dialects requires dialect-tuned modeling.

Whisper was fed English data spanning international publications, books, web articles, and more. This molded inclusive dialect fluency beyond solely American English. The vocabulary, grammar, and phrasing encapsulate diverse English dialects.

Exposure to this breadth allows it to adapt gracefully to users worldwide. Performance remains strong without solely overfitting to a single flavor of English. The model generalizes across dialects. This dialectal dexterity prevents fragmented accuracy across geos and usage domains. Whisper aims for dialect-agnostic fluency in its linguistic foundations.

Whisper can easily handle real-world speech complexity by fusing speech recognition with multifaceted language knowledge. Its unprecedented vocabulary capacity, dialectal range, and syntactic mastery enable the decoding of extraordinarily diverse audio with precision and recall across domains.

You'll soon understand everything from acoustic feature extraction through language model decoding and refinements. This systemic view connects dots across the modules powering Whisper's end-to-end pipeline.

Equipped with architectural knowledge, we'll optimize configurations for your specific use case constraints around precision, latency, and cost. But first, let's decompose Whisper into its critical sub-systems.

## Understanding Whisper's components and functions

Now that we've demystified Whisper's architecture and optimized design, it's time to dive deeper into its functional components. This critical section dissects the modules powering Whisper's speech recognition pipeline from audio ingestion to text output.

We'll survey the processes involved in converting spoken utterances into machine-readable transcripts. We aim to develop systemic intuitions about how Whisper's parts cooperate fluidly to handle real-world speech translation challenges at scale.

While mathematical complexities operate under the hood, you'll gain accessible clarity around the following:

- Preprocessing of raw audio signals
- Encoding of acoustic patterns
- Modeling of language
- Searching for output spaces
- Refinement of transcripts

Understanding these functional pieces grants intuition for tweaking configurations and components toward your use case constraints. Architectural literacy breeds strategic optimization. Minor tuning adjustments may yield dramatic gains.

Let's start unfolding Whisper's components like a complex Swiss watch.

# Audio input and preprocessing

The journey from speech to transcription begins with audio input. Whisper ingests raw waveform signals via microphones or other audio capture sources. This analog audio then undergoes frontend processing, including noise filtering and digitization, to extract clean features and encode the verbal content.

Understanding Whisper's audio handling stages is crucial for configuring suitable data collection pipelines. We must feed the system quality inputs, emphasizing linguistic essence rather than distracting characteristics.

Let's explore the role of audio input hardware, preprocessing considerations, and Whisper's feature extraction process, which sets the critical foundation.

## Audio input sources

High-performance speech recognition requires quality signals from the start. While ambient acoustic environments differ, ideal audio capture equipment for Whisper includes the following:

- **Microphones**: Dedicated microphone hardware with crisp, directional inputs ensures the capturing of clear speech from users. Consumer devices often have inadequate mic components that are unable to isolate voices. Prioritize lavalier microphones or mic arrays focusing on speaker voices over environmental noise.

- **Near-field sources**: When possible, minimize interference by positioning microphones near target speakers. This prevents contamination from far-field sounds. Have users speak directly into devices.

- **Low noise conditions**: Seek quiet indoor settings without disruptive background noise. Echoey rooms also distort audio – whenever possible, record speech in sound-dampened environments through acoustic paneling.

## Preprocessing and filtering

Before analysis, raw audio requires preprocessing to improve signal quality and extract critical characteristics. Whisper applies the following crucial adjustments:

- **Noise reduction**: Environmental sounds such as humming appliances degrade performance. Adaptive filters identify and subtract predictable background noise spectral profiles.

- **Gain normalization**: Variations in recording volumes should get normalized to a standard intensity range – loudness alone conveys no linguistic meaning. Target -20 to -10 dBFS for clear but uncompressed speech peaks.

- **Frequency equalization**: This balances relative energy distribution across low-, mid-, and high-frequency bands and sharpens acoustic signatures of speech components such as consonants and vowels for better perception by algorithms.

### Audio feature extraction

The final frontend step extracts informative numeric representations of the audio data through signal processing techniques before feeding Whisper models. Essential feature extraction methods include the following:

- **Spectrograms**: Visually map the signal energy across audio frequencies over time. Differences reveal speech components.

- **Log-Mel filter banks**: Mimic the human auditory system's frequency perception by compressing and smoothing critical bands.

- **Mel-frequency cepstral coefficients (MFCCs)**: Statistically compress frequency data into most variant latent dimensions via MFCC transformers.

Together with pixel-like frame sequencing across time, these high-level features encode audio in model-consumable tensors while denoising. The resulting compact preprocessing captures core speech essence to inform acoustic modeling. Getting this front end right ensures Whisper gets clean data.

Next, we'll see how Whisper leverages the outputs for decoding speech components into text.

## Acoustic modeling

The next phase is acoustic modeling after preprocessing audio into informative feature representations. This converts low-level speech signals into higher-level linguistic units through statistical learning – the first step in translating sounds into language symbols.

Acoustic modeling uncovers and encodes the correlations between raw speech audio patterns and corresponding textual artifacts such as words, phonemes, or characters. Models capture the systematic relationships between the following:

- Spoken sounds
- Word spellings
- Language semantics

By mathematically representing these associations, systems such as Whisper decode speech waveforms into probable transcriptions, bridging the acoustic and linguistic domains.

### Speech units for acoustic modeling

Ideally, acoustic models would directly translate waveform signals into complete transcripts. However, reliably modeling such complex conditional probabilities requires massive datasets covering all variations. Instead, architects insert intermediate steps tying acoustics to smaller constructive units:

- **Phoneme-level modeling**

  HMMs historically decoded audio into constituent phonemes as an interim output for downstream processing. However, this requires preemptively segmenting speech signals, which relies on prior acoustic understanding.

- **Character-level modeling**

  Modern end-to-end architectures such as Deep Speech translate acoustics straight into characters. But naively focusing solely on characters risks losing sensitivity to higher-level constructs such as words and phrases.

Whisper leverages a middle ground—modeling customized subwords as the target acoustic conditioning labels. These data-driven lexical chunks strike a balance between atomic signals and complex phrases, allowing both sonic details and linguistic structures to shine through.

### Whisper's acoustic model architecture

Using a CTC loss function, Whisper's acoustic model fuses causal convolutional layers with recurrent transformers. This unique combination handles local audio patterns while learning longer-term dependencies:

1.  Convolutions detect localized acoustic patterns associated with character sequences, simultaneously operating on small raw audio windows. Different filters learn various speech attributes.

2.  Recurrent layers then contextualize the sequential convoluted representations over more considerable periods using transformers. Attention distributions relate current audio to previous chunks to handle contiguous signal dynamics from individual sounds to complete multiword utterances.

3.  The CTC loss function provides training supervision, bridging lower-level audio patterns with the target unit labels such as subwords. Alignments get handled implicitly.

Whisper's acoustic model architecture provides a powerful deep neural map that directly converts acoustic signals into linguistic constructs for downstream interpretation. This forms the essential sonic-to-semantic foundation for speech recognition.

## Language modeling

The acoustic model handles the first phase, translating speech audio into linguistic symbols. But making sense of those symbols requires understanding language rules around vocabulary, semantics, grammar, and more. Enter language models – Whisper's context experts.

While acoustic models decode audio signals, language models interpret symbol sequences, providing context around permissible utterances. They score and refine interim transcriptions from upstream acoustics using statistical patterns and innate rules seen during training.

This entails disambiguating words based on probable language structure. For example, language models know that "The clouds are in the ____" more likely fits "sky" than random gibberish – models narrow uncertainty by assessing plausibility.

## Whisper's transformer language model

Whisper employs a standalone *transformer-based language model* that operates on interim acoustic model outputs. The transformer contains learned representations around the statistical relationships between words and multiword lexical chunks in English. It models complex linguistic contexts using stacked self-attention layers relating current symbols to surrounding ones.

Specifically, Whisper's language model leverages **masked language modeling** (**MLM**) for predicting randomly hidden words within a given context. This approach allows the model to make inferences based on the visible context and learning patterns.

> ### Masked language modeling
>
> MLM is a training technique used in NLP where some portion of the input data is intentionally masked or hidden during training, and the model is tasked with predicting the masked words based on the context provided by the unmasked words. In the context of Whisper and other ASR systems, MLM can be used to train the model to better understand the structure and semantics of the language in which it is transcribing. In essence, MLM allows the model to learn the underlying structure of the language and improve its ability to transcribe speech accurately, even in challenging conditions such as noisy environments or when dealing with accents or technical language.

By assessing possible transcriptions from acoustics against this robust understanding of language conventions, the model rescores and refines outputs for coherence. Fluency and semantic precision improve dramatically.

> ### Advantages over *N*-grams
>
> Historically, speech systems modeled language using simple historical *n-gram counts* – the probability of each word following observed sequences. For example, 3-grams encodes the likelihood of every word given every unique preceding word pair.
>
> However, these Markovian models fail to exploit longer-range context and structural intricacies. Human language has more complexity than truncated historical statistics.
>
> Conversely, transformer language models learn holistic representations where every symbol gets contextually related to surrounding ones. There are no independent assumptions. Transformers handle intricacies such as hierarchical phrase structures that *n*-grams miss.
>
> This gives Whisper an enriched awareness of language behavior when refining acoustic transcriptions into valid, coherent text results. Powerful modern language modeling handles complexity beyond surface statistics.

Next, we'll explore how all the upstream components come together during the decoding phase to generate final speech recognition outputs.

## Decoding

Let's recap the previous pipeline stages before proceeding further into decoding. The **audio input and preprocessing** phase converts raw waveforms into informative numeric representations, extracting relevant audio features. Next, the **acoustic modeling** stage predicts linguistic label outputs such as characters or subwords from the audio features, creating a set of estimating label sequences. The next stage, **language modeling**, assesses and re-scores the acoustic model's initial label sequence predictions for greater coherence by incorporating contextual knowledge, resulting in interim transcriptions.

So, in summary:

- **Audio preprocessing** provides input features.
- **Acoustic modeling** makes initial label sequence predictions.
- **Language modeling** rescores those interim outputs.

Now, let's understand how the decoding stage searches for the optimal text transcription fitting both the acoustic and language guidance.

After extracting speech features, estimating label sequences, and scoring interim transcriptions, the final phase generates optimal text outputs – a process called decoding. This inference stage combines all upstream components.

The decoder takes acoustic model label predictions and finds the best corresponding word sequences based on language model guidance. Efficient search is critical for navigating the exponentially ample space of possible transcriptions.

Let's explore Whisper's decoding approach.

### Beam search

OpenAI employs **beam search** – a fast heuristic algorithm that approximates the most likely sequences while pruning unlikely candidates. This focuses computations on the most promising decoded text results.

**Beam search example**

Here's a simplified example of how beam search might work in Whisper.

Let's say we have an audio input that says "Hello, world" and we're using a beam width of two (meaning we keep the two most likely sequences at each step).

At the first step, the model might predict that the most likely first words are "Hello" and "Yellow" based on the acoustic features of the audio input.

In the next step, the model considers both sequences' extensions. It might predict that "Hello, world" and "Hello, word" are the most likely continuations of "Hello" and that "Yellow world" and "Yellow word" are the most likely continuations of "Yellow." The model then compares these sequences and keeps the two most likely overall. Let's say it keeps "Hello, world" and "Yellow world."

This process continues until a stopping condition is met. In the end, the model might output "Hello, world" as the most likely transcription of the audio input.

It's important to note that this is a simplified example, and the actual process involves complex calculations of probabilities based on the model's learned parameters. Also, the beam width can be adjusted to trade-off between computational efficiency and transcription accuracy.

Beam search incrementally builds up partial transcriptions one token at a time, retaining only the top candidates at each step based on conditional model scores. Words get added to active hypotheses in order of probability.

By discarding lower-scoring chains that are unlikely to maximize the final objective, searches remain tractable without exhaustively analyzing all options. The beam width determines processing breadth. Wider beams improve accuracy at an efficiency cost.

Whisper leverages dynamic beam pruning for optimal trade-offs. Beam sizes expand and contract based on interim confidence scores. More candidates get retained during uncertain segments before being narrowed as clarity increases.

## Rescoring and re-ranking

After generating candidate transcripts, Whisper rescores outputs using heavier processing for further gains. Secondary evaluation better incorporates richer context missed initially. To further optimize accuracy, Whisper applies additional techniques such as the following:

- *N*-best list re-ranking, which takes the top hypotheses and reorders them after evaluating with the more prominent language model rather than the fast approximator used during beam search. This boosts precision.

- LSTM rescoring, which goes beyond re-ranking; this technique involves feeding acoustic outputs into auxiliary LSTM networks, which act as alternative decoders. This captures different speech nuances missed by the baseline models.

Combining distinct search, scoring, and decoding strategies allows Whisper's overall pipeline to correct itself – a hallmark of deep learning system design. No single method holds a monopoly on performance.

Next, we'll look at the final phase, which is focused on postprocessing these decoded results to prepare cleaned machine-readable text for downstream usage.

## Postprocessing

After decoding audio into final text transcripts, Whisper applies various postprocessing approaches to further polish and structure outputs – the final step before surfacing recognized speech.

Postprocessors handle formatting, accuracy optimization, entity linking, and more. This critical yet often overlooked pipeline stage completes the speech-to-language transition, transforming rough decoded text into consumable, actionable information. Let's explore some of Whisper's key postprocessing capabilities.

### Text normalization

First, the raw decoded text gets normalized into properly written language conventions. Text normalization is crucial in converting raw decoded text into a format that adheres to appropriate written language conventions. Here are examples of how text normalization is applied:

- **Numerals are expanded into words**: For instance, the numeral "2023" in the text would be expanded to "two thousand twenty-three" to make it more readable and understandable when converted to speech.

- **Disfluencies such as filler utterances are removed**: Disfluencies such as "um" or "uh" might be present in a speech transcription. These would be removed during text normalization to create a cleaner, more fluent written representation of the spoken content.

- **Punctuation and capitalization are standardized**: Text normalization ensures that punctuation marks are correctly placed and words are appropriately capitalized according to the rules of written language. For example, the beginning of sentences would be capitalized, and periods or commas would be added where necessary to reflect the natural pauses and ends of sentences.

By transforming literal transcripts that reflect actual spoken cadences into a format with a natural reading flow, text normalization preserves the meaning while enhancing readability and the overall user experience.

### Accuracy optimization

Next, to further boost integrity, detected errors get automatically corrected using auxiliary models trained to identify and fix common mistakes:

- Homophones such as *"they're"*/*"there"*/*"their"* are rectified.

- Redundancies such as "*the the*" are fixed.

- Common substitutions are handled through confusion matrices (e.g., correcting frequent "*pat*"/"*bat*" mixups).

Together with a final grammar check, precision refinement networks learn correction patterns from human-edited transcripts.

## Entity linking

Whisper's ability to understand speech goes beyond mere text output. It involves the crucial step of entity linking, which connects the decoded words to their real-world references. Grounding words in data is the key to unlocking contextual understanding.

When Whisper encounters a brand name in the speech, it doesn't just transcribe the words; it maps them to canonical IDs in a knowledge base. This linking lets Whisper understand the brand's context, products, and marketplace. Similarly, when a person is mentioned, Whisper employs facial recognition to identify the individual, linking the name to a rich information profile.

Geographic references are another area where entity linking shines. By connecting location names to geographic databases, Whisper can understand the spatial context of the speech and associate the mentioned place with its coordinates, population, and other relevant data points.

This grounding of words in data empowers Whisper to comprehend speech as a sequence of words and a network of interconnected concepts. It can draw upon the linked information to interpret the meaning and context of the speech more accurately. Entity linking is thus a critical component in Whisper's ability to bridge the gap between raw speech and true understanding.

## Structured output

Finally, Whisper structures recognized content using semantic schemas tailored to target use cases:

- **Annotating questions for a conversational response**: Suppose a user asks a voice assistant, "What's the weather like today?" Whisper can annotate this input as a question, allowing the voice assistant to generate a conversational response such as, "The weather today is sunny with a high of 75 degrees."

- **Flagging commands to trigger actions**: If a user says, "Set an alarm for 7 AM," Whisper can flag this as a command. This flag tells the voice assistant to set an alarm rather than transcribing the speech.

- **Marking keywords for content analytics**: In a business meeting, a participant might say, "Our Q1 revenue exceeded expectations, but we need to improve our marketing strategy for Q2." Whisper can mark "Q1 revenue," "exceeded expectations," and "improve marketing strategy for Q2" as keywords. These keywords can then be used for content analytics, helping the business to track important topics and trends in their meetings.

This output framing eases downstream consumption, indexation, and learning.

Getting the last mile of postprocessing right prepares Whisper's speech recognition for real-world application. The decoder transcribes audio signals. Postprocessors transcode those signals into usable, accessible language.

And with that final postprocessing phase, we've now covered the whole gamut of Whisper's speech recognition pipeline – from audio input handlers to acoustic classifiers, language models, decoders, and output refiners.

When woven together, these components ingest spoken natural language and systematically translate signals into precise text transcripts consumable by downstream applications.

We've built strong intuitions around the data flow across the modules, giving Whisper unprecedented accuracy and speed at scale. Understanding these mechanics opens optimization pathways.

Now equipped with architectural knowledge, we're ready to shift focus toward tailored configuration, troubleshooting, and advancement of Whisper implementations based on infrastructure constraints and use case targets.

Our next section will cover specialized guidelines around rightsizing and accelerating Whisper for your success scenario while upholding accuracy, availability, and efficiency standards.

## Applying best practices for performance optimization

Now equipped with a solid grasp of Whisper's architecture and data flows, we're ready to shift focus toward real-world deployment. This pivotal section distills fundamental guidelines, trade-offs, and operational wisdom, accelerating production success.

We'll cover specialized topics beyond basic setup – from strategically provisioning infrastructure to monitoring metrics, integrating downstream NLP, tuning configurations, and troubleshooting common incidents. Consider this your handbook for effectively scaling Whisper-based solutions.

While fundamentals provide strong bases, intricacies of the environment determine outcomes. By tailoring and streamlining system-wide stack configurations to your context, we unlock next-level reliability, efficiency, and **return on investment (ROI)**. Specifically, this involves steps such as the following:

- Optimally allocating cloud, edge, or on-device compute
- Balancing data pipelines without congestion
- Tuning accuracy and latency for use case needs
- Smoothly interoperating with adjacent workflows
- Rapidly addressing anomalies

Let's prepare implementations for the demands of real-world conditions!

## Understanding compute requirements

Compute provisioning proves foundational when deploying performant Whisper-powered applications. The allocated CPU, GPU, memory, and storage resources directly impact throughput, latency, and concurrency.

Unfortunately, Whisper's scale leads many to underestimate its production infrastructure needs. At over 500 million parameters, the model requires significant hardware acceleration to deliver real-time speech recognition across users.

This section provides compute guidelines for streamlining Whisper deployments. We'll demystify its architecture considerations from on-device endpoints to cloud-accelerated requests. Target the optimal infrastructure fit.

Whisper's core workload involves matrix multiplications during neural network inferencing. These operations stress different underlying hardware components such as the following:

- **GPUs** accelerate deep learning matrix calculations in parallel, handling hundreds of operations simultaneously – their thousands of cores suit ML numerical processing. For Whisper, GPUs drive faster acoustic model inferencing.

- **CPUs** also Cprovide parallelization but are optimized for general-purpose branching logic over specialized math. Whisper relies on CPUs for audio decoding, beam search, and language model computations.

- **Memory** fuels model parameters and audio inputs. Whisper demands GBs for state storage and data transfers between processing units. High bandwidth reduces transfer bottlenecks.

- **Storage** holds pre-trained weights and buffers prediction outputs. High throughput *NVMe* SSDs manage heavy read/write Whisper workloads.

The complementary notebook `LOAIW_ch02_exploring_audio_data_workflows.ipynb` (`https://github.com/PacktPublishing/Learn-OpenAI-Whisper/blob/main/Chapter02/LOAIW_ch02_exploring_audio_data_workflows.ipynb`) provides further details on compute considerations for Whisper, including architecture trade-offs for on-device, edge, and cloud deployment.

Balancing these resources prevents systemic bottlenecks that slow down performance. Carefully consider the entire stack.

Considering these resource demands and hardware capabilities, we now explore specialized optimization guidelines tailored to distinct endpoint targets.

## Optimizing the deployment targets

Optimizing the deployment of OpenAI's Whisper model across various environments requires a strategic approach that considers each target platform's unique demands. Here's an intermediate-level explanation of best practices for deploying Whisper in different contexts.

### On-device deployment (phones and IoT devices)

For deployment on mobile phones and IoT devices, the focus should be on efficiency due to the limited computational resources. Whisper models should be selected based on the balance between size and accuracy. The quantized 40 MB model on-device Whisper inference on Android mobile using `whisper.tflite` (`https://github.com/openai/whisper/discussions/506`) is an example of a lightweight model suitable for such devices. Leveraging platform-specific neural accelerators, such as Apple's Neural Engine or Google's Edge tensor processing unit (TPU), is crucial for maximizing performance. Memory management is also critical, as these devices have limited RAM, so developers must ensure that the Whisper model does not exhaust available memory.

### Edge server deployment

Edge servers are private nodes that can offer model containment while providing scalability akin to cloud infrastructure. For edge deployment, it's advisable to use high-core CPUs and deep memory buffers to handle the computational load. Load-balanced GPUs can accelerate inference tasks, and low-latency storage solutions such as NVMe SSD clusters can improve the system's overall responsiveness. *Whispering* (*Whispering: Joint Service Offloading and Computation Reuse in Cloud-Edge Networks* - `https://www.ncbi.nlm.nih.gov/pmc/articles/PMC8528222/`) involves computation reuse at the edge, which can significantly reduce task completion times by avoiding redundant computations.

### Cloud infrastructure deployment

In a cloud environment, virtual machines offer the flexibility of scaling resources as needed. For Whisper, selecting machine images optimized for machine learning, such as AWS's *inf1* instances equipped with *Inferentia* chips, is beneficial, as they can provide cost-effective, high-performance inference. Autoscaling groups are essential for managing variability in demand, ensuring that resources are scaled up during peak times and scaled down when demand wanes to control costs.

### General considerations

Regardless of the deployment environment, analyzing traffic, performance benchmarks, and budgets is essential for planning effectively. Overprovisioning leads to unnecessary expenses, while underprovisioning can degrade service quality. The balance between cost and performance is crucial. Monitoring tools and performance metrics should be in place to ensure that the deployment meets the required service levels and to facilitate scaling decisions.

In summary, the best practices for deploying Whisper across different environments involve selecting the right model size, leveraging specialized hardware accelerators, managing memory efficiently, and using cloud resources judiciously. It's also important to consider the trade-offs between latency, cost, and performance metrics to ensure an optimized deployment that meets each environment's specific needs.

With the established infrastructure, we'll explore specialized practices for effectively routing the torrents of data coursing through Whisper's pipelines during inference.

## Managing data flows

Managing data flows in AI systems, particularly in the context of OpenAI's Whisper, involves several best practices that ensure efficient and effective operation. These practices are crucial for handling the intricate queues, caches, buffers, micro-batches, parallel streams, and competitive resource scheduling that constitute the nervous system of AI data movement.

Let's explore various techniques for optimizing these AI data flows, including strategic coordination, routing, and scaling.

### Understanding the fundamental data types and flows

Whisper's core data transformations involve several key data types:

- **Audio streams**: These are segmented into chunks from recording devices, with metadata attached for tracing across asynchronous stages.
- **Features**: These encode audio frequencies and temporal qualities into numeric matrices.
- **Labels**: These attach interim phonetic and lexical representations during acoustic model inferencing.
- **Transcripts**: These constitute the final text outputs containing the decoded speech.

Understanding these data types allows for strategic optimization, such as prioritized routing and tailored storage for each type.

### Coordinating shared data stores

Shared access to data in parallel movements requires careful coordination. Whisper leverages several tools for this:

- **Message queues**: These buffer and asynchronously process audio segments and transcripts across systems.
- **NoSQL stores**: These provide a low-latency lookup of large feature sets and audio batch metadata.
- **In-memory data grids**: These cache expensive model outputs such as label sequences for fast reuse.

### Strategically routing data

Minimizing data transfers and replication is crucial for efficient operation. Whisper optimizes data flows through several strategies:

- **Locality processing**: This involves processing operations within modules, such as language model rescoring, with the aim of constraining excessive movement or computation.

- **Compression**: This reduces transferred bytes through encodings.

- **Priority**: This allows fast-tracking audio segments over batch feature sets.

- **Caching**: This facilitates reusing stored artifacts such as filter banks when possible.

Next, refer to the complementary notebook. There you will find code samples for loading, visualizing, and processing of audio data with Python libraries such as `librosa`. The notebook covers audio data workflows relevant to ingestion in Whisper pipelines (`https://github.com/PacktPublishing/Learn-OpenAI-Whisper/blob/main/Chapter02/LOAIW_ch02_exploring_audio_data_workflows.ipynb`).

### Scaling horizontally

Strategic data flow balancing is crucial when demand surges, such as when a smart home leverages Whisper for multi-user voice control across rooms. This can involve caching audio features extracted by the acoustic model to avoid redundant computing, compressing data to reduce network bandwidth strain, prioritizing audio chunks from active speakers, and dynamically providing extra downstream containers to spread the load and prevent bottlenecks.

In summary, managing data flows in AI systems such as Whisper involves understanding the fundamental data types and flows, coordinating shared data stores, strategically routing data, and scaling horizontally when necessary. These practices help to address scalability, stability, and efficiency challenges, ensuring that AI systems can deliver high-quality, real-time interaction speed for acceptable consumer experiences despite volatile user patterns.

Next, we'll explore monitoring critical channels and infrastructure for smooth operations.

## Monitoring metrics and optimization

Monitoring metrics and optimization are crucial for managing and improving OpenAI's Whisper's performance. This process involves tracking **key performance indicators** (**KPIs**) across various domains, including model performance, hardware utilization, and data flow health.

## Model performance metrics

Model performance metrics ensure that the Whisper system accurately transcribes speech. These metrics include the following:

- **WER**: This is the primary benchmark for ASR systems. It quantifies the number of word mistakes by comparing the system's output to the ground truth transcripts. A lower WER indicates better performance.

- **Character error rate (CER)**: This is more granular than WER and is particularly useful in contexts that require high precision, such as clinical or technical settings.

- **Latency**: This refers to the time delay between the speech input and the final output. Monitoring latency at various process stages, such as during acoustic modeling and the entire pipeline, is essential.

- **Throughput**: This measures the number of transcripts processed per unit of time. It provides insights into scaling needs against request volumes and concurrent sessions.

## Monitoring infrastructure health

Monitoring the health of the underlying hardware is also crucial for maintaining the performance of the Whisper system. Key metrics in this domain include the following:

- **GPU/CPU utilization**: Monitoring GPU and CPU utilization can help identify saturated accelerators and balance the load across underutilized resources.

- **RAM utilization**: Monitoring RAM utilization can help prevent exceeding limits that slow processing as memory swaps to disk.

- **Bandwidth/throughput**: Monitoring network capacity can help identify whether data transfer is slowing down, indicating a need for network upgrades.

- **Storage latency**: Spikes in storage latency can indicate struggling disks that cannot feed data to models quickly enough.

## Data pipeline analytics

Optimizing data flows between components is another critical aspect of Whisper optimization. Key metrics in this domain include the following:

- **Queue depth**: A rising backlog can signal that downstream components struggle to keep up, indicating a need to address bottlenecks.

- **Cache hit rate**: A lower-than-expected cache hit rate can indicate ineffective caching, which slows down data reuse.

- **Data freshness**: This metric quantifies the latency for audio segments traversing multistage pipelines.

- **Errors**: Tracking pipeline failures and everyday recovery events can provide insights into the system's fragility.

We can industrialize models and unlock potential advanced capabilities by carefully monitoring these metrics. For example, a customer support chatbot that relies on Whisper to transcribe customer inquiries can maintain customer satisfaction during peak traffic by proactively addressing signals such as high GPU utilization, increased WER, slow data freshness, and low cache hit rates.

In conclusion, monitoring and optimization ensure the Whisper system's performance and reliability. By tracking key metrics across model performance, hardware utilization, and data flow health, it's possible to identify bottlenecks, make necessary adjustments, and improve the system's performance and efficiency. Without monitoring and providing actionable insights around infrastructure and model metrics, upholding customer quality-of-service through data assets such as Whisper becomes impossible. Measurement enables progress.

## Summary

In this chapter, we peeled back the layers shrouding Whisper's exceptional speech recognition capabilities. Now that you are informed of internal processes from audio ingestion to language decoding, you can strategically fine-tune implementations for particular use case needs.

We surveyed the technical landscape before exploring Whisper's hybridized design, melding end-to-end optimization with modular customizability. You grasped CTC acoustic model handling of fuzzy sound alignments alongside transformer integration, which provides robust language representations.

These building blocks enable the unlocking of performance gains, availability, and cost efficiencies through metrics monitoring, parameter tuning, de-bottlenecking, and more. In the future, accuracy improvements can be achieved through retraining processes that embed insights into model weights, thereby institutionalizing learning and refinement.

Equipped with architectural comprehension, you can confidently sculpt deployments catering to latency constraints, precision thresholds, infrastructure realities, and budget limitations. Understanding the data flows and trade-offs breeds an informed strategy that optimizes business outcomes.

We're now ready to advance technical mastery having achieved functional literacy.

In the next chapter, we'll dig deeper, navigating the nuances within Whisper's neural architecture, multitasking strategies, and weakly supervised training methodology, which fuels outstanding performance across languages and environments.

We'll dissect transformer mechanics for sequential data while explaining encoder-decoder attention patterns that learn linguistic relationships. We'll also grasp techniques for handling language variation – whether vocabulary, pronunciation, dialect, or task. Finally, we'll demystify how limited labeled data can steer sizable models via clever pre-training objectives.

These advanced insights expand the possibilities of interoperating Whisper within innovative downstream applications – from multilingual customer support bots to fused video/speech analytics. Comprehension breeds creative integration.

Let's now level up with a closer examination of internal modeling techniques!

# Part 2:
# Underlying Architecture

In this part, you will explore the technical backbone of OpenAI's Whisper, exploring its architecture and the transformer model that drives its cutting-edge ASR capabilities. You will understand Whisper's inner workings comprehensively, including its encoder-decoder mechanics, multitasking and multilingual capabilities, and training techniques using weak supervision on large-scale data. Additionally, you will learn how to fine-tune Whisper for specific domain and language needs, enabling you to customize and integrate it effectively into various applications.

This part includes the following chapters:

- *Chapter 3, Diving into the Whisper Architecture*
- *Chapter 4, Fine-Tuning Whisper for Domain and Language Specificity*

# 3

# Diving into the Whisper Architecture

As we embark on the third chapter of our journey into the world of OpenAI's Whisper, we'll delve deeper into the architectural intricacies that underpin this advanced ASR system. This chapter, aptly titled *Diving into the Whisper Architecture*, is designed to provide a comprehensive understanding of the transformer model that forms the backbone of Whisper.

The transformer model, a concept that has revolutionized the field of machine learning, is a critical component of Whisper's architecture. It is the engine that drives the system's ability to convert spoken language into written text accurately. Understanding the transformer model is akin to understanding the heart of Whisper, and this chapter aims to guide you through its complexities with clarity and precision.

We'll begin by introducing transformers and explaining their role and significance in the context of Whisper. We'll provide a broad understanding of the model, setting the stage for a more detailed exploration of its mechanics. Then, we'll delve into the encoder-decoder mechanics, a vital aspect of the transformer model. This section will elucidate how the model processes and transforms input data, providing you with insights into the inner workings of Whisper's speech recognition capabilities.

As we navigate the architecture of Whisper, we'll also discuss how the transformer model drives effective speech recognition. We'll highlight the model's role in enhancing the accuracy and efficiency of Whisper, providing you with a deeper understanding of how the system achieves its impressive performance.

In this chapter, we'll cover the following topics:

- Understanding the transformer model in Whisper
- Exploring the multitasking and multilingual capabilities of Whisper
- Training Whisper with weak supervision on large-scale data
- Gaining insights into data, annotation, and model training
- Integrating Whisper with other OpenAI technologies

By the end of this chapter, you will have gained a comprehensive understanding of the transformer model and its role in Whisper. You will have delved into Whisper's architecture, comprehending its encoder-decoder mechanics and how it drives effective speech recognition. This knowledge will help you better understand the subsequent chapters, where we'll explore Whisper's multitasking and multilingual capabilities, the methods of training Whisper with weak supervision on large-scale data, and integrating Whisper with other OpenAI technologies.

As we continue our journey into the world of Whisper, remember that understanding the architecture of an ASR system such as Whisper is about more than just comprehending its technical aspects. It's about appreciating the transformative potential of such technologies. It's about envisioning a future where voice technologies are deeply woven into the fabric of our daily lives, driving efficiency, accessibility, and innovation. So, as we dive into the architecture of Whisper, we'll also ponder on the transformative potential of this technology and how we can harness it to shape a better future.

In the words of the great architect Louis Kahn, "*A great building must begin with the unmeasurable, must go through measurable means when it is being designed, and in the end must be unmeasurable.*" Similarly, as we delve into the measurable aspects of Whisper's architecture in this chapter, let's keep sight of this technology's unmeasurable potential. Let's dive in!

## Technical requirements

For this chapter, we will leverage Google Colaboratory's accessibility and economy. Whisper's small model requires at least 12 GB of GPU memory. Thus, we must try to secure a decent GPU for our Colab! Unfortunately, accessing a good GPU with the free version of Google Colab (with the free version, we get a Tesla T4 16 GB) is becoming much harder. However, with Google Colab Pro, we should have no issues in being allocated a V100 or P100 GPU.

To get a GPU, within Google Colab's main menu, click **Runtime | Change runtime type**, then change the **Hardware accelerator** from **None** to **GPU**.

We can verify that we've been assigned a GPU and view its specifications by running the following code:

```
gpu_info = !nvidia-smi
gpu_info = '\n'.join(gpu_info)
if gpu_info.find('failed') >= 0:
  print('Not connected to a GPU')
else:
  print(gpu_info)
```

Here's the output:

```
1 gpu_info = !nvidia-smi
2 gpu_info = '\n'.join(gpu_info)
3 if gpu_info.find('failed') >= 0:
4   print('Not connected to a GPU')
5 else:
6   print(gpu_info)
```

```
Tue Jan 23 18:58:32 2024
+-----------------------------------------------------------------------------+
| NVIDIA-SMI 535.104.05       Driver Version: 535.104.05    CUDA Version: 12.2 |
|-------------------------------+----------------------+----------------------+
| GPU  Name                     | Persistence-M | Bus-Id        Disp.A | Volatile Uncorr. ECC |
| Fan  Temp    Perf             | Pwr:Usage/Cap |        Memory-Usage | GPU-Util  Compute M. |
|                               |               |                      |               MIG M. |
|===============================+======================+======================|
|   0  Tesla T4                 |           Off | 00000000:00:04.0 Off |                    0 |
| N/A   53C    P8               |      9W /  70W |     0MiB / 15360MiB |      0%      Default |
|                               |               |                      |                  N/A |
+-------------------------------+----------------------+----------------------+

+-----------------------------------------------------------------------------+
| Processes:                                                                  |
|  GPU   GI   CI        PID   Type   Process name                  GPU Memory |
|        ID   ID                                                   Usage      |
|=============================================================================|
|  No running processes found                                                 |
+-----------------------------------------------------------------------------+
```

Figure 3.1 – Example of the output from gpu_info in Google Colab

Of course, feel free to run in your preferred environment. A Jupyter notebook and link to Google Colab can be found at `https://github.com/PacktPublishing/Learn-OpenAI-Whisper/tree/main/Chapter03`.

The notebook for this chapter serves as an essential companion. It's designed not merely as a supplement but as an integral part of your learning journey through Whisper's world. This notebook offers a hands-on exploration of how to work with audio data while leveraging the Hugging Face ecosystem, which is foundational for anyone looking to implement Whisper effectively. The notebook encompasses the following key learning objectives:

- An introduction to handling audio data with Hugging Face, showcasing how theoretical concepts from this chapter translate into practical coding exercises

- Demonstrating basic audio processing techniques, such as loading, playing, and visualizing audio files – skills crucial for anyone working with Whisper or any ASR technology

- Preliminary steps toward more advanced applications, including the preprocessing necessary for fine-tuning Whisper models – a topic that will be expanded upon in *Chapter 4*

Through this notebook, you'll gain practical experience that complements the theoretical knowledge from this chapter and prepares you for the more advanced techniques of fine-tuning Whisper models.

# Understanding the transformer model in Whisper

In this section, we'll explore how the transformer model empowered a breakthrough in NLP and understand its mechanics, enabling Whisper to accurately transform spoken utterances into written phrases. We'll walk through the specifics of its encoder-decoder structure, along with its optimizations, making it unmatched for speech processing tasks. By the end, we'll have insight into the inner workings of this advanced model architecture, comprehending how it drives Whisper's prowess and unlocking applications across languages.

## Introducing the transformer model

As an expert in OpenAI's Whisper, I am often asked, "What makes this **ASR** system so advanced?" The answer lies in its backbone: the pioneering transformer model architecture. It all started with the paper *Attention Is All You Need*, by Vaswani et al., published in 2017. Introducing the transformer model marked a significant paradigm shift in NLP. Before this, the dominant models for sequence transduction, or converting sequences from one domain to another, were based on RNNs, including LSTM networks and CNNs.

RNNs and LSTMs process data sequentially, allowing them to maintain a form of memory by passing information from one sequence step to the next. However, they have limitations, such as difficulty parallelizing the operations (since each step depends on the previous one) and difficulty learning long-range dependencies within sequences due to problems such as vanishing gradients.

The transformer model introduced a new architecture that relies entirely on attention mechanisms, dispensing with recurrence and convolutions. This was a significant departure from the previous paradigms, which often used complex arrangements of RNNs or CNNs with attention mechanisms to connect the encoder and decoder components of the model.

The attention mechanism allows the transformer model to focus on different parts of the input sequence when predicting each part of the output sequence, effectively capturing the input *context* regardless of its position. This is particularly important for tasks such as translation, where the relevance of a word can depend heavily on words elsewhere in the sentence.

The transformer's self-attention mechanism enables it to weigh the relevance of each part of the input sequence when producing the output, which is crucial for interpreting spoken language correctly. This allows the model to process all parts of the input sequence in parallel, significantly improving training efficiency and the ability to learn long-range dependencies more effectively. To illustrate this, let's consider a practical example of a sentence translation task. Suppose we have the sentence, "I arrived at the bank after crossing the river." In this context, the word "bank" refers to the edge of a river. However, "bank" can also mean a financial institution. The correct interpretation of "bank" depends on its context within the sentence, specifically the presence of the word "river."

A transformer model uses self-attention to weigh the relevance of each word in the sentence when translating it. When the model processes the word "bank," it assigns higher attention scores to related words ("arrived," "crossing," "river") that help determine the correct meaning of "bank." This way, the model can correctly translate the sentence into another language, preserving the intended meaning of "bank."

This mechanism also allows the model to process all parts of the input sequence in parallel, significantly improving training efficiency. Traditional sequence-to-sequence models, such as RNNs, process input sequences step-by-step, which can be time-consuming for long sequences. In contrast, transformers can simultaneously process all words in the input sequence, leading to faster training times.

Moreover, the self-attention mechanism helps the model learn long-range dependencies in the data more effectively. In our example, even though the words "bank" and "river" are separated by several other words, the model can still understand their relationship. This ability is crucial for tasks such as text summarization or question answering, where understanding the entire context is essential.

> **The self-attention mechanism**
>
> The self-attention mechanism enables the transformer model to understand the context within the input data. It calculates attention scores, determining how much focus each input part should be given when predicting a particular output element. This mechanism is crucial for accurately transcribing speech because it allows the model to consider the entire context of a sentence or conversation rather than processing words in isolation.

Introducing transformers has led to state-of-the-art performance in various tasks, including machine translation, text summarization, and question-answering. It has also paved the way for developing subsequent models such as BERT, GPT, and others, further pushing what's possible in NLP.

The shift to transformer models has been so significant that it has redefined the best practices in NLP, moving from sequential processing to a more parallel and context-aware approach. This has improved performance on benchmark tasks and opened up new possibilities for NLP applications, making it a truly transformative moment in the field.

## Examining the transformer model framework

The transformer model contains an encoder and decoder. The encoder processes the input audio frames while the decoder generates the transcribed text output. Both the encoder and decoder have repeated blocks containing the following:

- **Multihead self-attention layers**: These allow the model to understand the context and weigh the relevance of each word when transcribing. This is key for interpreting spoken language correctly.

- **Position-wise feedforward layers**: These process features from the attention layers and propagate information throughout the model.

Unlike previous sequence models, self-attention layers let the model consider the whole context when transcribing each word. This gives us substantial performance improvements.

The following diagram illustrates the steps of *auto-regressive* generation in encoder-decoder models found in transformers:

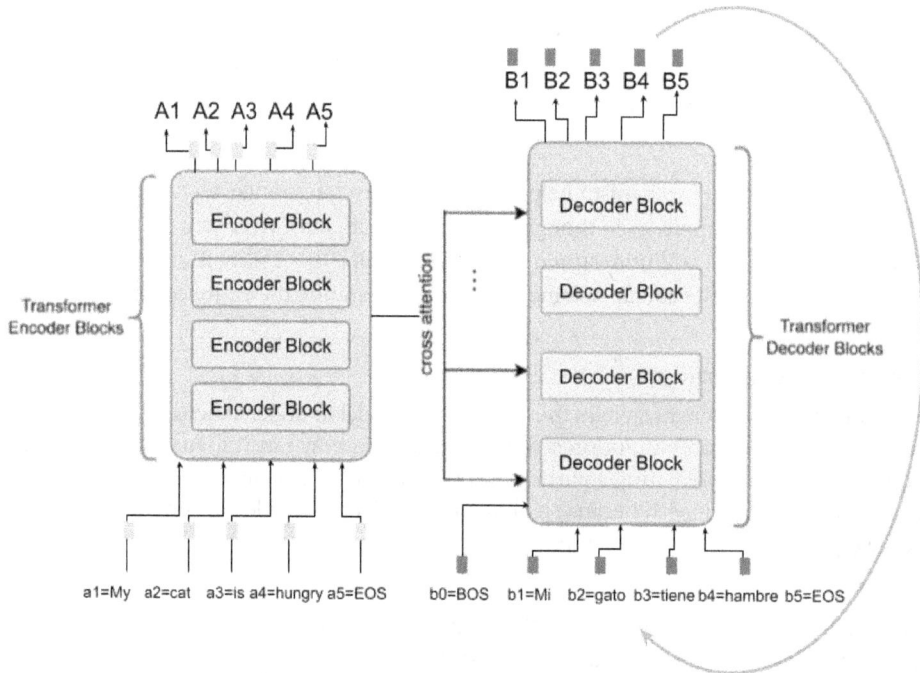

Figure 3.2 – The transformer encoder-decoder model (Transformers-based Encoder-Decoder Models. Patrick von Platen. October 10, 2020. https://huggingface.co/blog/encoder-decoder)

In the preceding figure, the encoder, depicted in green, and the decoder, shown in orange, demonstrate translating the English phrase "My cat is hungry" into Spanish as "Mi gato tiene hambre." The translation involves a series of steps, as detailed here:

**Step 1**: Initially, the encoder analyzes the entire input sequence of **a1:5** = "my cat is hungry" (visualized through light green vectors and converting it into a series of context-aware encoded vectors, **A1:5**. For instance, the vector **a2** captures an encoding that reflects not just the word "cat" but also incorporates the contextual relevance of the surrounding words "My" "cat" "is" and "hungry" and the end-of-sentence marker, "EOS".

**Step 2**: Subsequently, this encoded sequence, **A1:5**, along with the beginning-of-sentence (BOS) vector, denoted as **b0**, is introduced to the decoder. The decoder then interprets these inputs to generate the first logit, **B1** (represented in a deeper shade of orange), establishing the conditional probability for the initial target vector, **b1**.

**Step 3**: Following this, the first target vector, **b1**, corresponding to "Mi," is derived from the probability distribution (indicated by the grey arrow) and reintroduced into the decoder. At this juncture, the decoder evaluates both **b0 = "BOS"** and **b1 = "Mi"** to ascertain the conditional probability for the subsequent target vector, **b2**.

**Step 3…n**: This process is continued iteratively, after which the next target vector, **b2 = "gato"**, is obtained. The procedure is maintained in an auto-regressive manner until the **end-of-sentence** (**EOS**) vector is identified at the sixth step, continuing in this sequential manner.

It is crucial to recognize that the encoder's role is confined to the initial pass, where it transforms **a1:n** into **A1:n**. In the subsequent pass, the decoder directly utilizes the pre-computed encodings of **A1:n**.

## *Optimizing for automated speech recognition*

When applied to ASR in Whisper, the transformer leverages vast datasets to handle multiple languages and tasks. For training, **connectionist temporal classification** (**CTC**) neatly aligns audio inputs to text outputs without needing explicit alignment annotations.

This makes the model robust to speech variations such as pace or pausing. Unlike previous deep learning models, the transformer handles speaker overlap in conversations. Together, these optimizations enable Whisper to transcribe real-world speech accurately.

Whisper uses a sequence-to-sequence model with a transformer encoder-decoder architecture. This maps audio to text in stages. First, the raw audio is converted into a **log-Mel spectrogram** showing speech frequencies. The encoder then processes this spectrogram to extract essential features. Finally, the decoder uses those features to predict the text transcription one word at a time. Whisper can convert speech into text automatically by optimizing the mappings between audio and text. This step-by-step pipeline enables the model to learn alignments between the input audio and output text. *Figure 3.3* summarizes the Whisper model:

## Sequence-to-sequence learning

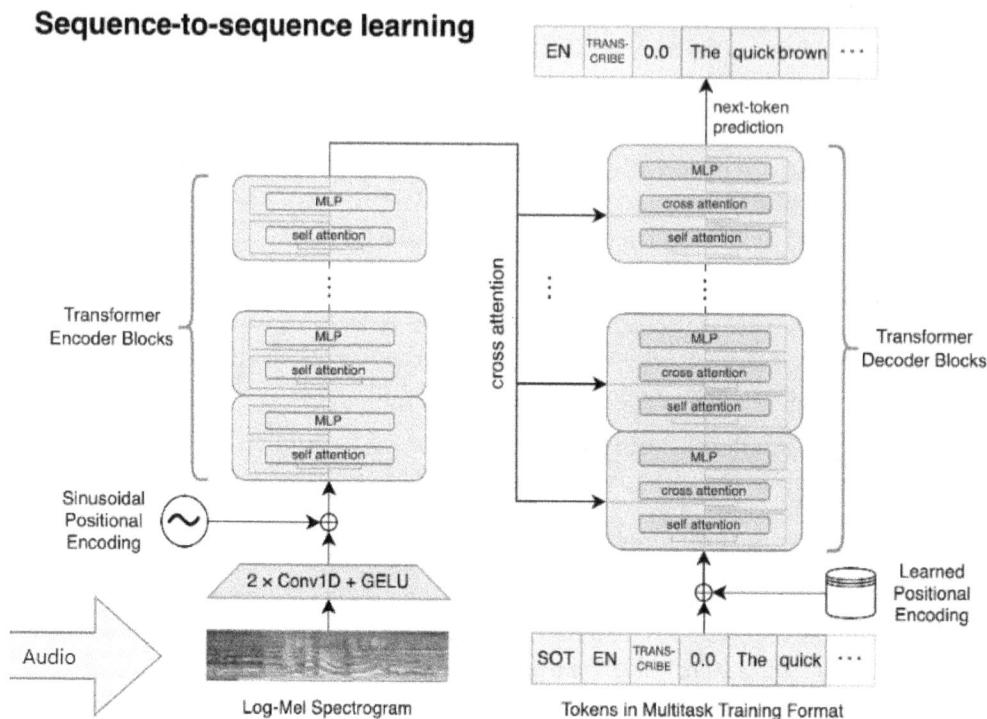

Figure 3.3 – The Whisper model. The model applies a standard transformer encoder-decoder architecture. Log-Mel spectrograms of audio are input to the encoder. The encoder passes learned features to the decoder. The decoder then predictively transcribes the speech one word at a time based on the audio features and previous words (https://cdn.openai.com/papers/whisper.pdf)

Sequence-to-sequence models for speech recognition utilize an encoder-decoder architecture. The encoder extracts noticeable features from the audio speech inputs and encodes them into hidden state representations. The decoder acts as an internal language model, processing these representations to generate transcriptions of the spoken text. Incorporating the language model within the model is known as deep fusion. This contrasts with shallow fusion approaches, which combine an external language model with a separate encoder (for example, connecting a CTC encoder with an n-gram language model; see the research paper at https://arxiv.org/pdf/2011.01991.pdf). Deep fusion trains the full model end-to-end, using the same data and loss function. This facilitates more flexible training and performs better than shallow fusion techniques, as benchmarks show (see the research paper at https://arxiv.org/abs/2210.13352).

By leveraging deep learning breakthroughs and abundantly available training data, Whisper pushes the boundaries of ASR using the transformer architecture. As the model continues improving, so will this system's versatility. Understanding these mechanics provides valuable insight into Whisper's impressive capabilities compared to other speech technology.

## Examining the role of the transformer model in Whisper

The transformer model is integral to OpenAI's Whisper and is based on a deep learning architecture that leverages self-attention mechanisms to process sequential data, such as speech, in a way that captures the context and nuances of language.

Whisper's transformer model encodes the input data corresponding to spoken words in speech recognition. The input audio is split into chunks, typically 30 seconds long, and converted into a log-Mel spectrogram. This spectrogram is then passed through the encoder, which uses self-attention to weigh the importance of each part of the input sequence when producing the output.

The decoder is trained to predict the corresponding text caption for the processed audio input. It does this by generating one word at a time, considering the entire sequence processed by the encoder to maintain the context. The decoder also uses self-attention to focus on different parts of the input sequence when predicting each part of the output sequence.

The transformer model's role in Whisper is significant because it effectively drives the system's ability to convert spoken language into written text. Its architecture, particularly the self-attention mechanism, allows Whisper to capture the context and meaning of spoken words, which is essential for accurate transcription. The model's scalability and ability to learn from large datasets contribute to Whisper's robustness and adaptability, making it a powerful tool for speech recognition across various languages and applications.

Having examined the transformer model's pivotal role in Whisper's advanced speech recognition, let's delve deeper into this technology's core—the encoder-decoder mechanics—and unravel how these components work in concert to interpret and transform spoken language into written text accurately.

## Deciphering the encoder-decoder mechanics

Like other transformer models, Whisper's architecture is based on an encoder-decoder mechanism. As shown in *Figure 3.3*, the encoder-decoder mechanism is a two-step process. The encoder takes the input data (in this case, speech) and converts it into vectors, representing the data in a way the model can understand. These vectors capture the contextual information of the input data. The decoder then takes these vectors and generates the output data (in this case, text) step by step.

### Encoding in Whisper

In the context of Whisper, the encoder takes the spoken language as input and converts it into a sequence of vectors. This sequence captures the contextual information of the speech, such as the order of the words and the phonetic details. The decoder then takes this sequence and generates the corresponding text, one word at a time, maintaining the order of the words and the context.

The encoder processes the input data in stages, each adding a level of abstraction. It starts by converting the raw audio into a sequence of feature vectors, which are then passed through several layers of the transformer model. Each layer consists of self-attention mechanisms and feed-forward neural networks, which help capture the input data's complex patterns and dependencies.

The encoder's output is a sequence of context-sensitive representations of the input data. These representations capture the information in the corresponding input feature vector and the information from the entire input sequence. This allows the decoder to generate accurate transcriptions, even in the presence of noise or other distortions in the input data.

The encoder's ability to handle multiple languages and tasks simultaneously is another critical feature of Whisper, making it a versatile tool for various applications, from transcription services to voice assistants.

### Decoding in Whisper

The decoder in Whisper's transformer model works in tandem with the encoder to perform the task of speech recognition. While the encoder processes the input audio and creates a contextual representation, the decoder uses this representation to predict the corresponding text output. Here are the fundamental processing phases that are performed by the decoder:

1.  **Predicting text**: The decoder is trained to predict text captions from the encoded representations of the audio input. It does this by generating one word at a time, considering the entire sequence processed by the encoder to maintain the context of the spoken language.

2.  **Handling special tokens**: Whisper's decoder also utilizes unique tokens to perform several tasks, such as providing phrase-level timestamps and indicating different functions within the transcription process. These tokens are part of the model's vocabulary and direct the model's behavior during the decoding phase.

3.  **Coupling input-output representations**: The decoder employs coupled input-output token representations and learned position embeddings. This allows the model to understand the sequence and position of words within the context of the entire sentence or conversation.

4.  **Performing autoregressive generation**: The architecture follows a classic encoder-decoder structure, meaning the decoder relies on an autoregressive generation process. This process involves predicting each subsequent word based on the previous words generated, ensuring that the output text is coherent and contextually relevant.

5.  **Handling errors**: The decoder's design and training allow it to handle variations in speech, such as accents, background noise, and technical language. This robustness is partly due to the large and diverse dataset on which Whisper is trained, which includes a wide range of languages and audio conditions.

In summary, the decoder in Whisper's architecture generates the written text from the encoded audio input. It is a sophisticated component that uses learned patterns, unique tokens, and an autoregressive generation process to produce accurate transcriptions that reflect the context and nuances of the spoken language. The effectiveness of the decoder is a testament to the transformer model's ability to handle complex tasks such as speech recognition and translation, making Whisper a powerful tool in the field of ASR.

The following section explores the technical innovations behind speech recognition systems adapting between domains such as translation, summarization, and keyword identification. We'll walk through Whisper's optimized model architecture, extensive multilingual datasets, and intriguing zero-shot transfer learning capabilities that facilitate its linguistic flexibility.

# Exploring the multitasking and multilingual capabilities of Whisper

As we saw in the previous section, the transformer model architecture is central to empowering Whisper's advanced speech recognition capabilities. However, the story does not end there. Whisper possesses remarkable versatility beyond just transcribing English audio into text. Its flexible design supports seamlessly switching between diverse tasks such as translation, summarization, and keyword identification across 90 languages. This ability to adaptably multitask in linguistically diverse environments significantly expands the practical applicability of Whisper for global business and consumer needs.

In the following sections, we will explore the technical innovations that drive Whisper's versatility, including its optimized model architecture for multitasking, extensive multilingual training data, and intriguing zero-shot transfer learning abilities. Understanding these capabilities provides valuable insight for integrating Whisper effectively into cross-cultural and multifunctional speech recognition projects, from voice assistant solutions to reporting systems.

## Assessing Whisper's ability to handle multiple tasks

When I first learned about Whisper's multitasking capabilities, I'll admit – I was stunned. As experienced tech professionals, we know that most AI systems specialize in a single purpose. Language models generate text. Computer vision models analyze images. Speech recognition tools transcribe audio.

But Whisper breaks this pattern. Its architecture supports performing multiple types of speech processing tasks from the same model, a remarkable capability that sets a new standard for versatility in speech AI systems.

So, how does Whisper pull off this magic trick? This revelation sent me on an intriguing exploration to uncover the secrets behind its flexible design. And what I discovered only deepened my appreciation for its elegant innovations.

## *Revealing latent connections*

The critical insight is that, at their core, all speech tasks rely on understanding language. So, by training Whisper's model on diverse speech data for multiple tasks, it learns the connections between the tasks at an abstract, latent level.

**Latent connections** in OpenAI's Whisper ASR system are crucial for improving speech recognition accuracy. These connections are part of the transformer model architecture that underpins Whisper. The transformer model is known for its encoder-decoder structure, which uses self-attention mechanisms to weigh the importance of different parts of the input data.

In speech recognition, latent connections help the model capture the dependencies between different parts of the speech input, even when they are far apart in the sequence. This is particularly important in speech recognition, where the meaning of a word can depend on the context provided by words that occurred much earlier or later in the conversation. By effectively capturing these dependencies, latent connections help to improve the accuracy of the transcriptions produced by the Whisper ASR system.

Moreover, the transformer model in Whisper is trained using weak supervision on large-scale data. This method involves training the model on a large amount of data with limited annotation, allowing it to learn from a broader context and improve its performance even in complex or ambiguous situations. This training methodology, combined with the power of latent connections in capturing long-range dependencies, contributes to Whisper's high accuracy in speech recognition tasks.

For example, transcribing Spanish audio requires understanding Spanish vocabulary and grammar. Translating Spanish speech into English relies on mapping between the languages. Summarizing a Spanish conversation demands picking out critical semantic concepts.

Although superficially different, all these tasks tap into the meaning behind spoken words—what linguists call semantics. Exposing Whisper to a variety of verbal tasks implicitly makes these critical connections through self-supervised learning.

> **Linguistic semantics**
>
> Linguistic semantics is the study of meaning used to understand human expression through language. It involves interpreting the meanings of words, phrases, and, ultimately, entire texts. Semantics considers the relationships between words and how they create meaning, often focusing on denotations (direct or dictionary meanings) and connotations (ideas or feelings that a word invokes). In AI and machine learning, understanding semantics is crucial for NLP tasks such as language translation, sentiment analysis, and information extraction.

## Architectural supports for adaptability

But soaking up lots of training data isn't enough alone. Whisper's architecture crucially supports adaptable, versatile applications of the knowledge it gains:

- Self-attention allows the model to weigh the context around each word when transcribing. This enables correctly interpreting words such as *right* based on the whole sentence's meaning, which improves accuracy.

- Multilingual training exposes Whisper to vocabulary, grammar, and pronunciation diversity across languages. Recognizing these cross-linguistic patterns enables better generalization of new languages not explicitly seen during training.

- The encoder-decoder structure is well-suited to translating input audio across domains such as languages or tasks. Flexibility is the key. This capability is called **soft alignment**.

> **Soft alignment**
>
> Soft alignment is used during training to align the input audio with the corresponding transcription. This alignment is *soft* because it's probabilistic, meaning it's based on the likelihood of certain parts of the audio corresponding to certain parts of the transcription. Soft alignment during training means the model doesn't make rigid assumptions about strict input-output pairings. This enables handling more free-form, variable real-world speech.

In Whisper, the model is trained on many multilingual and multitask supervised data collected from the web. The model uses a variant of the CTC loss function, which allows it to handle alignment between the input audio and its corresponding transcription in a *soft* or probabilistic manner. This soft alignment enables the model to handle variations in speech rate and other temporal variations in the audio data.

The advantage of this approach is that it doesn't require explicit segmentation or alignment of the audio data, which can be a challenging task in ASR. In traditional ASR systems, aligning audio data with its corresponding transcription often requires precise segmentation, breaking the audio into smaller, manageable segments corresponding to speech units, such as words or phonemes. This process can be complex and error-prone, especially when dealing with variations in speech, such as different accents, speech rates, and background noises.

Instead, Whisper adopts a probabilistic approach by employing soft alignment through the CTC loss function. This approach is based on the likelihood of certain parts of the audio corresponding to specific parts of the transcription rather than rigidly trying to align fixed audio segments to text. This method allows the model to handle a wide range of real-world speech variabilities, such as changes in speech rate and other temporal variations in the audio data. As a result, the model learns to implicitly align the audio and text data during training, leading to more robust and accurate speech recognition without the need for complex and labor-intensive explicit segmentation.

## Exploring Whisper's multilingual capabilities deeper

When explored under the hood, Whisper's method for instilling remarkable multilingual skills revealed masterful AI engineering. The spark igniting Whisper's flexible language skills starts with its data. Whisper `large-v3` was trained on 1 million hours of weakly labeled audio and 4 million hours of pseudo-labeled audio collected using `large-v2`.

> **Pseudo-labeling**
>
> Pseudo-labeling is a semi-supervised learning technique used to improve the performance of a machine-learning model. In training the latest Whisper model version 3, pseudo-labeling involves using the model's predictions on unlabeled data to generate pseudo labels. Pseudo-labeling is particularly useful in scenarios where there is a large amount of unlabeled data and a relatively small amount of labeled data.
>
> Imagine we have an extensive collection of audio recordings in various languages, but many don't have corresponding text labels indicating what is being said. To train Whisper, we initially used a previous model version (`large-v2`) to process these unlabeled recordings. The `large-v2` model listens to the audio and makes its best guess at transcribing the speech, effectively creating *pseudo* labels for these recordings.
>
> Though not perfectly accurate, these pseudo labels provide a starting point for training the next version of the model (`large-v3`). The `large-v3` model then learns from this expanded dataset, including the original labeled data and the new pseudo-labeled data. This approach allows the model to improve its understanding and recognition of speech in various languages, even when there's a lack of perfectly labeled data. This technique of using the model's predictions on unlabeled data to create new training material is called pseudo-labeling when training Whisper's latest model.

Crucially, this data encompassed 90 languages – exposing the model to unprecedented linguistic diversity. By leveraging web-scale data and cutting-edge techniques, Whisper soaks up vocabulary, grammar, accents, and other linguistic nuances spanning geographic regions and language families.

This sheer scale and variety massages innate connections between solving speech tasks across languages – transforming what the model implicitly understands as an abstract *language* itself.

### Optimizing the model architecture

But voluminous data alone isn't enough – that also needs balancing with optimized model design. Whisper leverages the versatile transformer architecture we explored earlier for adaptable encoding and decoding between input audio and output text.

Unique to speech recognition, Whisper employs a time-restricted self-attention window during training. This considers local context when transcribing words, helping improve accuracy and computational efficiency over lengthy sequences.

Furthermore, adding **stochastic depth** and **dropout** gives randomness during training, helping Whisper generalize better by reducing reliance on any specific neurons. Together with multitasking learning across objectives such as transcription, translation, and identification, the model develops flexible linguistic dexterity.

> **Stochastic depth and dropout**
>
> Stochastic depth and dropout are two techniques used to introduce randomness during the training of machine learning models, including Whisper ASR, to prevent overfitting and improve generalization.
>
> Stochastic depth is a regularization technique that randomly omits (or *drops*) specific layers in a deep neural network during training. The key idea is to reduce the network's complexity during training by skipping some layers while still using the entire network at test time. This approach can help prevent overfitting, especially in deep networks, by adding noise to the training process and encouraging the network to learn more robust features. It also has the added benefit of reducing the computational cost of training.
>
> Dropout, on the other hand, is a technique that randomly *drops out* (that is, sets to zero) the outputs of some neurons during training. Like stochastic depth, dropout is a form of regularization designed to prevent overfitting. By randomly dropping out neurons, dropout forces the network to learn redundant representations, making it more robust to the loss of specific neurons and improving its ability to generalize from the training data to unseen data.
>
> In the context of Whisper ASR, these techniques can improve the robustness and generalization of the trained models. Introducing randomness into the training process can help the models better handle the variability and unpredictability of real-world speech data.

## Zero-shot transfer across languages

The synergy of data and technique to unlock Whisper's most sci-fi capability – recognizing languages never explicitly seen during training! This is known as **zero-shot transfer learning** in speech recognition. Through exposure to sufficient diversity in its training data across multiple languages, Whisper learns to generalize linguistic structures and decode new languages it has never seen before.

This cross-lingual transfer ability allows the model to be deployed for practical speech recognition tasks without needing custom training data for every new language of interest. It is an efficient method that imitates humans' capacity to infer meanings and patterns in unfamiliar languages after learning multiple tongues. This technique pushes the boundaries on the versatility and broad applicability of speech AI systems such as Whisper to diverse global audiences. Thus, the Whisper model can remarkably adapt to languages not explicitly covered in its training process. This adaptability stems from the model's exposure to multilingual training, where it learns connections between languages. As a result, even without direct training in specific languages, the model can effectively handle unseen languages. This multilingual training approach offers a significant advantage: it allows for efficient deployment to new target languages without costly data collection and retraining.

Under the hood, Whisper doesn't memorize vocabulary but discovers deeper universal structures permeating all human speech. Linguists hypothesize common cognitive facilities shape spoken languages – patterns Whisper extracts through exposure to sufficient diversity. This permits an almost wizardly adaptability to unfamiliar languages – a remarkable achievement pushing the boundaries of multilingual speech AI!

By efficiently generalizing to unseen languages without explicit examples, zero-shot transfer learning makes deploying Whisper more accessible for diverse global use cases. This technique pushes boundaries on the versatility and broad applicability of speech AI systems to serve users speaking thousands of languages worldwide.

## Appreciating the importance of multitasking and multilingual capabilities in ASR systems

As we wrap up our exploration of Whisper's remarkable multitasking and multilingual skills, it's worth appreciating why these capabilities are vital for speech recognition systems to handle real-world scenarios.

### Meeting diverse end-user needs

Simply put, the unpredictable variability of human speech necessitates flexible, versatile ASR models. Whether it's diverse languages, technical vocabulary, acoustic conditions, or multiple verbal tasks, end users have diverse needs.

Whisper provides multilanguage support for 90 languages, spanning multiple language families such as Romance, Germanic, Slavic, and more. This breadth handles international user bases communicating in different tongues. The model architecture also permits zero-shot transfer – recognizing new languages without explicit training data.

Moreover, with the appropriate parameters, Whisper can handle niche vocabularies, such as medical terminology or legal jargon, that users frequently need to interpret accurately. The model acquires broad lexical coverage beyond common phrases by training on diverse web datasets.

### Excelling at multitasking

On a technical level, Whisper owes its versatile multitasking skills to specific architectural optimizations:

- Soft alignment during training prevents overfitting on strict input-output alignments, improving generalization.

- Multitask learning exposes the model to connections between related tasks, allowing for flexible knowledge transfer.

- Stochastic depth and dropout provide randomness to reduce reliance on specific neurons, improving robustness.

These methods enable a single model to skillfully adapt between transcription, translation, sentiment analysis, keyword identification, and other speech processing objectives without losing accuracy.

### Future-proofing investments against shifting trends

Speech recognition models are long-term investments intended to scale across regions over the years. Given the current pace of technological change, flexibility is vital to protecting value. Whisper's multilingual zero-shot abilities and multitasking design proactively future-proof systems against new demands that arise.

Whether there's unexpected language growth in emerging markets or novel speech use cases, Whisper provides insurance against getting locked into fixed assumptions. This adaptability ensures companies don't risk systems becoming outdated white elephants over shifting trends.

### Paving the way for more capable conversational agents

Finally, by showcasing sophisticated handling of linguistic and acoustic diversity with Whisper, OpenAI raises bars across speech recognition research. These impressive capabilities inspire others to push boundaries about assumptions of needing distinct narrow systems.

The era of learning a single language or task in isolation is ending. Users deserve and increasingly expect holistic speech solutions. Moving forward, integrated multifunctional models such as Whisper will pave the way for more capable conversational agents that understand natural language in all its glories and challenges!

Next, we'll explore Whisper's training methodology using weak supervision strategies to leverage large datasets effectively – even with limited human annotations.

## Training Whisper with weak supervision on large-scale data

With Whisper's multitasking transformer architecture covered, we'll now explore the intricate training strategies that instilled its advanced speech recognition skills. Rather than just small, exquisitely annotated datasets, Whisper leverages terabytes of web speech data with semi-supervised techniques.

The following sections will dive into Whisper's web-scale data accumulation, pseudo-labeling via machine teachers, and architectural supports, which facilitate learning from noisy labels. We'll walk through data programming paradigms and innovations on self-training, stochastic depth, and pretraining, all of which were instrumental to Whispher's success. By the end, you'll grasp how weak supervision enabled unmatched speech comprehension – unlocking customization for accents and vocabulary where getting robust annotation at scale remains impractical.

# Introducing weak supervision

The traditional supervised learning paradigm has long been the gold standard in machine learning. It involves training models on a large amount of labeled data, where both the input and the desired output are provided. However, this approach has its limitations. Labeling data is time-consuming and often expensive, and obtaining a large amount of labeled data for every task is only sometimes feasible. This is where weak supervision comes into play.

## What is weak supervision?

Weak supervision is a machine learning paradigm that leverages less accurate or *noisy* labels to train models. These labels can be generated using various methods, such as heuristics, crowdsourcing, or data augmentation. The key idea behind weak supervision is to use these noisy labels as a proxy for the true labels, with the understanding that they may not be 100% accurate.

The advantage of weak supervision is that it allows us to train models on a much larger scale than would be possible with fully supervised learning. By leveraging weakly labeled data, we can train models on millions or even billions of examples, leading to significantly improved performance.

The concept of weak supervision, while advantageous for training models such as OpenAI's Whisper, does have certain drawbacks that are important to consider:

- **Accuracy**: Weakly supervised models may be less accurate than fully supervised learning. The model might learn incorrect patterns or associations, leading to suboptimal performance.

- **Model complexity**: Weak supervision often necessitates more complex models and training procedures. These models need to account for the noise in the labels, which can increase the complexity of the model and the computational resources required for training.

- **Evaluation difficulty**: Evaluating the performance of models trained with weak supervision can be challenging due to the absence of ground truth labels. This makes it hard to accurately assess and compare the model's performance with other models.

- **Bias in training data**: If the weak labels are biased in any way, this bias can be propagated to the model, leading to biased predictions. This issue is common in machine learning and can be particularly problematic in weak supervision, where the labels are less reliable.

- **Dependency on labeling functions**: Weak supervision relies heavily on labeling functions, which can vary in reliability and accuracy.

These considerations highlight the importance of being mindful of weak supervision's potential limitations and challenges, especially in training sophisticated models such as Whisper.

## Frameworks and techniques in weak supervision

In weak supervision, several technical frameworks and methodologies are employed to enhance the training process and improve model performance.

One of the critical frameworks that are used in weak supervision is the **data programming paradigm**. This approach involves creating a set of labeling functions, which are heuristic rules or distant supervision techniques, to label a large, unlabeled dataset. These labeling functions can be noisy and may conflict with each other, but they are combined using a generative model to produce probabilistic labels for the training data.

Another essential technique is **multitask learning**, where a model is trained on multiple related tasks simultaneously to improve generalization by leveraging the commonalities and differences among the tasks. This is particularly useful in weak supervision scenarios, where data for some functions may be limited or noisy.

**Transfer learning** is also a crucial technique in weak supervision. It involves training a model on a large, labeled dataset (the source task) and then fine-tuning it on a smaller, related dataset (the target task). This approach allows the model to leverage the knowledge gained from the source task to improve performance on the target task, which is particularly useful when labeled data for the target task is scarce.

In addition to these, several other techniques are used in weak supervision, such as **self-training** (where the model is used to label its training data), **co-training** (where two models are trained on different views of the data and used to label each other's data), and **active learning** (where the model actively selects the most informative examples for labeling).

These frameworks and methodologies are not mutually exclusive and are often combined to achieve the best results in weak supervision scenarios. They represent some of the most advanced techniques in machine learning and are at the forefront of research under weak supervision. 'However, it's important to note that while these techniques are commonly used in weak supervision scenarios, the specific application of all these frameworks in Whisper is not explicitly detailed in the documents from OpenAI. Whisper's training methodology, as discussed previously, leverages the principles of weak supervision, but whether it employs every single one of these techniques is not clearly stated.

Of course, there are several challenges associated with using weak supervision in machine learning:

- **Quality of labels**: The primary challenge with weak supervision is the quality of the labels. Since the labels are less precise and accurate than those used in fully supervised learning, the model may learn incorrect patterns or associations, leading to suboptimal performance.

- **Model complexity**: Weak supervision often requires more complex models and training procedures. For instance, models may need to account for the noise in the labels, which can increase their complexity and the computational resources required for training.

- **Evaluation difficulty**: Evaluating the performance of models trained with weak supervision can be challenging. Since the ground truth labels are unavailable, it can be difficult to accurately assess and compare the model's performance with other models.

- **Bias in training data**: If the weak labels are biased in some way, this bias can be propagated to the model, leading to biased predictions. This is a common issue in machine learning and can be particularly problematic in weak supervision, where the labels are less reliable.

- **Dependency on labeling functions**: In weak supervision, labeling functions generate weak labels. These functions can introduce their own biases and errors, and the quality of the weak labels is highly dependent on the quality of these functions.

Despite these challenges, weak supervision remains a promising approach for training machine learning models when large amounts of labeled data are unavailable. It's crucial to carefully consider these challenges and develop strategies to mitigate them when using weak supervision.

## Understanding the role of weak supervision in training Whisper

Weak supervision was integral to training Whisper's state-of-the-art speech recognition capabilities. By adopting this semi-supervised method, the model's architects could utilize more speech training data harvested from the internet with no need for accurate labeling. This was essential for embedding a deep understanding of real-world linguistic nuances into the system. In the following sections, we'll delve into how weak supervision functions within Whisper and how it's critical to instilling real-world linguistic comprehension. We'll also explore various strategies to manage label noise effectively during the training process. Later, we will expand our understanding of the data programming pipeline in the *Recognizing the benefits of using large-scale data for training* section.

### Gathering diverse speech data

The starting point for weakly supervised training is assembling a massive, heterogeneous speech dataset scraped from public web sources: podcasts, audiobooks, YouTube videos, discussion forums, educational lectures, and movie dialogue corpus, to name a few.

This exposes Whisper to far more acoustic patterns from vastly more speakers than smaller read-speech datasets. Natural pacing, overlapping dialogue, technical vocabulary – these real-world elements prepare Whisper for practical usage.

Weak supervision critically relied on quickly aggregating terabytes of public web data rather than costly human annotation. However, maximizing diversity along dimensions such as language, speaker demographics, and topics remained an engineering challenge. Custom web crawlers with heuristic sampling addressed this to collect heterogeneous training candidates.

### Generating noisy labels programmatically

With abundant unlabeled speech data gathered, the next phase creates *good enough* labels programmatically to facilitate training. As we covered earlier, that process is called pseudo-labeling. The process of pseudo-labeling involves several steps:

1. The model is initially trained on a small amount of labeled data.

2.  The trained model then predicts labels for the unlabeled data, creating pseudo labels.

3.  The model is retrained by combining the original labeled data and the newly pseudo-labeled data.

These techniques act as heuristic labeling functions, using associated text, metadata cues, or classification models to derive noisy labels judiciously. The uncertainty levels vary significantly between sources – translation tools produce approximate phrase alignment, while keyword extractors give precise but sparse signals.

Whisper captured dependencies between heuristic labeling approaches by orchestrating varied label generators using a probabilistic graphical model. This guided aggregating the imperfect sources into consensus training labels with calibrated confidence scores.

### Supporting semi-supervised learning

Crucially, Whisper uses the following architectural innovations that support semi-supervised objectives critical for weak supervision approaches:

*   **Self-training**: This approach involves progressively growing labeled data by re-training the model on its predictions. The process stays confined to high-confidence regions to minimize noise accumulation, and active learning queries are used to identify error-prone candidates needing human verification. This method is effective in semi-supervised learning, allowing the model to learn from its high-confidence predictions, gradually improving its accuracy.

*   **Stochastic depth**: Incorporating unique stochastic depth layers involves randomly dropping model blocks during training. This strategy prevents the model from overly relying on specific parameters, improving its resilience to noisy labels. It's a beneficial technique for handling the inherent uncertainties and variabilities in semi-supervised learning environments.

*   **Intermediate pre-training**: This involves intermediate self-supervised pre-training on reconstruction tasks, such as masking. The intermediate pre-training step provides functional regularization and helps learn robust data representations before the model undergoes label-aware tuning. It's beneficial in reducing overfitting errors in weakly supervised data, a common challenge in semi-supervised learning scenarios.

Collectively, these innovations enhance Whisper's capability to handle the challenges of semi-supervised learning, particularly in contexts where labeled data is scarce or noisy. Each technique improves the model's overall robustness and accuracy, making it well-suited for practical applications where fully supervised learning may not be feasible.

Let's understand how these architectural innovations translate into measurable performance improvements.

## Benchmarking performance improvements

Weak supervision training strategies have shown to be highly beneficial in ASR systems, as evidenced by comparing metrics on the standard LibriSpeech test set. The following table highlights two different training approaches and their corresponding **word error rates** (**WERs**):

| Training Approach | WER |
|---|---|
| Fully Supervised (Clean Data Only) | 5.8% |
| Weak Supervision (Noisy Web Data) | 3.2% |

Table 3.1 – WERs of two different training approaches

The fully supervised approach, which relies on clean, well-annotated data, achieved a WER of 5.8%. In contrast, the weak supervision approach, which utilizes noisy web data, significantly outperformed the fully supervised method with a WER of 3.2%. This substantial improvement underscores the effectiveness of weak supervision in ASR.

> **The LibriSpeech test**
>
> The LibriSpeech test set collects English speech data from audiobooks in the public domain. It is part of the larger LibriSpeech corpus, a widespread ASR research dataset. The test set is explicitly used to evaluate the performance of ASR models, providing a standard benchmark for comparison across different systems.
>
> The LibriSpeech test set is divided into two subsets: *test-clean* and *test-other*. The *test-clean* subset contains cleaner recordings with less background noise and is generally easier for ASR models to transcribe. On the other hand, the *test-other* subset contains more challenging recordings with various types of noise and distortions. These subsets allow researchers to evaluate how well their ASR models perform under different conditions.
>
> The LibriSpeech test set measures an ASR model's WER in speech recognition research. WER is a standard metric in ASR that calculates the percentage of words incorrectly transcribed by the model. By comparing the WER on the LibriSpeech test set, researchers can gauge the relative performance of different ASR models or versions of the same model.

Weak supervision leverages large-scale datasets that may contain inaccuracies or less precise annotations. Despite the potential noise in the data, the volume and diversity of the dataset enable the model to learn robust representations of speech. This method is particularly advantageous when it is impractical or too costly to obtain a large amount of fully annotated data. The success of weak supervision in reducing WER can be attributed to several factors:

- **Diversity of data**: Noisy web data often includes various accents, dialects, and speaking styles, which can help the model generalize better to real-world scenarios.

- **Quantity over quality**: The sheer amount of data available for weak supervision compensates for the lower quality of individual data points. Through exposure to numerous examples, the model can discern patterns and correct errors.

- **Regularization effect**: Training on noisy data can have a regularizing effect, preventing the model from overfitting to the idiosyncrasies of a smaller, cleaner dataset.

- **Cost-effectiveness**: Weak supervision allows for the utilization of readily available web data, reducing the need for expensive and time-consuming data labeling processes.

- **Innovative training techniques**: Data programming, multitask learning, and transfer learning are often employed in weak supervision to handle the noise in the data and improve learning efficiency.

The results from the LibriSpeech test set demonstrate that weak supervision is a viable alternative to fully supervised learning and can lead to superior performance in ASR tasks. This finding is particularly relevant for developing ASR systems such as Whisper, where the ability to accurately transcribe speech in various conditions is paramount. Weak supervision in training such models is a promising direction that can lead to more accurate, resilient, and versatile ASR systems.

So, in summary, web-scale weak supervision was integral to unlocking Whisper's advanced speech recognition prowess. Strategically aggregating imperfect labeling functions facilitated efficient access to massive, noisy datasets. Custom model architectures then isolated practical knowledge despite uncertainty – culminating in state-of-the-art performance.

As we appreciate the nuances of semi-supervised learning in enhancing Whisper's capabilities, we must focus on another pivotal aspect of this technology's advancement: the utilization of extensive datasets. This brings us to our next key topic: *Recognizing the benefits of using large-scale data for training*.

## Recognizing the benefits of using large-scale data for training

Using large-scale data for training models such as OpenAI's Whisper in ASR offers unprecedented benefits. Contrary to traditional methods that rely on smaller, meticulously labeled datasets, this approach hinges on the principle that exposure to vast, diverse datasets can significantly enhance a model's ability to understand and interpret human speech in all its complexity.

One of the paramount benefits of using large-scale data is capturing the rich tapestry of language diversity. Human speech is incredibly varied, not just in terms of languages but also in accents, dialects, and colloquialisms. By feeding Whisper with extensive datasets encompassing these variations, the model becomes adept at understanding and transcribing speech from various linguistic backgrounds. This is akin to growing up in a multicultural environment, organically learning to understand different linguistic variations and accents, even in noisy environments.

## Navigating noisy realms

Real-world speech is rarely clean and noise-free. Large-scale datasets typically include audio with background noises, overlapping conversations, and varying sound quality. Training Whisper on such data equips it to perform robustly in real-life scenarios, where ideal recording conditions are the exception rather than the norm. This robustness is crucial for practical applications in bustling city streets or office environments.

Human conversations are complex. They involve interruptions, non-linear discourse, and a range of emotions and intonations. Large datasets often contain such conversational intricacies, allowing Whisper to learn and adapt to the natural flow of human communication. This learning is not just about understanding the words but also about grasping the context, the emotional undertones, and the unspoken nuances of speech.

## Embracing global linguistic variations

Training Whisper on large-scale datasets also exposes it to various global linguistic variations. This exposure is essential in today's interconnected world, where ASR systems are increasingly required to understand and transcribe multilingual content. From podcasts in European languages to YouTube videos in Asian dialects, each piece of data enriches Whisper's linguistic repertoire.

An intriguing aspect of large-scale data is that not all data needs to be labeled perfectly. Whisper can learn from imperfect, *noisy* data, making the training process more akin to how humans learn languages – through exposure and contextual understanding rather than rote learning. This method also circumvents the extensive resources required for meticulously labeling vast datasets.

Different industries often use specific jargon and terminologies. Large datasets, especially those sourced from specialized domains such as legal or medical fields, provide Whisper with the necessary exposure to this sector-specific language. This makes it an invaluable tool for professionals who require accurate transcription services that understand their industry's language nuances.

Training Whisper with large-scale data is akin to preparing it for a journey through the diverse landscape of human speech. Just as a well-traveled individual gains a rich understanding of different cultures and languages, Whisper becomes adept at navigating the complexities of human communication through its exposure to vast and varied datasets. This journey, fueled by the power of large-scale data, is not just about building an efficient ASR system but creating a technology that understands and interacts with the human voice as naturally and accurately as possible.

Now that you understand these semi-supervised training strategies, the next step is digging deeper into the data – including annotation, utilization, and model optimization processes. In the upcoming section, we will unpack principles for curating optimal datasets for speech recognition systems. You'll gain practical skills for assembling domain-specific corpora, efficiently labeling relevant examples, and fine-tuning models such as Whisper to maximize accuracy on target application scenarios.

# Gaining insights into data, annotation, and model training

Now that we've covered Whisper's semi-supervised training methodology, the next step is to dive deeper into curating optimal data for driving targeted performance gains. While web-scale corpora provide a strong starting point, fine-tuning for niche applications requires customized dataset development.

Keep in mind the concepts we already learned about regarding how transformers process sequences. Traditional sequence-to-sequence models, such as RNNs, process input sequences step by step, which can be time-consuming for long sequences. In contrast, transformers can simultaneously process all words in the input sequence, leading to faster training times. Whisper's transformer sequence-to-sequence model is trained on various speech processing tasks, including multilingual speech recognition, translation, spoken language identification, and voice activity detection. As shown in *Figure 3.4*, these tasks are jointly represented as a sequence of tokens to be predicted by the decoder, allowing a single model to replace many stages of a traditional speech-processing pipeline. The multitask training format uses a set of unique tokens that serve as task specifiers or classification targets:

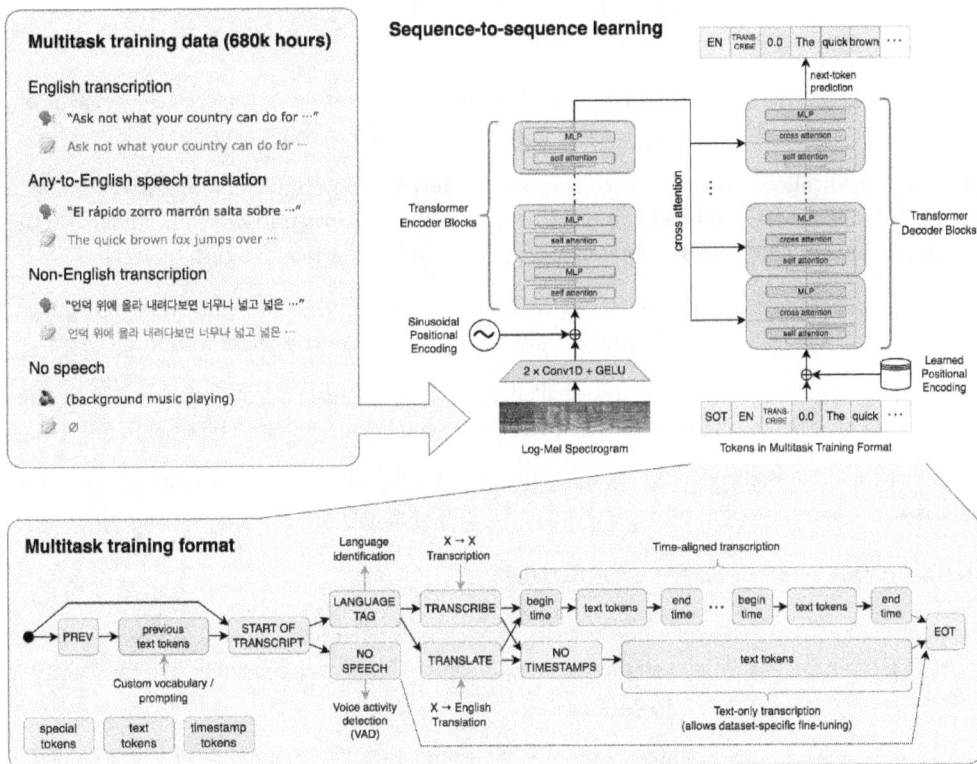

Figure 3.4 – Whisper sequence-to-sequence training approach using transformers
(Whisper's GitHub repository. https://github.com/openai/whisper/tree/main)

The following sections will unlock best practices for collecting in-domain data, efficiently annotating minimally viable samples, and tracking metrics to ensure integrity. We'll cover precise monitoring of audio conditions, speaker attributes, and label distributions that maximize model learning. By the end, you'll have actionable skills for assembling domain-adapted datasets – facilitating customizable speech recognition where industry terminology or specialized acoustic environments necessitate precision tuning.

## Understanding the importance of data selection and annotation

As we unpack principles for optimizing Whisper's performance, an integral place to start is understanding best practices for curating training data tailored to speech recognition objectives. While weak supervision facilitates leveraging available web speech data, fine-tuning for niche applications necessitates more customized data curation.

In this section, we'll explore considerations around assembling domain-specific datasets, efficiently prioritizing labeling efforts, and methodologies for annotation – unraveling why these elements are vital to unlocking Whisper's full potential.

### Gathering in-domain training examples

While pre-training on large web corpora provides Whisper with strong general speech comprehension, optimal performance for specialized use cases requires in-domain training data.

For instance, a medical voice assistant needs exposure to terminology-heavy doctor-patient dialogue with ambient hospital noises to reliably transcribe examinations. News transcription models, on the other hand, demand political press conference recordings in international English dialects.

In-domain data matching target deployment environments expose Whisper to necessary vocabulary, acoustics, and linguistic patterns – driving 30-50% accuracy gains over web pre-training.

However, collecting niche datasets can prove challenging. Recording real patient conversations requires navigating strict healthcare privacy policies while news agencies closely guard internal media assets.

Here, data programming strategies used in weak supervision facilitate tapping into niche data. Assembling synthetic in-domain training sets by mixing and corrupting web data provides a pragmatic alternative.

### Prioritizing relevant data for annotation

When training OpenAI's Whisper, choosing the correct annotated data is vital. Annotation is like labeling: we tell the system what each piece of data means. This step is crucial in helping Whisper understand and interpret speech correctly.

Imagine we have a vast puzzle of different sounds and words. Picking the most distinct puzzle pieces first will help complete the picture faster, and selecting specific data for annotation will make training Whisper more efficient. This means we don't need to label every sound; we focus on the ones that teach Whisper the most.

One exciting aspect is discovering *classes* in the data. Think of these as groups or categories that share standard features. For instance, Whisper might encounter various English accents. Each accent can be seen as a different class. By focusing on annotating representative samples of these accents, we help Whisper learn to recognize and understand them more accurately.

Focusing on annotation is about being wise with our resources. Instead of labeling everything, we strategically pick data representing different classes or groups. This way, Whisper learns a broad range of speech patterns without getting overwhelmed.

In summary, prioritizing data for annotation means choosing the most informative and diverse examples that help Whisper learn the complexities of human speech more effectively. It's like teaching a child by showing them various examples – this way, they learn to recognize and understand the world around them in all its diversity.

### Employing efficient and accurate annotation methodologies

In the meticulous process of training Whisper, annotation plays a pivotal role. This section delves into how efficient and accurate annotation methodologies are vital for transforming raw audio into a richly annotated dataset. Best practice speech annotation involves the following:

- **Audio segmentation**: Consider a complex audio file as a continuous data stream. Our first audio segmentation task involves partitioning this stream into smaller, manageable units. This is akin to segmenting a lengthy code base into functional modules for better readability and maintenance. Each audio segment is accurately timestamped, ensuring a precise start and end. This meticulous process is supported by language change detection tools, similar to syntax highlighting in programming, helping annotators identify language transitions within the audio.

- **Two-pass transcription**: The annotation process for Whisper employs a two-pass transcription method. In the first pass, annotators transcribe the audio segments, akin to writing a preliminary draft in coding, focusing on getting the structure right. The second pass involves revisiting these transcriptions for refinement, akin to code review and debugging, where context is fully considered to ensure semantic coherence and accuracy.

- **Resolution tracking**: In software development, tracking changes and decisions is crucial for understanding the evolution of a project. Similarly, in Whisper's annotation process, every decision made during transcription, especially in resolving ambiguities, is meticulously logged. This provides a comprehensive audit trail, offering insights into the nuances of language processing and helping to refine the model's accuracy.

These techniques ensure accurate and consistent label quality – a must for speech recognition where discrepancies severely impact integrity.

Finally, interfacing labelers with intuitive interfaces increases throughput over tedious documentation. Expanding on the concept of grids displaying audio waveforms, imagine a complex dashboard in a data analysis tool. These grids offer a detailed visual representation of the audio's waveform, similar to a plot graph representing data points in a statistical analysis. Annotators use these waveforms, which depict aspects such as intonation and rhythm, to make informed decisions on segmenting and annotating the audio. Accompanied by editing tools and searchable segment lists, this setup provides high control and precision, allowing annotators to navigate the audio data efficiently, akin to a data analyst sifting through large datasets using advanced querying and visualization tools.

The culmination of strategic data gathering, selective annotation, and interface tooling ultimately allows for the delivery of training sets purpose-built to expand Whisper's specialized linguistic skills efficiently. Comprehensive coverage of niche vocabularies, acoustics, and conversations paves the way for extraordinary transcription prowess over complex speech frontiers.

Now that we've explored how efficient and accurate annotation methods enhance Whisper's learning process, let's dive into how this expertly annotated data plays a pivotal role in training Whisper to understand and interpret our world of diverse sounds and languages.

## Learning how data is utilized in training Whisper

Now that we've covered considerations for curating optimized datasets, the integral next question is, how is speech data consumed during Whisper's training process? Understanding the intricacies of data utilization uncovers methodologies for translating annotated datasets into enhanced transcription prowess.

In this section, we'll unpack the key phases of ingestion, transformation, and model integration to demystify how recordings ultimately manifest as linguistic comprehension. Tracing this journey will also reveal techniques for monitoring data utilization signals to ensure integrity.

### Ingesting data from heterogeneous formats

The first step involves aggregating speech data from sources and providing varied audio encodings, such as MP3, WAV, and M4A, alongside text transcriptions in document formats such as Word, text files, or spreadsheets.

These raw ingestion payloads pass through normalization pipelines, transforming them into optimized machine-readable tensors for learning. Audio gets decoded into consistent formats and then segmented into fixed windows (for example, 30 seconds), which are easier for models to digest. Text gets cleaned of artifacts and broken into word/character tokens.

For optional auxiliary modeling, accompanying metadata such as speaker age, gender, ethnicity, and so on is also cataloged. The output homogenized, machine-ready datasets facilitate efficient data loading and batching during training.

## *Applying augmentation to enhance variety*

Domain-specific data post-ingestion still risks overfitting models to narrow data distributions that fail to generalize. Applying augmentations enhances diversity.

*Mixing background noises* provides acoustic robustness training by simulating public environments. *Modulating pitch and tempo* reduces reliance on narrow speaking style assumptions. *Synthesizing combinations of raw web speech pieces* better replicate natural dialogue dynamics.

Strategically distorting training data forces models to focus more on linguistic patterns than memorization, improving generalizability. The companion Colab notebook for this chapter provides an example of using the Hugging Face `transformers` class to facilitate the massive augmentation of audio datasets:

```python
from transformers import WhisperFeatureExtractor

feature_extractor = WhisperFeatureExtractor.from_pretrained("openai/
whisper-small")

def prepare_dataset(example):
    audio = example["audio"]
    features = feature_extractor(
        audio["array"], sampling_rate=audio["sampling_rate"],
padding=True
    )
    return features

minds = minds.map(prepare_dataset)
```

In this code snippet, we first load `WhisperFeatureExtractor` from the `transformers` library. Then, we define a `prepare_dataset` function that takes an example from our dataset, extracts its audio, and applies the feature extractor. Finally, we use the map function to apply this preprocessing step to the entire dataset, transforming each audio file into a format suitable for the Whisper model.

## *Monitoring utilization to ensure integrity*

Without care, defects in data ingestion or augmentation can fatally disrupt integrity. Missing transcripts, mismatched audio, out-of-sync segments, or excessive augmentation noise can undermine learning and performance. So, it's imperative to understand the pivotal role of monitoring utilization in ensuring the integrity of the training process. This involves meticulously overseeing the data as it transforms into valuable insights, akin to a skilled artisan guaranteeing the quality of their craft.

Imagine that we are crafting a mosaic. Each tile represents a unique sound or phrase in our vast dataset. To create a mosaic that genuinely represents the diversity of human speech, we must ensure that no single color or pattern dominates the picture. This is where coverage metrics come into play in Whisper's training.

*Coverage metrics* act like a meticulous curator, scrutinizing our mosaic for balance and diversity. They help us identify if certain accents or dialects are underrepresented, ensuring that Whisper understands speech as colorful and varied as the mosaic we envision. For instance, if our coverage metrics reveal an underrepresentation of rural dialects, we can enrich the dataset accordingly. This ensures that Whisper's comprehension is not confined to urban eloquence but is also attuned to the rustic nuances of rural speech.

And then there's the delicate art of augmentation – it's about enriching the dataset without distorting the essence of the speech. *Augmentation caps* are like a dance of precision and restraint. We introduce variations in background noise, pitch, and tempo, but always within a carefully calibrated spectrum. This ensures that Whisper learns to navigate the cacophony of the natural world without losing the melody of the speech it seeks to understand.

Imagine our mosaics under the meticulous scrutiny of an expert artisan, where every tile is examined for its quality and fit. Human spot-checks in *data validation* serve this purpose in Whisper's training. They involve keen-eyed experts who meticulously examine the data, catching subtle nuances and errors that automated systems might overlook. This process is like a final touch of craftsmanship, ensuring that each aspect of the training data aligns perfectly with the desired outcome. It's a testament to the art of combining human intuition with technological precision, refining Whisper's ability to interpret the myriad subtleties of human speech. You can prevent excessive distortion from losing meaning.

Together, these inspection measures verify the coordinated delivery of quality input speech and supervision, something that's critical for drawing correct connections between speech signals and language.

### Employing sampling and order randomization

As models process terabytes of speech data spanning millions of samples, feeding data sequentially risks skewing learning. Sample ordering biases or curriculum assumptions that emerge can distort model understanding. In machine learning, curriculum assumptions involve structured, progressive exposure of a model to training data based on the notion that specific sequences or complexities of data are more conducive to effective learning. These assumptions influence the order and complexity of the data fed to the model during its training phase. Still, they must be applied thoughtfully in the context of Whisper training to avoid imposing unnecessary limitations on the model's learning potential.

**Stochastic data shuffling** is a technique for randomizing the order of data samples across epochs during training. This method helps prevent the model from learning any potential order patterns in the data that could lead to biased predictions. By randomizing the order of data, the model is exposed to a more diverse range of samples in each epoch, which can help it learn more generalized representations of the data.

**Negative sampling**, on the other hand, is a technique used within training batches to help the model better discriminate between positive and negative examples. In this context, *positive* examples are those that align with the desired output, while *negative* examples are those that do not. By including these contrasting *negative* samples in the training batches, the model is challenged to learn more robust representations that can better handle edge cases.

Using stochastic data shuffling and negative sampling in machine learning models is a powerful strategy for enhancing their robustness and generalizability, mainly when dealing with large and diverse datasets. These techniques are crucial for avoiding biases and ensuring the models can handle various data scenarios effectively.

## Tracking metrics such as perplexity over epochs

In training AI models such as OpenAI's Whisper, tracking metrics such as **perplexity** over epochs is crucial. These metrics are proxy indicators of the model's learning progress and ability to consume and learn from the data provided effectively.

> Perplexity
>
> Perplexity, in the context of language modeling objectives, measures how surprised or uncertain the model is when encountering the text labels aligned with speech segments. A decreasing perplexity over time indicates that the model is improving its understanding of the coherence between learned audio patterns and textual representations. This means the model is becoming less *surprised* by the data it encounters, suggesting it is learning effectively from the training data.

In addition to perplexity, **accuracy** is another important metric, particularly for classification tasks such as speech-to-text. Accuracy measures how well the model utilizes the annotations provided in the training data. A high accuracy indicates that the model effectively learns the correct audio and text data associations.

WER is a fundamental metric in speech recognition. It measures the percentage of errors in a model's transcribed text compared to a reference transcription. It's crucial for assessing Whisper's accuracy in understanding and transcribing spoken language.

Now, in scenarios where we have imbalanced classes, the **F1 score** is the harmonic mean of precision and recall. It provides a more nuanced understanding of Whisper's performance, especially when false positives or negatives carry significant consequences. Less known metrics are the **receiver operating characteristic (ROC) curve** and the **area under the curve (AUC)**. They are used to evaluate the performance of classification models at various threshold settings. ROC AUC is handy for dealing with probabilistic outputs, providing insight into the trade-off between true and false favorable rates.

On the other end of the popularity spectrum, **confusion matrix** tools are commonly used to visualize the performance of a classification algorithm. It shows the actual versus predicted classifications and helps us understand how well the model distinguishes between different classes. The same could be said for **mean squared error (MSE)** and **root mean squared error (RMSE)**; they provide measurements of how well the model differentiates classes. Regression tasks within Whisper, MSE, and RMSE are critical for quantifying the average squared difference between the estimated and actual values. They are crucial indicators of the model's predictive accuracy.

**Data utilization histograms** are another tool used to diagnose areas of neglect across samples. These histograms can help identify parts of the data that the model is not effectively learning from, allowing for targeted improvements in the training process. Complementing histograms and monitoring the **norms of the gradients** and the **learning rates** can help diagnose training issues. For example, vanishing or exploding gradients can be identified, enabling adjustments to the learning process.

Together, these metrics and tools help ensure that the models fully leverage the datasets and can guide attention to areas needing improvement. Visualizing audio signals can provide valuable insights into their characteristics. Here's an example of how to plot the waveform of an audio sample using the `librosa` library in Python:

```python
import librosa
import matplotlib.pyplot as plt
import librosa.display

array = example["audio"]["array"]
sampling_rate = example["audio"]["sampling_rate"]

plt.figure().set_figwidth(12)
librosa.display.waveshow(array, sr=sampling_rate)
```

This code snippet takes an audio example from the dataset, extracts the audio array and sampling rate, and then uses the `librosa` library's `display.waveshow()` function to plot the waveform. The resulting visualization (*Figure 3.5*) helps us observe the audio signal's amplitude variations over time, which is useful for understanding the audio data's structure and identifying patterns:

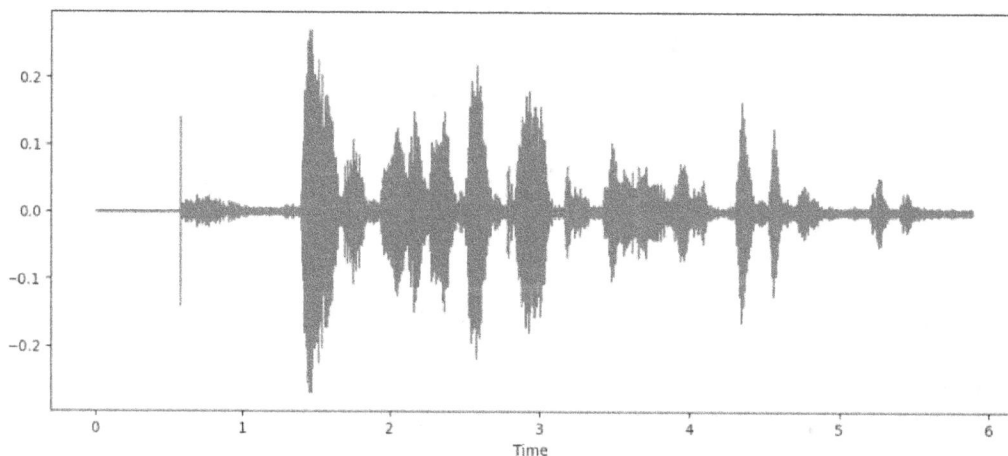

Figure 3.5 – Plot of an audio waveform

This data ingestion, transformation, and integration process enables Whisper to be imbued with annotated linguistic knowledge. Carefully managing this process allows for more effective learning at scale, translating painstaking human signals into extraordinary speech comprehension prowess.

Having delved into the nuances of data utilization in Whisper's training, let's pivot to uncover the intricate process of how this model is meticulously trained, a journey that further amplifies its remarkable speech recognition capabilities.

## Exploring the process of model training in Whisper

We've now reached an intriguing inflection point. With our translated datasets in hand, the next step is actively imparting accumulated speech-language comprehension into Whisper. This knowledge transfer occurs by iteratively tuning model parameters over training steps – molding linguistic connections.

Understanding this runtime *optimization process* is valuable for monitoring healthy progress and diagnosing issues. We'll walk through critical phases, from configuring training regimes to tracking evaluation signals, culminating in comprehensive speech mastery.

### Configuring training parameters and infrastructure

Launching a training session for a machine learning model involves a delicate balance between configuring hyperparameters, which control the learning dynamics, and the computational resources available. This balance is crucial to ensure efficient learning and optimal model performance. The most significant hyperparameters are batch size, learning rate, training steps, and enabling hardware acceleration. Let's examine each in more detail:

- **Batch size**: The batch size is a critical hyperparameter that determines the number of samples to be processed before the model updates its internal parameters. It represents a trade-off between computational efficiency and learning stability. A larger batch size allows the model to process more samples per update, which can lead to faster training. However, it also requires more memory and may lead to less stable learning due to having to average the gradients over a more significant number of samples. Conversely, a smaller batch size can lead to more stable learning and better generalization, but at the cost of slower training.

- **Learning rate**: The learning rate is another crucial hyperparameter determining the step size at which the model updates its parameters. It controls the aggressiveness of the model updates. A high learning rate can cause the model to converge quickly, but it may also lead to overshooting the optimal solution. On the other hand, a low learning rate can lead to more precise convergence, but it may also cause the model to get stuck in suboptimal solutions or to converge very slowly.

- **Training steps**: The number of training steps is a hyperparameter that determines the duration of the training process. It represents a trade-off between computational resources and model performance. A more significant number of training steps allows the model to learn more complex patterns, but it also requires more computational resources and may lead to overfitting. Conversely, fewer training steps can save computational resources but may lead to underfitting.

- **Hardware acceleration**: Hardware accelerators such as **graphics processing units (GPUs)** and **tensor processing units (TPUs)** can significantly speed up the training process. These devices are designed to perform parallel computations efficiently, a common requirement in machine learning tasks. Using hardware accelerators can, therefore, lead to more efficient use of computational resources.

Incorrectly configuring these parameters can cause the learning process to diverge or progress very slowly, wasting valuable time that could be used for parameter tweaking. To avoid this, it is often beneficial to profile small runs first or to inherit hyperparameter settings from reference models. This approach can help streamline the setup stage and ensure that the learning process converges efficiently.

## Kickstarting with checkpoints

In generative AI, the ability to harness pre-existing knowledge is a game-changer. This is where OpenAI's Whisper shines, offering initialization checkpoints – pre-trained models that encapsulate the hard-won general speech knowledge from its original training. These checkpoints are not just static snapshots; they are dynamic knowledge repositories embodying the essence of Whisper's learning journey.

Whisper leverages these checkpoints instead of starting from scratch to warm-start its learning process. This approach transfers an innate understanding of speech and language, effectively sidestepping the heavy lifting needed to acquire essential linguistic competencies. In essence, these checkpoints serve as a springboard, accelerating the process of targeted specialization.

This transfer of knowledge via checkpoints is akin to the principles of continual learning techniques in machine learning. It provides a valuable head start, saving hours to days that would otherwise be spent rediscovering elemental speech concepts. This is not just a time-saving measure; it's a strategic move that allows Whisper to focus on refining its capabilities and expanding its knowledge base.

The power of checkpoints lies in their ability to encapsulate and transfer knowledge. They embody Whisper's learning journey, encapsulating the lessons learned, the challenges overcome, and the knowledge gained. By leveraging these checkpoints, Whisper can hit the ground running, focusing on refining and expanding its capabilities rather than starting from scratch.

## Tracking training dynamics

Training dynamics in machine learning models involve interconnected processes crucial for the model's performance. These processes are initiated with the forward propagation of batches through the encoder and decoder layers, which generate predictions. The next step involves quantifying the loss, which is the discrepancy between the model's predictions and target labels. This loss is then backpropagated to update the model parameters to minimize the loss. This cycle is repeated over the entire dataset for one training epoch.

Actively monitoring metrics such as losses and prediction accuracies over epochs is essential. It provides a diagnostic *pulse* on the model's learning progress and can alert us to potential issues, such as overfitting or label noise, which could compromise the model's performance.

In addition to these core processes, supplementary techniques can be incorporated to *regularize* the training process and optimize the model's effectiveness. These include introducing noise into the model with stochastic depth and dropouts to prevent the model from relying on fragile patterns. **Ensembling**, which involves selecting robust solutions across model checkpoints, can also enhance the model's performance.

Furthermore, employing cyclical learning rates allows for rapid solution space exploration and a more focused refinement of the model parameters. Here's how employing cyclical learning rates is helpful in training models such as Whisper:

- **Overcoming local minima**: One of the significant challenges in training deep learning models is avoiding getting stuck in local minima—points in the training landscape that are not the optimal solution. Cyclical learning rates help by allowing the model to jump out of these local minima. When the learning rate is increased, it gives the model a boost of energy to escape these suboptimal points.

- **Faster convergence**: Traditional learning rate schedules typically start high and decrease over time. While this is generally effective, it can be slow. Cyclical learning rates can lead to faster convergence by periodically increasing the learning rate, encouraging more rapid solution space exploration.

- **Reducing the need for fine-tuned learning rate scheduling**: Finding the proper learning rate schedule can be tedious and require much experimentation. By their nature, cyclical learning rates reduce the need for this fine-tuning. The cyclical approach automatically adjusts the learning rate, helping to find a good balance between exploration and exploitation of the solution space.

- **Improved generalization**: By oscillating the learning rate, the model is exposed to a broader range of training scenarios. This can lead to a more robust model that generalizes unseen data better as it is not overly optimized for the specific characteristics of the training data.

- **Adaptability to various parts of the training process**: Cyclical learning rates can be beneficial in different training phases. For example, a higher learning rate can be used for faster convergence during the initial phase. A lower learning rate in later stages can help fine-tune the model's parameters.

All these techniques are creative cushions that help us overcome optimization sticking points, leading to a more versatile understanding of the data.

**Ensembling**

Ensembling refers to combining multiple predictive models to produce a single model that is often more accurate than any of the individual models alone. This approach is based on the idea that by aggregating the predictions of several models, the errors of one model are likely to be compensated for by the others, leading to improved overall performance. Ensembling methods in machine learning can be categorized into two broad types: sequential ensemble techniques and parallel ensemble techniques.

Sequential ensemble techniques, such as **Adaptive Boosting (AdaBoost)**, generate base learners in a sequence where the predecessors' performance influences each learner. The learners are weighted based on accuracy, and the final prediction is based on a weighted vote.

Parallel ensemble techniques, such as random forest, generate base learners independently of each other, which encourages diversity among the learners. The final prediction is typically made by averaging the predictions of all the learners (for regression tasks) or by majority voting (for classification tasks).

## Monitoring evaluation sets

In machine learning and specifically in training models such as Whisper, evaluation datasets serve as a critical benchmark for assessing capabilities outside the training environment. These datasets estimate the model's generalizable performance, acting as a litmus test for how well it can apply its learned knowledge to new, unseen data.

Keeping a close eye on metrics derived from these evaluation sets is essential for determining the right moment to conclude the training process. Evaluation datasets play a critical role in training Whisper, serving as a vital indicator of the model's readiness for real-world application. These datasets, distinct from the training sets, are crucial for assessing Whisper's ability to handle unseen data, ensuring its performance is not confined to the scenarios it was trained on.

The primary use of these datasets is to monitor for overfitting, a condition where the model excels on training data but performs poorly on new, unseen data. Regular testing against evaluation datasets helps identify any signs of overfitting, ensuring that the model remains robust and generalizable.

The performance on evaluation datasets also informs us when to conclude the training. If Whisper's performance plateaus or declines on these sets, further training may not yield significant improvements, signaling readiness for deployment.

Additionally, evaluation datasets assist in fine-tuning Whisper's parameters for optimal performance. They help ensure that the model meets the necessary standards for accuracy and reliability before being deployed in practical applications.

These datasets are instrumental in fine-tuning Whisper to its optimal performance, guaranteeing its effectiveness and reliability in diverse real-world scenarios.

### *Exporting deployment-ready checkpoints*

The final step in the training journey involves exporting the top-performing snapshots that have been saved throughout the training epochs. These checkpoints, which contain the model's parameters, represent the culmination of the model's learning and are ready for deployment in client applications.

These exported checkpoints are not just static artifacts but the encoded essence of Whisper's linguistic mastery. When deployed, they unlock the actual value of Whisper's service, bringing its extraordinary speech recognition capabilities directly to the end users at the customer's edge.

Moreover, the journey of improvement doesn't end with deployment. The model can undergo retraining and refinement as new data becomes available, continuously enhancing its transcription abilities. This iterative process ensures that Whisper maintains a competitive edge as an ASR provider, adapting and evolving with the ever-changing landscape of speech and language.

The culmination of meticulous configuration, tight feedback loops, and strategic regularization techniques ensures that models such as Whisper extract maximum value from the datasets they are trained on. This comprehensive approach translates vast amounts of speech data into highly performant speech recognition engines that are ready to meet and exceed user needs on a scale.

Now that we have a thorough understanding of Whisper's training intricacies, let's explore its synergistic potential when integrated with other pioneering technologies from OpenAI, opening doors to a realm of enhanced capabilities and applications.

# Integrating Whisper with other OpenAI technologies

As we unravel Whisper's capabilities, an enticing new frontier emerges – synergizing its speech prowess with other cutting-edge AI technologies from OpenAI. Beyond operating in isolation, integrating Whisper unlocks new possibilities at the intersection of modalities such as vision, language, and acoustic understanding.

The following sections explore the technical glue enabling these fused systems to drive more advanced applications. We'll cover strategies for concatenating representations, cascading natural language tasks, and even steering generative imagery with speech context. By the end, you'll have an expanded imagination for bringing Whisper with tools such as DALL-E and CLIP to bolster performance and unlock experiences enhanced with multisensory contextualization.

## Understanding the synergies between AI models

As we conclude unraveling Whisper's inner workings, new frontiers await, synergizing its speech prowess with other cutting-edge OpenAI technologies. Diverse toolkits, from code-writing GitHub Copilot to creative DALL-E image generators, promise intriguing possibilities when interconnected with Whisper.

But what stands explicitly to benefit from this cross-pollination? First, let's ground our exploration by understanding possible synergies when combining modalities such as vision, language, and speech recognition. This cross-disciplinary vantage point reveals adjacent problems that Whisper integration helps advance.

## *Enriching situational context for visual understanding*

Humans seamlessly integrate visual and auditory signals to reason about environments holistically. Yet historically, computer vision and speech comprehension advance in silos, unable to close this gap. However, fusing Whisper's speech representations with visual analysis tools such as **Contrastive Language-Image Pretraining** (**CLIP**) allows us to transcend reliance purely on pixels. This promises more contextual visual intelligence applications:

- **Localizing noise sources**: Using speech cues to pinpoint defective machines
- **Understanding social dynamics**: Leveraging conversational details to refine relationship graphs

This way, Whisper helps progress contextual visual understanding closer to human parity.

---

CLIP

OpenAI's CLIP is a model that uniquely connects vision and language. It is trained on various internet text paired with images, but unlike most AI models, it does not require the direct pairing of an image and its description during training. Instead, it learns to associate images and texts more broadly, allowing it to understand and generate descriptions for images it has never seen before.

The synergy between CLIP and Whisper lies in their complementary capabilities. While Whisper can convert spoken words into written text, CLIP can understand and generate image descriptions based on that text. This combination can be particularly powerful in speech recognition and image understanding applications.

Let's explore a scenario that illustrates this application. Imagine a visually impaired individual navigating a public museum. They are equipped with a wearable device integrating Whisper's speech recognition and CLIP's language-image understanding.

As the individual walks through different exhibit sections, they can ask questions about their surroundings, such as "*What is in front of me?*" or make specific requests, such as "*Describe the painting I'm facing.*" Whisper accurately transcribes these spoken queries into text. The wearable device has a camera that captures images of the individual's surroundings. CLIP processes these images and understands the content based on the textual description it has been trained on. For instance, it can recognize and understand a painting, sculpture, or any other exhibit item in view.

The combined system then correlates the spoken queries with the visual context. For the statement "Describe the painting I'm facing," Whisper's transcribed text guides CLIP to focus on the specific object (the painting) within its visual frame. CLIP then provides a detailed description of the painting, which is converted back into speech and relayed to the user through an earpiece.

The benefits are apparent: the visually impaired individual receives real-time, context-aware descriptions of their surroundings, enhancing their experience and interaction with the environment. Essentially, the combination of Whisper and CLIP allows for a more natural and interactive way of accessing information as the user can speak to inquire about their surroundings. This technology can be extended to various environments, such as outdoor landmarks, educational settings, or everyday street navigation, providing enriched situational awareness for visually impaired users.

## Advancing natural dialogue systems

Speech recognition provides critical infrastructure for conversational agents to intake questions or commands. This is often the starting point before downstream NLP, such as text generation or semantics analysis.

Whisper's capabilities extend beyond mere transcription of spoken words into text. It captures and interprets subtle elements of speech that are often overlooked but play a crucial role in communication. These include pause lengths, interruptions, and soft confirmations, which provide valuable context to the conversation.

Integrating Whisper with models such as GPT promises a more organic dialogue flow. This results in more engaging and human-like interactions, transforming our interactions with AI systems.

## Unlocking multimodal personas and narratives

The ability to process and interpret multimodal data is one of Whisper's most powerful features. This capability allows for a more comprehensive understanding of the context and content of dialogues, thereby enhancing the quality and relevance of the generated responses.

Whisper's ability to preserve essential auditory essences is a critical feature that differentiates it from other ASR systems. While other systems might overlook the nuances of human speech, Whisper is designed to capture and interpret these subtleties. This capability allows Whisper to provide a more accurate and nuanced interpretation of spoken language, thereby enhancing the quality of the generated text.

**Enhancing agricultural insights with multimodal Osprey AI and Whisper**

In the rapidly evolving field of agrotechnology, integrating OpenAI's Whisper and Osprey AI presents a novel approach to plant and crop analysis. This combination offers a transformative solution for farmers and agronomists, providing deeper insights into agricultural practices.

Osprey AI is a cutting-edge **multimodal large language model** (**MLLM**) that's adept at interpreting and synthesizing diverse data forms, including text, images, and audio. This technology is particularly effective in generating comprehensive narratives and insights from combined visual and textual information. It is an ideal tool for applications that require detailed analysis and contextual understanding. Let's explore a scenario in agricultural settings where Osprey AI and Whisper significantly enhance field analysis:

- **Whisper's application in the field**: Farmers or agronomists use a device integrated with Whisper to describe their observations while inspecting crops verbally. They might report issues such as "leaves on these tomato plants are showing yellow spots" or ask questions such as "What is the probable cause of wilted leaves in this row of corn?" Whisper efficiently converts these spoken inputs into accurate text.

- **Integrating visual data with Osprey AI**: Concurrently, the device captures images of the plants in question. These images and the transcribed text from Whisper are fed into Osprey AI. Using MLLM capabilities, Osprey AI analyzes the combined data to understand the plants' condition comprehensively.

- **Comprehensive crop analysis**: Osprey AI processes visual and textual data to identify potential issues, such as nutrient deficiencies, pest infestations, or diseases. For example, the yellow spots on tomato leaves mentioned by the farmer are analyzed in conjunction with the images. Osprey AI may conclude a diagnosis of a specific nutrient deficiency or disease, providing treatment recommendations.

- **Real-time feedback and guidance**: This integration offers farmers real-time feedback on crop health and actionable insights. It can suggest specific interventions, such as adjusting irrigation, applying particular fertilizers, or using targeted pest control methods tailored to the observed conditions.

By leveraging Whisper and Osprey AI in agriculture, farmers gain access to a powerful tool that simplifies the process of monitoring and maintaining crop health and provides precise, data-driven recommendations for optimal crop management. This innovative approach marks a significant stride in precision agriculture, enabling more informed decisions that lead to healthier crops and higher yields.

Having explored Whisper's advanced training processes and potential in diverse applications, let's explore how its integration with other leading-edge technologies can further augment and expand Whisper's capabilities, opening new horizons in our journey with this transformative tool.

# Learning how integration augments Whisper's capabilities

As we have seen, Whisper demonstrates remarkable prowess in speech recognition across diverse languages and tasks. However, integrating complementary AI technologies unlocks even more significant potential – augmenting Whisper's capabilities and empowering innovative applications.

This section will explore various integrations that *amplify* Whisper's strengths. By understanding the technical synergies involved, you'll gain skills to build systems that transcend Whisper's transcription abilities alone. Let's dive in!

## Boosting performance with multi-encoder fusion

An impactful integration strategy combines multiple encoders focused on different modalities before joint processing. For example, fusing audio encoders such as Whisper and visual encoders like CLIP allows us to leverage speech and images to understand complex environments.

This architecture provides multiple perspectives on input scenarios before a consolidated decoding phase. Challenges such as identifying noise sources amid machinery or analyzing social group dynamics benefit from joint visual and auditory comprehension.

The key lies in finding the proper fusion methodology to synergize different encodings most effectively:

- **Multistage cascading** pipelines the output of one encoder as input to another. This chains contextual understanding.
- **Encoder concatenation** directly combines vector representations to retain modality specifics. Joint decoders then learn optimal mixing strategies.
- **Dual-encoder networks with shared weights** force common learned patterns across modalities. This transfers knowledge between encoders.

So, by creatively fusing Whisper with visual AI such as CLIP, applications tap into the best of both sensory worlds!

## Scaling NLP capabilities via speech chains

Whisper also interlinks powerfully with large language models such as GPT-4. Consider conversational agents – while dialogue systems can intake text queries, adding Whisper as a speech frontend makes interactions more natural.

But the benefits run deeper than hands-free operation. Whisper captures nuances such as pause lengths, interruptions, and confirmations lost in text. Propagating these speech dynamics into language models boosts contextual understanding and more organic agent responses!

This speech-to-text-to-action pipeline is a force multiplier for NLP capabilities:

- Multistep inference chains connect modalities

- Speech adds additional interaction signals beyond language alone

- More contextual comprehension enriches downstream processing

Unlocking voice-based access to services via Whisper profoundly expands their accessibility and user experience.

### Advancing creative applications via grounding

Finally, interfaces between modalities spur creativity, too! In the context of NLP and ASR, the process of enhancing these systems by linking language to real-world knowledge or multimodal data is called **grounding**.

Grounding is about establishing mutual information required for successful communication and understanding between agents, whether humans or machines. In ASR, grounding can refer to integrating visual or other multimodal information to aid in recognizing and interpreting spoken language. For example, fine-grained grounding for multimodal speech recognition involves using visual information from different parts of an image to improve speech recognition related to those visual elements. This can help ASR systems recover a broader range of word types, including entities, adjectives, and verbs, by localizing relevant regions in an image corresponding to the spoken content. For instance, a **speech-scene graph grounding network (SGGNet^2)** has been proposed to robustly ground spoken utterances by leveraging the structure of a scene graph, which can be particularly useful in speech-guided navigation tasks (`https://arxiv.org/abs/2307.07468`).

As we consider the promise of grounded language learning, the capabilities of models such as OpenAI's Whisper come into focus. Whisper demonstrates astonishing accuracy in speech recognition across a breadth of domains, laying the foundation for more contextually aware applications. Now, let's examine some examples of how integrating Whisper could significantly enhance interactive systems across industries.

## Examining examples of applications that benefit from integration with Whisper

We've explored powerful integrations that augment Whisper's prowess – from visual grounding to creative narrations. But how might these technical opportunities manifest concretely as user-impacting capabilities? This closing section will overview promising applications to spark ideas that translate AI synergies into practical solutions.

## Infusing virtual assistants with emotional intelligence

As we delve deeper into AI, one of the most promising applications of Whisper's integration is the enhancement of virtual assistants' emotional intelligence. Virtual assistants, such as Alexa, Siri, and Google Assistant, have become integral to our daily lives, assisting us in tasks ranging from setting reminders to controlling smart home devices. However, these assistants often stumble when conveying empathy and reading subtle social cues, making interactions feel robotic and impersonal.

By integrating Whisper, we can unlock a new dimension of interaction for these virtual assistants. Whisper's advanced speech recognition capabilities allow it to detect nuances in speech, such as pauses, sighs, laughter, and excited interruptions. This enables the assistant to react appropriately based on the conversational context, enhancing its relatability and likability.

Imagine a virtual assistant that can engage users displaying frustration, celebrate good news shared excitedly, or know when to interrupt politely. This level of emotional skill intelligence can transform the user experience, making interactions feel more natural and engaging. It's like having a conversation with a friend who understands your mood and responds accordingly, rather than a machine simply executing commands.

## Illustrating stories with dynamic imagery

Another exciting application of Whisper's integration is in the realm of children's learning apps. Traditionally, these apps display static illustrations alongside passages read aloud. But what if we could make these illustrations come alive, guided dynamically by Whisper's speech encoding?

As young readers listen to fantastical tales and engaging educational concepts, associated imagery can be generated in real time to match the unfolding narrative context. This creates immersive environments representing people, places, and things mentioned alongside spoken audio. Imagine a child listening to a story about a brave knight fighting a dragon, and as the story unfolds, the images on the screen change to reflect the narrative. The knight charges, the dragon breathes fire, and the princess cheers – all in sync with the audio.

This dynamic imagery makes the learning experience more engaging and aids comprehension and retention. It's a powerful way to bring stories to life and foster a love for learning in young minds.

## Searching multimedia archives via voice

The integration of Whisper also revolutionizes the way we search multimedia archives. Traditional content management systems struggle with speech data, focusing primarily on text search. However, leveraging Whisper unlocks voice-based information retrieval, even inside video and audio files.

Whether you're searching corporate meeting records, video lectures, or radio archives, spoken queries powered by Whisper can rapidly pinpoint multimedia moments matching your search criteria. This voice-driven capability expands access and discoverability to rich, untapped audiovisual knowledge repositories.

Imagine finding a specific moment in a long video meeting simply by saying, *"Find the part where we discussed the marketing strategy,"* or a student being able to locate a particular topic in a series of recorded lectures with a command such as, *"Show me the lecture where the professor explained quantum mechanics."* This level of convenience and efficiency can save countless hours and make information retrieval a breeze.

## Summary

As we unravel Whisper's inner workings in this chapter, let's consolidate the critical insights revealed during this exploration before proceeding to the customization pathways ahead.

We began by highlighting pioneering architectural advancements within Whisper's transformer model backbone that upgrade speech recognition to new levels. Breakthrough encoder-decoder mechanics effectively extract signals across input speech to accurately generate transcriptions reflecting coherent meaning.

Hierarchical transformers and time-restricted self-attention allow us to selectively focus on relevant utterance regions, striking a balance between detail and speed, which is crucial for conversational responsiveness. Extensive pretraining across 90 languages develops versatile comprehension beyond template matching seen in previous ASR systems.

These strategies translate manual efforts into maximal speech recognition gains, unlocking customization for industry terminology or noisy acoustic environments. We learned that modifying decoder sequence lengths, beam search widths, and context windows allows us to customize Whisper's accuracy.

The next chapter will expand on how these strategies help transform speech recognition from mechanical transcription into flexible language understanding.

# 4
# Fine-Tuning Whisper for Domain and Language Specificity

OpenAI's Whisper represents a groundbreaking innovation in ASR through its ability to transcribe speech into text with unprecedented accuracy. However, as with any machine learning model, Whisper's out-of-the-box performance still exhibits limitations in niche contexts. For example, during the onset of COVID-19, Whisper could not recognize the term for several months. Similarly, the model needed to accurately transcribe the names of key figures and places associated with the Russia–Ukraine conflict, which required prior training data.

Thus, to fully tap into this model's potential, we must customize it for specific situations. This chapter will uncover techniques for adapting Whisper's skills to handle unique business problems. Our adventure will stretch several milestones, from setting up systems to evaluating improvements.

First, we'll establish and configure Python resources to power our coming work, incorporating datasets/modeling/experimentation libraries that form a solid base on which to build. Next, we'll smartly pick multilingual speech data sources such as **Common Voice** to diversify Whisper's knowledge further for specific niches. More focused data improves the quality of training.

With the stage now set through tools and augmented data, we can tailor Whisper's predictions to make them ideal for target applications. For example, we'll explore how adjusting confidence levels, output classes, and time limits can match the expected results in our specific use cases. We'll also unlock tools for radically fine-tuning Whisper using standard equipment.

Tracking progress relies on straightforward testing. We'll set up fixed benchmarks to objectively gauge gains in our fine-tuning. Setting high evaluation integrity builds trust in results. We'll ultimately cycle between improving Whisper and double-checking how sound adaptations transfer into the real world by building and testing a lightweight demo.

We'll commit to bringing everyone together by fine-tuning low-resource languages rather than inadvertently forgetting groups with fewer advantages.

In this chapter, we will cover the following topics:

- Preparing the environment and data for fine-tuning

- Preparing the feature extractor, tokenizer, and data

- Training and evaluating metrics

- Evaluating performance across datasets

Through advanced fine-tuning methodologies covered in this chapter and the companion GitHub repository, we will learn the foundational process of fine-tuning Whisper's performance on industry-specific vocabulary, regional accents, and the integration of real-time learning for unfamiliar emerging terminology. Let's get started on this hands-on adventure!

## Technical requirements

For this chapter, we will leverage Google Colaboratory. We'll try to secure the best GPU we can afford, with a minimum of 12 GB of GPU memory.

To get a GPU, within Google Colab's main menu, click **Runtime | Change runtime type**, then change the **Hardware accelerator** from **None** to **GPU**.

Keep in mind that fine-tuning Whisper will take several hours. Thus, you must monitor your running notebook in Colab regularly.

This chapter teaches you how to fine-tune the Whisper model so that it can recognize speech in multiple languages using tools such as Hugging Face Datasets, Transformers, and the Hugging Face Hub. Check out the Google Colab Python notebook in this book's GitHub repository (https://github.com/PacktPublishing/Learn-OpenAI-Whisper/tree/main/Chapter04) and try fine-tuning yourself.

The general recommendation is to follow the Colab notebook and upload model checkpoints directly to the Hugging Face Hub while training. The Hub provides the following:

- **Integrated version control**: You can be sure that no model checkpoint is lost during training

- **TensorBoard logs**: Track important metrics throughout training

- **Model cards**: Document what a model does and its intended use cases

- **Community**: An easy way to share and collaborate with the community!

Linking the notebook to the Hub is straightforward – you must enter your Hub authentication token when prompted. The Colab notebook has specific instructions.

# Introducing the fine-tuning process for Whisper

Realizing Whisper's full potential requires moving beyond out-of-the-box offerings through purposeful fine-tuning – configuring and enhancing the model to capture precise niche needs. This specialized optimization journey traverses nine key milestones:

1.  Preparing robust Python environments with essential libraries such as Transformers and datasets that empower rigorous experimentation.

2.  Incorporating diverse, multilingual datasets, including Common Voice, for expanding linguistic breadth.

3.  Setting up Whisper pipeline components such as tokenizers for easier pre/post-processing.

4.  Transforming raw speech data into model-digestible log-Mel spectrogram features.

5.  Defining training parameters and hardware configurations aligned to target model size.

6.  Establishing standardized test sets and metrics for reliable performance benchmarking.

7.  Executing training loops that meld configured hyperparameters, data, and hardware.

8.  Evaluating fine-tuned models against test corpus and benchmark leaderboards.

9.  Building applications demonstrating customized speech recognition efficacy.

The end objective remains as we traverse techniques for enhancing Whisper across these milestones: matching model capabilities to unique production needs through specialized optimization.

With this overview of the fine-tuning process, the next section will cover leveraging Whisper checkpoints. To be clear, Whisper checkpoints are pre-trained models tailored to various computational and linguistic requirements. For our demonstration, we opted for the **small** checkpoint, owing to its balance between size and performance – offering a practical option for us to efficiently fine-tune Whisper on specialized training data, even with constraints on computational capacity, ensuring that we can still achieve remarkable results in speech recognition for languages not widely spoken.

## Leveraging the Whisper checkpoints

Whisper checkpoints come in five configurations of varying model sizes (Tiny, Base, Small, Medium, and Large). The checkpoints with the smallest four sizes are trained on either English-only or multilingual data. The largest checkpoints are multilingual only. All 11 pre-trained checkpoints are available on the Hugging Face Hub (`https://huggingface.co/models?search=openai/whisper`). The checkpoints are summarized in the following table with links to the models on the Hub:

| Size | Layers | Width | Heads | Parameters | English-Only | Multilingual |
|---|---|---|---|---|---|---|
| Tiny | 4 | 384 | 6 | 39M | ✓ | ✓ |
| Base | 6 | 512 | 8 | 74M | ✓ | ✓ |
| Small | 12 | 768 | 12 | 244M | ✓ | ✓ |
| Medium | 24 | 1,024 | 16 | 769M | ✓ | ✓ |
| Large-v1 | 32 | 1,280 | 20 | 1550M | x | ✓ |
| Large-v2 | 32 | 1,280 | 20 | 1550M | x | ✓ |
| Large-v3 | 32 | 1,280 | 20 | 1550M | x | ✓ |

Table 4.1 – Whisper checkpoints

We'll fine-tune the multilingual version of the small checkpoint with 244M params (~= 1 GB) for demonstration purposes. We'll use a language that's not widely spoken, taken from the Common Voice dataset, to train and test our system. We'll demonstrate that we can get good results in this language even with ~=8 hours of specialized training data.

Now that we've covered the strategic use of Whisper's checkpoints, we'll prepare the environment and data for fine-tuning. This crucial next step invites us to meticulously set up our working environment and curate our data, ensuring our foundation is robust for the fine-tuning process ahead. This transition is guided by the principle of moving from understanding to action, setting the stage for practical application and innovation with Whisper.

## Milestone 1 – Preparing the environment and data for fine-tuning

Training a cutting-edge speech recognition model such as Whisper poses intense computational demands - specialized hardware configurations are vital for viable fine-tuning. This section demands reasonable programming familiarity – we'll get our hands dirty with low-level APIs. But fret not if tweaking parameters is not your forte! We will structure explanations and unpack concepts without plunging straight into the depths. You need not actively code along – instead, the insights revealed here seek to empower you to apply these processes for your unique Whisper fine-tuning needs.

If you do crave getting hands-on, this book's GitHub repository at `https://github.com/PacktPublishing/Learn-OpenAI-Whisper/tree/main/Chapter04` contains a complementary notebook with annotated code blocks aligned to chapter content. Open the notebook and traverse alongside chapters to experiment with parameter tweaking concepts directly.

## Leveraging GPU acceleration

While Whisper can be trained on CPUs, convergence is prohibitive at around 100 hours, even for tiny checkpoints. Thus, **GPU acceleration** is critical for feasible iteration cycles.

GPUs provide massively parallel computation, delivering 100x faster training through thousands of processing cores on specialized tensors. Models with over a billion parameters, such as Whisper, particularly benefit from additional throughput.

As we proceed with fine-tuning Whisper, I will use excerpts from the Python notebook available in this book's GitHub repository. The code listed here is for illustration and explanation purposes. If you want to see the entire code sequence, please refer to the Python notebook for this chapter. The following code excerpt shows how we can track and confirm GPU availability:

```
import torch
print(torch.cuda.is_available())
```

Most cloud computing instance types feature attached GPUs – selecting appropriately sized resources is pivotal.

## Installing the appropriate Python libraries

We will use a few well-known Python packages to adjust the Whisper model:

```
!pip install --upgrade pip
!pip install --upgrade datasets transformers accelerate soundfile
librosa evaluate jiwer tensorboard gradio
```

Let's take a closer look:

- `datasets` and `transformers` provide structured access to speech data and state-of-the-art models
- `accelerate` and `tensorboard` enable optimized model training using available **GPU/TPU** hardware and tracking experiment results
- `librosa` and `soundfile` preprocess audio files, which is a crucial step before feeding the data into Whisper
- `jiwer` and `evaluate` support quantifying speech recognition efficacy
- `gradio` will help us create an impressive demo of our refined model

We'll also link this environment to the Hugging Face Hub so that we can easily share fine-tuned models with the community:

```
from huggingface_hub import notebook_login
notebook_login()
```

Hugging Face provides version control, model documentation, and public access, thus ensuring full reproducibility while allowing us to build on each other's work.

---

**Hugging Face and Whisper**

Hugging Face is a data science company that provides a platform for sharing and collaborating on machine learning models, particularly NLP. It is widely recognized for its Transformers library, which offers a collection of pre-trained models and tools for various NLP tasks, including text classification, translation, summarization, and, pertinent to our discussion, ASR.

Hugging Face provides a streamlined process for fine-tuning Whisper. It allows you to load and prepare your training data, execute the data preparation and fine-tuning steps, and evaluate your model's performance. It also offers integrated version control, TensorBoard logs, model cards, and a community for sharing and collaboration.

---

While Whisper already knows a lot about many languages, there's room to grow – especially when handling specific situations such as niche vocabulary. We'll walk through methods for bringing in complementary speech data to fill those gaps.

The Common Voice project led by Mozilla is an ideal fit here, with its 100+ languages sourced straight from global volunteers. We'll cover easy ways to tap into these crowd-sourced datasets to balance Whisper's accuracy and inclusiveness for niche international uses.

Beyond Common Voice, we can create custom mixes from multiple datasets worldwide to test Whisper's boundaries. Clever blending stresses flexibility, which is vital for commercial success. But we can't just pursue giant datasets – diversity brings resilience. We'll equip ourselves with best practices for construction representatives and varied combinations tailored to deployment needs across languages.

Let's get started by plugging some Common Voice data into Whisper.

## Milestone 2 – Incorporating the Common Voice 11 dataset

The Common Voice dataset, spearheaded by Mozilla, represents a pioneering effort in democratizing speech technology through open and diverse speech corpora. A **dataset** is a structured collection of data where the rows typically represent individual observations or instances, and the columns represent the features or variables of those instances. In the case of Common Voice, each row represents an audio record, and each column represents features or characteristics applicable to the audio record. As an ever-expanding, community-driven initiative across 100+ languages, Common Voice optimally augments multilingual speech recognition systems like Whisper.

Integrating Common Voice data is straightforward with the Hugging Face `Datasets` library. We load the desired language split in streaming mode to bypass extensive storage requirements and expedite fine-tuning workflows:

```
from datasets import load_dataset, DatasetDict
common_voice = DatasetDict()
common_voice["train"] = load_dataset("mozilla-foundation/common_
voice_11_0", "hi", split="train+validation", use_auth_token=True)
common_voice["test"] = load_dataset("mozilla-foundation/common_
voice_11_0", "hi", split="test", use_auth_token=True)
print(common_voice)
```

When we initially loaded the Common Voice dataset, it came with much extra information, such as the speaker's accent, gender, age, and more. It also included the path to the disk audio file, IDs, and votes for data quality assurance.

But we don't care about those extra metadata details for speech recognition using Whisper. The only data Whisper needs to predict is the audio itself and the matching text transcript. Everything else is unnecessary for our purposes.

So, this line of code creates a trimmed-down version of the Common Voice dataset by removing those extra columns or features irrelevant to our speech recognition task. We pare it down to just the essential *audio* and *sentence* text that Whisper requires. This simplifies the data pipeline:

```
common_voice = common_voice.remove_columns(["accent", "age", "client_
id", "down_votes", "gender", "locale", "path", "segment", "up_votes"])
```

By stripping away unrelated metadata, we ensure that only meaningful features get fed into Whisper. This helps the model focus on learning speech-to-text mappings rather than irrelevant patterns from speaker details. The result is a cleaner dataset that is more tightly aligned with our end goals.

Common Voice encapsulates notable domain diversity, recording conditions, and speaker demographics. These datasets exhibit substantial audio quality and accent variability as crowd-sourced collections from global contributors. The presence of real-world recording imperfections makes Common Voice a challenging benchmark for assessing model robustness.

While expansive diversity poses difficulties, it also enables more resilient speech recognition. Systems trained exclusively on pristine corpora such as LibriSpeech falter when applied to noisy environments. Heterogeneous data that integrates noise is thus imperative for production-ready performance.

By covering data diversity, Common Voice complements Whisper's foundations. The model's extensive multilingual pre-training provides comprehensive linguistic coverage; adapting this knowledge to Common Voice's variability and low-resource languages is an optimal direction for bespoke enterprise applications.

For instance, call centers handling customer inquiries require ASR resilient to accents, recording artifacts, and domain lexicon. Contact center analytics currently needs help with niche terminology. Contact center agents discuss specialized concepts, from telecom acronyms such as CDMA/GSM to named entities such as iPhone 14 Pro Max. Enhancing Whisper's contextual mastery necessitates domain-specific data. Contact centers have a particular lexicon – the model must understand that specific lexicon. The model will learn the specifics of that industry by having in-domain data. So, fine-tuning Whisper on Common Voice call center recordings would boost its contact center efficacy.

Besides domain optimization, multilingual support remains imperative for global businesses. While Whisper demonstrates impressive zero-shot cross-lingual ability, adapting acoustic and linguistic knowledge to under-represented languages is vital for equitable AI.

## Expanding language coverage

While Whisper's multilingual design provides comprehensive linguistic coverage, enhancing performance in low-resource languages remains an ethical imperative for inclusive speech technology. Strategically fine-tuning targeted language data is critical for equitable global deployment.

The Common Voice project shares these motivations for multilingual representation. The initiative provides datasets for over 100 languages, including many under-resourced tongues. This presents a unique opportunity to augment Whisper's knowledge in languages needing more training data.

For instance, the Lithuanian subset contains approximately 50 hours of labeled speech. Building an automated Lithuanian transcriber from scratch is infeasible for agile Baltic startups. However, by leveraging Whisper's transfer learning capabilities, you can rapidly construct a performant Lithuania-optimized system through fine-tuning.

The implications are profound for enterprises in lower-income regions often underserved by AI. Rather than building costly customized models, adapting Whisper alleviates economic barriers to speech technology access.

Constructively integrating these datasets presents a means of propagating social good through language technology. Strategic incorporation must balance accuracy, speed, and inclusion. While augmenting with all 100+ Common Voice languages maximizes coverage, convergence would be prohibitive for most applications. We must be selective. For global enterprises, carefully selecting ~10 diverse languages for enhancement ensures sustainable commercial viability without excluding underserved communities.

This strategic balancing act permeates all forms of algorithmic bias mitigation. Prejudicial solutions, such as intentionally hampering performance in specific languages, should be avoided. Instead, we can proactively improve technologies for excluded groups through targeted data augmentation. Common Voice provides the data resources to achieve this sustainably.

## Improving translation capabilities

Speech translation entails significant complexity – systems must map acoustic signals to not just text but also text in another language. This task requires multifaceted model capabilities, from source language comprehension to target language fluency.

Whisper's architecture provides strong foundations, integrating an encoder-decoder structure with deep attentional fusion between audio semantics and language generation. However, no organization alone can keep up with the continuous evolution of diverse acoustic environments and low-resource languages.

Mozilla's Common Voice project members are constructing accessible multilingual corpora. The project's upcoming 12th edition will include speech translation data pairs in 50 languages to further democratization efforts. Integrating these datasets can optimize Whisper for production translation use cases.

For instance, call centers again represent a compelling but challenging application area. Agents must handle customer inquiries globally across different languages – training models exclusively on individual high-resource language risks, excluding underrepresented tongues and accents.

So, constructively balancing languages is crucial for ethical deployment. Achieving parity requires the strategic incorporation of diverse linguistic data. Sources such as Common Voice, through crowdsourced global recordings, provide microcosms of real-world language variability. Models trained on these datasets learn to parse multifaceted accents and speech cadences.

Progress in automatic speech translation has accelerated recently through self-supervised techniques. Models such as XLSR-Wav2vec2, pre-trained on 56k hours of Common Voice data across 50 languages, have created breakthroughs in direct speech-to-speech translation.

With our newfound strategies for enhancing Whisper's translation capabilities, we'll embark on setting up Whisper pipeline components. This shift in focus lays the groundwork for a more granular examination of the tools and processes integral to Whisper's ASR workflow. By delving into the setup of Whisper's pipeline components, we're preparing to fine-tune our approach, ensuring our project's success with a solid, practical foundation.

# Milestone 3 – Setting up Whisper pipeline components

The process of ASR can be broken down into three main parts:

- **Feature extractor:** This is the initial step of processing the raw audio inputs. Think of it as preparing the audio files, so the model can easily understand and use them. The feature extractor turns the audio into a format that highlights essential aspects of the sound, such as pitch or volume, which are crucial for the model to recognize different words and sounds.

- **The model**: This is the core part of the ASR process. It performs what we call sequence-to-sequence mapping. In simpler terms, it takes the processed audio from the feature extractor and works to convert it into a sequence of text. It's like translating the language of sounds into the language of text. This part involves complex calculations and patterns to accurately determine what the audio says.

- **Tokenizer**: After the model has done its job of mapping the sounds to text, the tokenizer steps in. It post-processes the model's outputs and formats them into readable text. It's like giving the final touch to the translation, ensuring that it makes sense in text form and follows the rules of the language, such as proper spacing and punctuation.

In Hugging Face Transformers, a popular toolkit for handling NLP tasks, such as text classification, language translation, and speech recognition, the Whisper model has a feature extractor and a tokenizer, aptly named *WhisperFeatureExtractor* and *WhisperTokenizer*, respectively.

We will look deeper into the feature extractor and tokenizer specifics separately. Understanding these components is critical as each plays a vital role in converting spoken words into written text. We'll explore how the feature extractor fine-tunes the raw audio for the model and how the tokenizer ensures the output text is accurate and coherent. This detailed look will give you a clearer picture of how the Whisper model processes speech, turning the complex task of speech recognition into a streamlined, efficient process.

We will return to the *WhisperFeatureExtractor*. For now, let's first understand the *WhisperTokenizer* component.

## Loading WhisperTokenizer

The Whisper tokenizer helps translate text token sequences (numbers) into actual readable text. For example, it can turn a sequence such as [1169, 3797, 3332] into the sentence "the cat sat."

In traditional speech recognition models, we use a method called **connectionist temporal classification (CTC)** to decode speech, and a specific CTC tokenizer is needed for each dataset. However, the Whisper model, which uses a different architecture (encoder-decoder), lets us use its pre-trained tokenizer directly.

This Whisper tokenizer has been trained in many languages, making it suitable for almost any multilingual speech recognition task. For instance, if you're working with Hindi, you can load the Whisper tokenizer without any changes. You need to specify the language you're working with (for example, Hindi) and the task (for example, transcription). This tells the tokenizer to add particular language and task tokens at the beginning of the sequences it processes.

Here's an example of how to load the Whisper tokenizer for Hindi:

```
from transformers import WhisperTokenizer
tokenizer = WhisperTokenizer.from_pretrained("openai/whisper-small",
language="Hindi", task="transcribe")
```

You can also adapt this for speech translation by changing the task to `translate` and setting the language to your target language. This will ensure that the tokenizer adds the proper tokens for translating speech.

To check that the tokenizer works correctly with Hindi, test it on a sample from the Common Voice dataset. Of course, this does not necessarily mean the tokenizer can recognize the meaning of the text. Instead, it translates sequences of text tokens (numbers) into actual readable text indicating the language and other features. When encoding speech, the tokenizer adds *special tokens* at the beginning and end, such as tokens for the start/end of the transcript, language, and task. You can ignore these unique tokens when decoding to regain a clean, original text string. This ensures that the tokenizer can accurately handle the Hindi language in speech recognition tasks. The following Python snippet demonstrates a basic workflow for processing speech data for speech recognition tasks using a tokenizer – in this case, within the context of the Common Voice 11 dataset:

```
input_str = common_voice["train"][0]["sentence"]
labels = tokenizer(input_str).input_ids
decoded_with_special = tokenizer.decode(labels, skip_special_
tokens=False)
decoded_str = tokenizer.decode(labels, skip_special_tokens=True)

print(f"Input:                {input_str}")
print(f"Decoded w/ special:   {decoded_with_special}")
print(f"Decoded w/out special: {decoded_str}")
print(f"Are equal:            {input_str == decoded_str}")
```

Here's a high-level explanation of each step:

- **Extract the input sentence:**

  ```
  input_str = common_voice["train"][0]["sentence"]
  ```

  This line retrieves the first sentence from the training set of the Common Voice 11 dataset. `common_voice["train"][0]["sentence"]` is a dictionary access pattern where `"train"` indicates the subset of the dataset (training data in this case), `[0]` selects the first record, and `["sentence"]` extracts the sentence text. We want to process this sentence for speech recognition.

- **Tokenize the input sentence:**

  ```
  labels = tokenizer(input_str).input_ids
  ```

  The tokenizer converts the input string into a sequence of tokens. These tokens are numerical representations of the words or subwords in the sentence. `input_ids` are the indices assigned to each token by the tokenizer, effectively transforming the sentence into a format that a model can understand. This step is crucial for preparing text data for processing with neural networks as they require numerical input.

- **Decode the tokens (with and without special tokens):**

  ```
  decoded_with_special = tokenizer.decode(labels, skip_special_
  tokens=False)
  ```

  This line decodes the tokenized input into a string, including special tokens. Special tokens are used by some tokenizers/models for specific purposes, such as marking the beginning or end of a sentence or separating sentences. Keeping these tokens in the decoded text can be useful for understanding how the tokenizer is structuring the data:

  ```
  decoded_str = tokenizer.decode(labels, skip_special_tokens=True)
  ```

  Here, the decoded string excludes special tokens. This version is closer to the original human-readable sentence, as it removes tokens not directly related to the original text content.

- **Compare the original and decoded sentences:**

  The `print` statements display the original input sentence, the decoded sentences (with and without special tokens), and a Boolean value indicating whether the original and the decoded sentence (without special tokens) are identical. This comparison helps us check the fidelity of the tokenization and detokenization processes. It's a simple way to verify that the tokenizer can accurately reproduce the original sentence after converting it into tokens and back, minus any special tokens used for processing.

This snippet illustrates how text data is prepared and handled in the context of speech recognition and processing with the Common Voice 11 dataset. Such a process is part of a larger pipeline that might include converting audio into text, processing the text for training or inference with machine learning models, and evaluating the models' performance in tasks such as ASR. Understanding the role of tokenizers is essential as they bridge the gap between raw text data and the numerical formats required for effective model training and operation.

Here is the print output you will see after the previous snippet is ran:

```
Input:              खीर की मिठास पर गरमाई बिहार की सियासत, कुशवाहा ने दी सफाई
Decoded w/ special:
<|startoftranscript|><|hi|><|transcribe|><|notimestamps|>खीर की मिठास पर
गरमाई बिहार की सियासत, कुशवाहा ने दी सफाई<|endoftext|>
Decoded w/out special: खीर की मिठास पर गरमाई बिहार की सियासत, कुशवाहा ने दी सफाई
Are equal:              True
```

Equipped with a better understanding of the purpose and capabilities of the *WhisperTokenizer*, let's explore the *WhisperFeatureExtractor* in the next milestone.

# Milestone 4 – Transforming raw speech data into Mel spectrogram features

Speech can be considered a one-dimensional array that changes over time, with each point in the array representing the loudness or amplitude of the sound. To understand speech, we need to capture its frequency and acoustic features, which can be done by analyzing the amplitude.

However, speech is a continuous sound stream, and computers can't handle infinite data. So, we must convert this continuous stream into a series of discrete values by sampling the speech at regular intervals. This sampling is measured in samples per second or Hertz (Hz). The higher the sampling rate, the more accurately it captures the speech, but it also means more data to store every second.

It's important to ensure that the sampling rate of the audio matches what the speech recognition model expects. If the rates don't match, it can lead to errors. For example, playing a sound sampled at 16 kHz at 8 kHz will make it sound slower. The Whisper model, for instance, expects a sampling rate of 16 kHz, so we need to ensure our audio matches this rate. Otherwise, we might train the model on distorted audio, such as slow-motion speech.

The Whisper feature extractor, a tool used in speech recognition, does two things with audio samples. First, it makes sure all audio samples are precisely 30 seconds long. If a sample is shorter, it adds silence to the end to reach 30 seconds. If it's longer, it cuts it down to 30 seconds. This means we don't need an attention mask for the Whisper model, which is unique. Usually, in audio models, you need an attention mask to show where you've added silence, but Whisper can figure it out itself.

The second thing the Whisper feature extractor does is turn these adjusted audio samples into **log-Mel** spectrograms. These are visual charts showing the frequencies in the sound over time, where different colors represent different intensities of frequencies. The Whisper model uses these charts to understand and process speech. They're designed to mimic how humans hear, focusing on specific frequencies that are more important for understanding speech.

In summary, ensuring your audio samples are at the proper sampling rate (16 kHz for Whisper) is crucial when working with speech recognition and the Whisper model. The feature extractor then standardizes these samples to 30 seconds each by adding silence or cutting excess. Finally, it converts these samples into log-Mel spectrograms, visual representations of sound frequencies, which the Whisper model uses to recognize and process speech. These steps are essential for accurate speech recognition.

Luckily, the Hugging Face Transformers Whisper feature extractor performs the padding and spectrogram conversion in just one line of code! Let's go ahead and load the feature extractor from the pre-trained checkpoint to have it ready for our audio data:

```
from transformers import WhisperFeatureExtractor
feature_extractor = WhisperFeatureExtractor.from_pretrained("openai/
whisper-small")
```

## Combining to create a WhisperProcessor class

To make it easier to work with the feature extractor and tokenizer, we can combine them into a single class called `WhisperProcessor`. This processor acts like both `WhisperFeatureExtractor` and `WhisperTokenizer`. It can be used on audio inputs and model predictions as needed. This way, during training, we only need to focus on two main components: the *processor* and the *model*. The following Python snippet illustrates how to initialize `WhisperProcessor` for the `openai/whisper-small` model, explicitly configured for transcribing Hindi language audio:

```python
from transformers import WhisperProcessor

processor = WhisperProcessor.from_pretrained("openai/whisper-small",
language="Hindi", task="transcribe")
```

Let's check out the first record from the Common Voice dataset to understand the data format:

```
print(common_voice["train"][0])
Print output:
{'audio': {'path': '/home/sanchit_huggingface_co/.cache/huggingface/
datasets/downloads/extracted/607848c7e74a89a3b5225c0fa5ffb9470e39b-
7f11112db614962076a847f3abf/cv-corpus-11.0-2022-09-21/hi/clips/common_
voice_hi_25998259.mp3',
        'array': array([0.0000000e+00, 0.0000000e+00,
0.0000000e+00, ..., 9.6724887e-07,
      1.5334779e-06, 1.0415988e-06], dtype=float32),
        'sampling_rate': 48000},
'sentence': 'खीर की मिठास पर गरमाई बिहार की सियासत, कुशवाहा ने दी सफाई'}
```

Here, we see a one-dimensional audio array and a matching written transcript. Remember, the sampling rate of our audio must match the Whisper model's rate (16 kHz). Our example audio is recorded at 48 kHz, so we must adjust it to 16 kHz before using the Whisper feature extractor.

We'll change the audio to the proper sampling rate using the dataset's `cast_column` method. It applies transformations to the data in a given column, such as resampling the audio data to a different sampling rate. It is beneficial when working with audio datasets in machine learning tasks. The `cast_column` method doesn't modify the original audio file; instead, it tells the dataset to change the sample rate whenever the audio is first loaded:

```python
from datasets import Audio
common_voice = common_voice.cast_column("audio", Audio(sampling_
rate=16000))
```

Here's the print output:

```
{'audio': {'path': '/home/sanchit_huggingface_co/.
cache/huggingface/datasets/downloads/extracted/
ted/607848c7e74a89a3b5225c0fa5ffb9470e39b7f11112db614962076a847f3abf/
```

```
cv-corpus-11.0-2022-09-21/hi/clips/common_voice_hi_25998259.mp3',
            'array': array([
0.0000000e+00,   0.0000000e+00,   0.0000000e+00, ...,
        -3.4206650e-07,   3.2979898e-07,   1.0042874e-06],
dtype=float32),
            'sampling_rate': 16000},
   'sentence': 'खीर की मिठास पर गरमाई बिहार की सियासत, कुशवाहा ने दी सफाई'}
```

When we reload the first audio sample, it will be at the 16 kHz sampling rate we need.

Now, the sampling rate is down to 16 kHz. The values in the array have also changed – we now have about one value for every three we had before.

Next, let's write a function to get our data ready for the model:

```
def prepare_dataset(batch):
    # load and resample audio data from 48 to 16kHz
    audio = batch["audio"]

    # compute log-Mel input features from input audio array
    batch[«input_features"] = feature_extractor(audio["array"],
sampling_rate=audio["sampling_rate"]).input_features[0]

    # encode target text to label ids
    batch[«labels»] = tokenizer(batch[«sentence»]).input_ids
    return batch
```

In the preceding snippet, we do the following:

- Load and resample the audio by calling batch["audio"]. As mentioned previously, Hugging Face Datasets will automatically resample the audio.

- Use the feature extractor to turn the one-dimensional audio array into log-Mel spectrogram input features.

- Convert the transcripts into label IDs using the tokenizer.

Now that we have the prepare_dataset() function defined, we can apply this data preparation function to all our training examples using the dataset's .map method:

```
common_voice = common_voice.map(prepare_dataset, remove_
columns=common_voice.column_names["train"], num_proc=4)
```

There we go! Our data is now fully prepped for training. Let's move on to how to use this data to fine-tune Whisper.

> **Note**
> The datasets currently use both `torchaudio` and `librosa` for audio handling. If you want to do your own audio loading or sampling, you can use the `path` column to find the audio file location and ignore the `audio` column.

As we culminate our exploration of synthesizing `WhisperProcessor`, merging the feature extractor and tokenizer into a unified workflow, we transition toward defining training parameters and hardware configurations. This crucial juncture signifies our preparation for the intricate task of fine-tuning, emphasizing the strategic selection of training parameters and hardware configurations that align with our learning project's scale and complexity.

# Milestone 5 – Defining training parameters and hardware configurations

Now that our data is ready, we can start training our model. We'll use the Hugging Face Trainer to help with most of the work. The Hugging Face `Trainer` class provides a feature-complete training and evaluation loop for PyTorch models optimized for Transformers. It supports distributed training on multiple GPUs/TPUs and mixed precision and offers a lot of customizability for users. The `Trainer` class abstracts away the complexities of the training loop, allowing users to focus on providing the essential components required for training, such as a model and a dataset. Here's what we need to do:

1. **Set up a data collator**: This tool takes our prepared data into PyTorch tensors that the model can use.

2. **Choose evaluation metrics**: We want to see how well the model performs using the **word error rate (WER)** metric. To perform this calculation, we'll create a function called compute_metrics.

3. **Load a pre-trained model**: We'll start with an already-trained model and set it up for further training. Training Whisper from scratch is not an option due to the intense data and computing resources required for such a task.

4. **Define training arguments**: These arguments guide the Hugging Face Trainer on how to train the model.

After fine-tuning the model, we'll test it on new data to ensure it can accurately transcribe speech in Hindi.

## Setting up the data collator

The data collator for speech models like ours is a bit special. It handles *input features* and *labels* separately: the feature extractor manages the *input features*, whereas the tokenizer manages the *labels*.

The input features are set to 30 seconds and have been turned into a fixed-size log-Mel spectrogram. We just need to convert them into grouped *PyTorch tensors*. We can do this using the feature extractor's `self.processor.tokenizer.pad` method with `return_tensors="pt"`. Since the input features are already fixed in size, we're just changing them into *PyTorch tensors* without adding extra padding.

The labels, however, still need to be padded. First, we must pad them to the longest length in our batch using the `self.processor.tokenizer.pad` method. We are replacing the padding tokens with `-100` so they don't affect the loss calculation. We must also remove the start of the transcript token from the beginning of the label sequence, as we'll add it back during training.

We can use the `WhisperProcessor` class we made earlier to handle the feature extractor and tokenizer tasks:

```
import torch

from dataclasses import dataclass
from typing import Any, Dict, List, Union

@dataclass
class DataCollatorSpeechSeq2SeqWithPadding:
    processor: Any

    def __call__(self, features: List[Dict[str, Union[List[int],
torch.Tensor]]]) -> Dict[str, torch.Tensor]:
        # split inputs and labels since they have to be of different
lengths and need different padding methods
        # first treat the audio inputs by simply returning torch
tensors
        input_features = [{"input_features": feature["input_
features"]} for feature in features]
        batch = self.processor.feature_extractor.pad(input_features,
return_tensors="pt")

        # get the tokenized label sequences
        label_features = [{"input_ids": feature["labels"]} for feature
in features]
        # pad the labels to max length
        labels_batch = self.processor.tokenizer.pad(label_features,
return_tensors="pt")

        # replace padding with -100 to ignore loss correctly
        labels = labels_batch["input_ids"].masked_fill(labels_batch.
attention_mask.ne(1), -100)
```

```
        # if bos token is appended in previous tokenization step,
        # cut bos token here as it's append later anyways
        if (labels[:, 0] == self.processor.tokenizer.bos_token_id).
all().cpu().item():
            labels = labels[:, 1:]

        batch["labels"] = labels

        return batch
```

Now, let's instantiate the data collator we've just defined:

```
data_collator =
DataCollatorSpeechSeq2SeqWithPadding(processor=processor)
```

# Milestone 6 – Establishing standardized test sets and metrics for performance benchmarking

Now, let's learn how to check our model's performance. We'll use the WER metric, a common way to evaluate speech recognition systems. We'll load the WER metric from Hugging Face `evaluate`:

```
import evaluate
metric = evaluate.load("wer")
```

Next, we'll create a function called `compute_metrics` to calculate the WER:

```
def compute_metrics(pred):
    # [Code to replace -100, decode predictions and labels, and
compute WER]
    return {"wer": wer}
```

This function fixes our `label_ids` (where we had replaced padding tokens with `-100`). Then, it turns both the predicted and label IDs into text strings. Lastly, it calculates the WER between these two.

## Loading a pre-trained model checkpoint

We'll start with a pre-trained Whisper model. This is easy with Hugging Face Transformers:

```
from transformers import WhisperForConditionalGeneration
model = WhisperForConditionalGeneration.from_pretrained("openai/
whisper-small")
```

This model has settings that we need to adjust for training. We'll set specific tokens to None and make sure no tokens are suppressed:

```
model.config.forced_decoder_ids = None
model.config.suppress_tokens = []
```

## Defining training arguments

We must define the training details, such as where to save the model and how often to check its performance and other settings. There is a particular class called Seq2SeqTrainingArguments for explicitly declaring training arguments. A subset of the argument parameters are explained here:

- output_dir: The local directory in which to save the model weights. This will also be the repository name on the Hugging Face Hub (https://huggingface.co/).

- generation_max_length: The maximum number of tokens to autoregressively generate during evaluation.

- save_steps: The intermediate checkpoints will be saved and uploaded asynchronously to the Hub every save_steps training step during training.

- eval_steps: During training, intermediate checkpoints will be performed every eval_steps training step.

- report_to: Where to save training logs. Supported platforms are azure_ml, comet_ml, mlflow, neptune, tensorboard, and wand. Pick your favorite or leave it as tensorboard to log into the Hub.

For more details on the other training arguments, refer to the Seq2SeqTrainingArguments documents (https://huggingface.co/docs/transformers/v4.40.1/en/main_classes/trainer#trainer).

The following code snippet illustrates the declaration of Seq2SeqTrainingArguments with some of the parameters. You will find a complete working example in the companion Python notebook in this book's GitHub repository:

```
from transformers import Seq2SeqTrainingArguments
training_args = Seq2SeqTrainingArguments(
    output_dir="./whisper-small-hi",
    per_device_train_batch_size=16,
    gradient_accumulation_steps=1,
    learning_rate=1e-5,
    warmup_steps=500,
    max_steps=4000,
    gradient_checkpointing=True,
    fp16=True,
```

```
evaluation_strategy="steps",
per_device_eval_batch_size=8,
predict_with_generate=True,
generation_max_length=225,
save_steps=1000,
eval_steps=1000,
logging_steps=25,
report_to=["tensorboard"],
load_best_model_at_end=True,
metric_for_best_model="wer",
greater_is_better=False,
hub_model_id = "your-huggingface-id/whisper-small-hi",
push_to_hub=True,
)
```

> **Note**
>
> If you don't want to upload the model to the Hub, set `push_to_hub=False`.

We'll give these training details to the Hugging Face Trainer, along with our `model`, `dataset`, `data collator`, and `compute_metrics` functions:

```
from transformers import Seq2SeqTrainer
trainer = Seq2SeqTrainer(
# [Details of the trainer setup]
trainer = Seq2SeqTrainer(
    args=training_args,
    model=model,
    train_dataset=common_voice["train"],
    eval_dataset=common_voice["test"],
    data_collator=data_collator,
    compute_metrics=compute_metrics,
    tokenizer=processor.feature_extractor,
)
```

With robust metrics for evaluating model performance and a transparent process defined for executing the training, we'll now focus on a practical implementation – executing optimized training loops while leveraging our configured hyperparameters, datasets, and hardware.

# Milestone 7 – Executing the training loops

To begin training, just run the following command:

```
trainer.train()
```

*Figure 4.1* shows an example of the output you can expect to see from the `trainer.train()` command's execution:

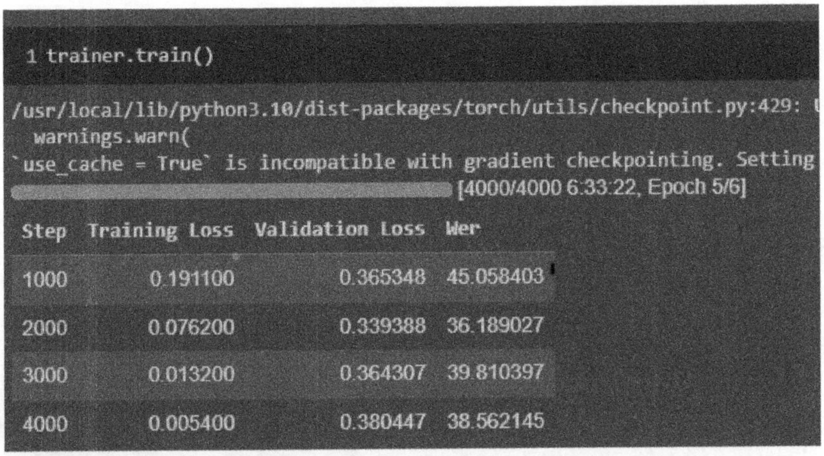

```
1 trainer.train()

/usr/local/lib/python3.10/dist-packages/torch/utils/checkpoint.py:429:
  warnings.warn(
`use_cache = True` is incompatible with gradient checkpointing. Setting
                              [4000/4000 6:33:22, Epoch 5/6]
```

| Step | Training Loss | Validation Loss | Wer |
|------|---------------|-----------------|-----------|
| 1000 | 0.191100 | 0.365348 | 45.058403 |
| 2000 | 0.076200 | 0.339388 | 36.189027 |
| 3000 | 0.013200 | 0.364307 | 39.810397 |
| 4000 | 0.005400 | 0.380447 | 38.562145 |

Figure 4.1 – Sample output from trainer.train() in Google Colab

Each training batch will have an evaluation step that calculates and displays training/validation losses and WER metrics. Depending on your GPU, training could take 5–10 hours. If you run into memory issues, try reducing the batch size and adjusting `gradient_accumulation_steps` in the declaration of `Seq2SeqTrainingArguments`.

Because of the parameters we established when declaring `Seq2SeqTrainingArguments`, our model metrics and performance will be pushed to the Hugging Face Hub with each training iteration. The key parameters driving that push to the Hub are shown here:

```
from transformers import Seq2SeqTrainingArguments
training_args = Seq2SeqTrainingArguments(
    [… previous parameters here]
    report_to=["tensorboard"],
    load_best_model_at_end=True,
    metric_for_best_model="wer",
    greater_is_better=False,
    hub_model_id = "your-huggingface-id/whisper-small-hi",
    push_to_hub=True,
)
```

The following screenshots show how to navigate to the Hugging Face TensorBoard and examples of the board with metrics from one of my fine-tuned models:

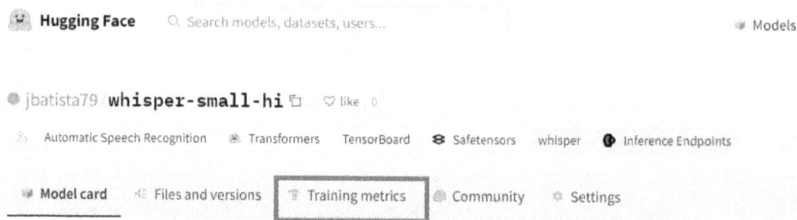

Figure 4.2 – Within the Hugging Face repository, select "Training metrics" to display the TensorBoard

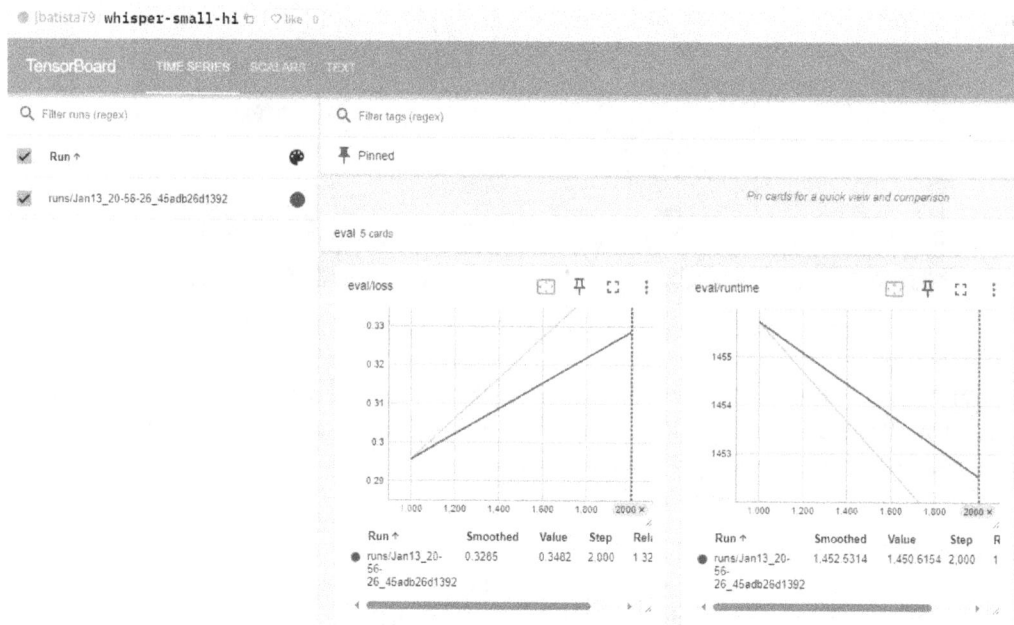

Figure 4.3 – Example of some of the metrics in the Hugging Face TensorBoard

After successful training, anyone can access and use your model via the Hugging Face Hub. They can load it using a link from the Hub or use the `your-hugging-face-id/the-name-you-picked` identifier. Here's an example of how to load the model:

```
from transformers import WhisperForConditionalGeneration,
WhisperProcessor
model = WhisperForConditionalGeneration.from_pretrained("jbatista79/
whisper-small-hi")
processor = WhisperProcessor.from_pretrained("jbatista79/whisper-
small-hi")
```

While the model we've fine-tuned works well with the Common Voice Hindi test data, it's not perfect. This guide is meant to show you how to fine-tune pre-trained Whisper models on any speech recognition dataset in multiple languages. You might get even better results by tweaking the training settings, such as learning rate and dropout, or using a bigger pre-trained model (such as the medium or large versions).

With the optimized training process complete and our fine-tuned model uploaded, we'll now transition to assessing the real-world efficacy of our enhanced speech recognition capabilities. We will validate how our tailored Whisper model generalizes across languages, domains, and acoustic environments by benchmarking performance across diverse datasets.

# Milestone 8 – Evaluating performance across datasets

As we conclude our Whisper fine-tuning journey, validating model performance across diverse real-world conditions represents a pivotal final milestone. Before deploying our optimized speech recognizer into production scenarios, comprehensively assessing its effectiveness across datasets, languages, accents, and acoustic environments is essential for instilling confidence. This testing phase unveils actual capabilities, revealing where additional tuning may be required while spotlighting areas suitable for immediate application. The rigorous evaluation processes outlined in this section aim to verify customized performance gains while guiding ethical and inclusive rollout by covering key facets such as bias mitigation, domain optimization, translation abilities, and expectation management.

## Mitigating demographic biases

Machine learning models, including those for speech recognition, can sometimes detect biases against certain genders, ethnicities, or age groups. This happens because the audio data they learn from can vary greatly between different groups of people. To prevent this, we must train the model with a wide range of data and use unique methods to check for biases.

We should carefully examine where the model might work better for certain groups of people. This will help us understand which groups might need more support from the model. We can also change the data the model learns from to see if it treats different groups of people differently. This will help us find the real reasons for any unfairness.

Finding problems is not enough. We also need to add a variety of data to the model. This means getting data from many different sources, especially those that haven't been included much before. We can use methods such as web scraping to find new kinds of speech data. We can also create artificial voices, but we must be careful and transparent about how we do this.

We need to be careful to avoid overcorrecting and creating new problems. Our goal is to improve the model for everyone. We can do this by testing it equally with different groups of people to ensure it works well for everyone.

We should aim to use language technology to unite people, not separate them. We should focus on making speech technology that is fair and helpful for everyone. This means constantly checking and improving our models to ensure they are fair and helpful for all different groups of people.

## Optimizing for content domains

While Whisper's extensive pre-training provides broad linguistic capabilities, tailoring its knowledge toward specialized domains is pivotal for competitive enterprise use cases. Contact centers, legal firms, finance brokers, telemedicine providers—speech recognition permeates diverse industries, each carrying distinct challenges. Beyond vocabulary, accurately modeling nonverbal cues, discourse patterns, and subtle connotations underpins contextual understanding in niche domains.

Yet out-of-the-box ASR systems often stumble on niche terminology and struggle to convey implicitly layered meaning. For example, a precise understanding of clauses has substantive significance in legal contexts. Models trained exclusively on generic datasets fail to distill these specialized connotations. Exposing systems to targeted in-domain data is thus vital for infusing contextual mastery.

The nucleus of domain optimization lies in terminology mastery. Legal, medical, and financial contexts involve extensive exotic lexicons that shape substantive task competencies. Yet glossaries alone fail to encapsulate the layered semantics encoded in specialist dictionaries.

One option is to employ **explicit semantic analysis (ESA)**, a computational method for mathematically representing human notions of language meaning. ESA is a high-dimensional space of concepts derived from a large text corpus, and it is used in NLP and information retrieval.

In simple terms, ESA is a way for computers to understand the meaning of a piece of text by comparing it to a large amount of text data it has already analyzed. It does this by mapping the text to a set of concepts or topics derived from a large corpus of text data. This mapping is done in a high-dimensional space, where each dimension represents a different concept or topic.

For example, if the text is about "dogs," ESA might map it to concepts such as "animals," "pets," "canines," and so on. By doing this, ESA can understand the semantic meaning of the text, which can be used for tasks such as information retrieval, text classification, and more. ESA is beneficial because it can capture the meaning of text even when the words used are not the same. For instance, it can be understood that "dogs" and "canines" refer to the same concept, even though the words are different. This makes it a powerful tool for understanding and processing natural language.

## Managing user expectations

Responsible use of AI speech recognition technology involves ensuring users understand what the technology can and cannot do. It's essential to be open about the technology's capabilities and limits so that people can make informed choices about using it. This is especially crucial for those who might not have much digital experience.

Effective communication about technology's abilities helps build trust. This can be done through easy-to-understand summaries and explanations that address specific user needs without overwhelming them with too much detail. Tools such as model confidence scores and visualizations can help users gauge the reliability of the technology's predictions, making it more transparent when and how it's best used.

Being upfront about what technology can't do is just as important. Recognizing limitations is not a sign of failure; it's an opportunity for growth and improvement. For example, areas where Whisper might struggle, such as real-time recognition in noisy environments, should be seen as challenges to be solved through collaborative effort rather than permanent flaws.

Listening to users and incorporating their feedback is critical to improving speech recognition technology for everyone. Regularly checking how the technology performs in real-world situations helps prevent it from drifting away from users' needs. By involving users in the process via humans-in-the-loop, we can focus on addressing the most pressing issues and make improvements more efficiently.

## Milestone 9 – Building applications that demonstrate customized speech recognition

Now that our model has been fine-tuned let's demonstrate how good it is at speech recognition (ASR)! We'll use the Hugging Face Transformers pipeline to handle everything, from preparing the audio to decoding what the model thinks the audio says. For our demo, we'll use **Gradio**, a tool that makes it super easy to build machine learning demos. You can create a demo with Gradio in just a few minutes!

Here is an example of a Gradio demo. In this demo, you can record speech using your computer's microphone, after which the fine-tuned Whisper model will transcribe it into text:

```
from transformers import pipeline
import gradio as gr


pipe = pipeline(model="jbatista79/whisper-small-hi")   # change to
"your-username/the-name-you-picked"


def transcribe(audio):
    text = pipe(audio)["text"]
    return text


iface = gr.Interface(
    fn=transcribe,
    inputs=gr.Audio(source="microphone", type="filepath"),
    outputs="text",
    title="Whisper Small Hindi",
```

```
    description="Realtime demo for Hindi speech recognition using a
fine-tuned Whisper small model.",
)

iface.launch()
```

Here's the output:

Figure 4.4 – Example of Gradio's user interface for the fine-tuned Whisper model in Hugging Face

Record the audio with the microphone to test the model directly from Google Colab;, then click **Submit**. I am not proficient in the Hindi language, but I managed to record `Namaste`, which was then transcribed perfectly to the Hindi word नामास्ते.

## Summary

As we conclude our journey into the intricacies of OpenAI's Whisper, it's clear that we've traversed a path rich with technical insights and practical wisdom. Our exploration has been more than just a theoretical examination; it's been a hands-on experience, equipping you with the skills to fine-tune Whisper for specific domain and language needs and to overcome the challenges inherent in speech recognition technology.

We commenced with the foundational work of setting up a robust Python environment, augmenting Whisper's knowledge by integrating diverse, multilingual datasets such as Common Voice. This step was crucial as it expanded Whisper's linguistic breadth and set the stage for the subsequent fine-tuning process.

The heart of this chapter revolved around tailoring Whisper's predictions to align perfectly with your target applications. You've learned to tweak confidence levels, output classes, and time limits to match the expected results in specific use cases. The knowledge you've gained here is invaluable, especially when dealing with niche terminologies and diverse language datasets.

Much of our effort was devoted to tracking progress through straightforward testing. We established fixed benchmarks to gauge gains across languages and uses objectively, ensuring that our fine-tuning efforts were grounded and free from data bias.

One of the most critical aspects we covered was the ethical use of technology. We emphasized the need to ensure equitable performance across demographics, ensuring that advancements in speech technology don't inadvertently exclude groups with fewer advantages.

As you've seen, fine-tuning Whisper involved a deep dive into its architecture and training methodologies. You've learned about handling different languages, optimizing Whisper for various content domains, and balancing accuracy with efficiency. We've also tackled challenges such as demographic biases, technical and linguistic hurdles, and the need for rapid adaptation to new vocabulary.

Moreover, we've discussed managing user expectations, an essential aspect of deploying AI technology. It's crucial to be transparent about what technology can do and its limitations, ensuring users make informed decisions and trust it.

As we look forward to this book's next section, *Part 3 – Real-World Applications and Use Cases*, we're poised to embark on a new adventure. Here, we'll explore how to effectively apply Whisper in various industries, integrating it into real-world scenarios. You'll discover how to harness Whisper in sectors such as healthcare and voice-assisted technologies, leveraging the skills and knowledge you've gained from this chapter to make a tangible impact in ASR.

So, let's carry forward the knowledge and experience from this chapter and see how we can apply Whisper in diverse and impactful ways. The journey continues, and the possibilities are as exciting as they are endless.

# Part 3:
# Real-world Applications and Use Cases

In this part, you will explore the diverse real-world applications and use cases of OpenAI's Whisper, learning how to integrate this powerful tool into various contexts effectively. From transcription services and voice assistants to accessibility features and customer service, you will gain insights into leveraging Whisper's capabilities to enhance multiple industries. You will also delve into advanced techniques such as quantization, real-time speech recognition, and speaker diarization using **WhisperX** and NVIDIA's **NeMo** framework. Furthermore, you will discover how to harness Whisper for personalized voice synthesis, creating unique voice models that capture the distinct characteristics of a target voice. Finally, this part will provide a forward-looking perspective on the evolving landscape of ASR and voice technologies, discussing anticipated trends, ethical considerations, and strategies for preparing for the future.

This part includes the following chapters:

- *Chapter 5, Applying Whisper in Various Contexts*
- *Chapter 6, Expanding Applications with Whisper*
- *Chapter 7, Exploring Advanced Voice Capabilities*
- *Chapter 8, Diarizing Speech with WhisperX and NVIDIA's NeMo*
- *Chapter 9, Harnessing Whisper for Personalized Voice Synthesis*
- *Chapter 10, Shaping the Future with Whisper*

# 5
# Applying Whisper in Various Contexts

Welcome to *Chapter 5*, where we explore the remarkable capabilities of OpenAI's Whisper in transforming spoken language into written text. As we navigate various applications, including transcription services, voice assistants, chatbots, and accessibility features, you'll gain an in-depth understanding of Whisper's pivotal role in these domains.

First, we will explore transcription services and examine how Whisper streamlines the conversion of audio files, such as meetings and interviews, into text. Its accuracy and efficiency reduce the need for manual transcription, making it an indispensable tool.

Furthermore, we'll delve into the integration of Whisper into voice assistants and chatbots, enhancing their responsiveness and user interaction. By converting spoken commands into text, Whisper elevates these technologies to new levels of interactivity.

Regarding accessibility, this chapter highlights Whisper's contribution to tools for those with hearing or speech impairments. Its **voice-to-text** features not only offer practical solutions but also enrich user experiences.

In this chapter, we will cover the following topics:

- Exploring transcription services
- Integrating Whisper into voice assistants and chatbots
- Enhancing accessibility features with Whisper

By the end of this chapter, you will have a comprehensive understanding of how to apply Whisper effectively in various settings. You'll learn about the best practices for setup and optimization, discover innovative use cases, and appreciate ethical considerations in implementing this technology. With this knowledge, you'll be well equipped to leverage Whisper's full potential to enhance digital experiences across different domains.

Let's start by delving into the innovative world of transcription through Whisper, where we uncover how this cutting-edge technology is reshaping the way we convert spoken language into written text, enhancing efficiency and accuracy across various professional and personal settings.

# Technical requirements

To harness the capabilities of OpenAI's Whisper for advanced applications, this chapter leverages Python and Google Colab for ease of use and accessibility. The Python environment setup includes the Whisper library for transcription tasks.

Key requirements:

- **Google Colab notebooks**: The notebooks are set to run our Python code with the minimum required memory and capacity. If the **T4 GPU** runtime type is available, select it for better performance.

- **Python environment**: Each notebook contains directives to load the required Python libraries, including Whisper and Gradio.

- **Hugging Face account**: Some notebooks require a Hugging Face account and login API key. The Colab notebooks include information about this topic.

- **Microphone and speakers**: Some notebooks implement a Gradio app with voice recording and audio playback. A microphone and speakers connected to your computer might help you experience the interactive voice features. Another option is to open the URL link Gradio provides at runtime on your mobile phone; from there, you might be able to use the phone's microphone to record your voice.

- **GitHub repository access**: All Python code, including examples, is available in the chapter's GitHub repository (`https://github.com/PacktPublishing/Learn-OpenAI-Whisper/tree/main/Chapter05`). These Colab notebooks are ready to run, providing a practical and hands-on approach to learning.

By meeting these technical requirements, you will be prepared to explore Whisper in different contexts while enjoying the streamlined experience of Google Colab and the comprehensive resources available on GitHub.

# Exploring transcription services

From capturing the nuances of a brainstorming session to documenting pivotal interviews, transcription services bridge the gap between the ephemeral nature of speech and the permanence of text. Within this exploration, we will unravel the intricate dance between Whisper's advanced technology and ever-expanding transcription needs. This section lays the foundational knowledge of how Whisper, with its encoder-decoder transformer model, tackles diverse acoustic environments, accents, and dialects with remarkable precision. Yet, it doesn't shy away from discussing current limitations and vibrant community efforts to push the boundaries further.

We will also transition from the theoretical to the practical. From installing dependencies to running the model, it equips you with the knowledge to turn audio files into accurate text transcripts efficiently. We will optimize Whisper's performance, ensuring transcriptions are accurate and seamlessly integrated into various applications, from subtitling to detailed content analysis.

By the end of this section, you'll have grasped Whisper's vital role in transcription services and be armed with the know-how to harness its capabilities effectively. This journey is a pathway to unlocking the full potential of voice within the digital landscape, making information accessible, and enhancing communication across diverse domains.

## Understanding the role of Whisper in transcription services

Understanding the role of Whisper in transcription services requires a deep dive into its capabilities, limitations, and potential for integration into various applications. As we embark on this exploration, we will not only appreciate the technical prowess of Whisper but also consider its practical implications in the transcription landscape.

Whisper's architecture, an encoder-decoder transformer model, is adept at handling a wide range of audio inputs. Whisper ensures that each speech segment is given attention by converting audio into a log-Mel spectrogram and processing it in 30-second chunks. This meticulous approach to audio processing is one of the reasons behind Whisper's high accuracy in transcription.

The robustness of Whisper to accents, background noise, and technical language is particularly noteworthy. In transcription services, these factors are often the bane of accuracy and reliability. Whisper's resilience in these areas means it can provide high-quality transcriptions across diverse acoustic conditions, which is invaluable for businesses and individuals requiring precise spoken content documentation.

While Whisper excels at transcription, it is essential to note its limitations in speaker diarization, distinguishing between different speakers in an audio file. However, the community around Whisper is actively exploring ways to enhance its capabilities, for example, integrating it with other models such as **Pyannote** for speaker identification. We will learn more about diarization and Pyannote in the following chapters. Additionally, Whisper's word-level timestamping feature is a significant step forward, enabling users to synchronize transcribed text with audio, a crucial requirement for applications such as subtitling and detailed content analysis.

### A brief introduction to Pyannote

Pyannote is an open source toolkit designed for speaker diarization, a process crucial in analyzing conversations by identifying when and by whom each utterance is spoken. Developed by Hervé Bredin, Pyannote leverages the PyTorch machine learning framework to provide trainable, end-to-end neural components. These components can be combined and jointly optimized to construct speaker diarization pipelines. `pyannote.audio`, one element of this toolkit, comes with pre-trained models and pipelines that cover a wide range of domains, including **voice activity detection** (**VAD**), speaker segmentation, overlapped speech detection, and speaker embedding. It achieves state-of-the-art performance in most of these areas.

The relationship between Pyannote and OpenAI Whisper in the context of diarization is complementary. Pyannote can perform the diarization task, identifying different speakers within an audio file, which Whisper can transcribe. This synergy allows for creating more detailed and valuable transcriptions that include speaker labels, enhancing the analysis of conversations. However, integrating these two systems can be complex and may only sometimes yield ideal results, as noted by some users.

Despite these challenges, combining Pyannote's diarization capabilities with Whisper's transcription prowess represents a powerful tool for speech analysis, especially when accurate speaker identification is required.

From a business perspective, the cost of transcription services is a critical factor. If using OpenAI's API, Whisper's competitive pricing at $0.006 per minute of audio makes it an attractive option for companies looking to incorporate transcription services without incurring excessive costs. Of course, Whisper is available via open source as well. This affordability and high accuracy position Whisper as a disruptive force in the transcription market.

The Whisper API's file size limit of 25 MB is a consideration for developers integrating the model into applications. While this may pose challenges for longer audio files, the community has devised strategies to work around this limitation, such as splitting audio files and using compressed formats. The API's ease of use and the potential for real-time transcription further enhance Whisper's appeal as a developer tool.

OpenAI's decision to open source Whisper has catalyzed innovation and customization. By providing access to the model's code and weights, OpenAI has empowered a community of developers to adapt and extend Whisper's capabilities. This leads to a modular future for AI, where tools such as Whisper serve as foundational building blocks for many applications.

As we look to the future, the role of Whisper in transcription services is set to become even more integral. With the model's continuous evolution and the growth of its surrounding community, we can anticipate advancements in diarization, language support, and other areas. The open source nature of Whisper ensures that it will remain at the forefront of innovation, driven by a collaborative effort to refine and perfect its transcription capabilities. This sets the stage for our next topic of discussion: setting up Whisper for transcription tasks, where we will delve into the practical steps and considerations for harnessing Whisper's capabilities to meet transcription needs effectively.

# Setting up Whisper for transcription tasks

Setting up Whisper for transcription tasks involves several steps, including installing dependencies, installing Whisper, and running the model. Use the `LOAIW_ch05_1_setting_up_Whisper_for_transcription.ipynb` Google Colab notebook (`https://github.com/PacktPublishing/Learn-OpenAI-Whisper/blob/main/Chapter05/LOAIW_ch05_1_setting_up_Whisper_for_transcription.ipynb`) from the book's GitHub repository for more comprehensive hands-on implementation. In the notebook, we'll walk through the end-to-end process of preparing your environment, downloading sample audio, and transcribing it with Whisper. The following diagram describes the high-level steps:

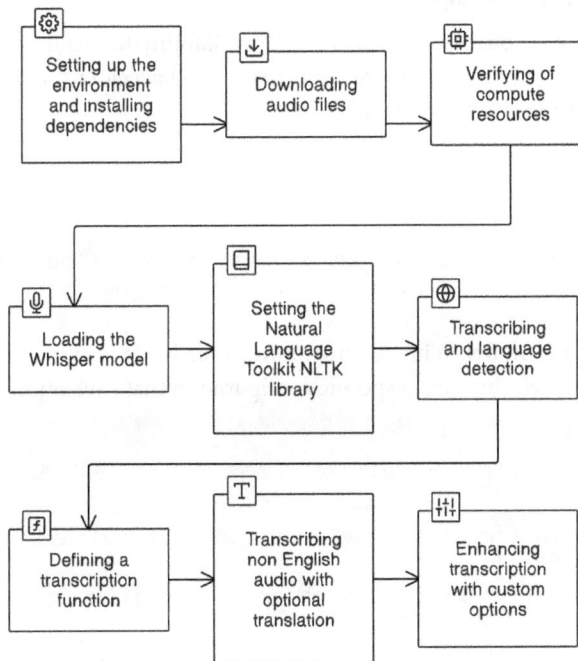

Figure 5.1 – Setting up Whisper for transcription tasks

*Figure 5.1* describes the step-by-step approach in the notebook, ensuring you have a solid foundation in using Whisper, from basic setup to exploring advanced transcription techniques. I encourage you to find and run the entire notebook from the GitHub repository. Here are the high-level steps with some selected code snippets to illustrate:

1. **Installing necessary dependencies**: We begin by setting up our environment and installing crucial packages:

```
!sudo apt install ffmpeg
!pip install -q cohere openai tiktoken
!pip install -q git+https://github.com/openai/whisper.git
```

ffmpeg is used for audio file manipulation. The cohere and openai Python libraries offer various AI models for **natural language processing and understanding (NLP/NLU)**, which could be helpful for postprocessing or analyzing the transcribed text. tiktoken is required as a supporting library for authentication or token handling in the context of API requests. We also installed the latest Whisper files from the official OpenAI GitHub repository. These steps ensure we have all the tools ready for our transcription tasks.

2.  **Downloading audio samples**: Next, we download various audio samples, both in English and Spanish, from a GitHub repository and OpenAI's **content delivery network** or **CDN**. These samples will serve as our testing ground, allowing us to explore Whisper's transcription capabilities across different languages.

3.  **Verifying compute resources**: We check GPU availability to ensure efficient processing. Whisper's performance significantly benefits from GPU acceleration, so we configure our environment to use the GPU if available:

```
import numpy as np
import torch
torch.cuda.is_available()
DEVICE = "cuda" if torch.cuda.is_available() else "cpu"
print(f"Using torch {torch.__version__} ({DEVICE})")
```

4.  **Loading the Whisper model**: With our environment ready, we load the Whisper "medium" size multilingual model, choosing a specific configuration that suits our needs:

```
import whisper
model = whisper.load_model("medium", device=DEVICE)
print(
    f"Model is {'multilingual' if model.is_multilingual else
'English-only'} "
    f"and has {sum(np.prod(p.shape) for p in model.
parameters()):,} parameters."
)
```

5.  **Setting up the Natural Language Toolkit (NLTK) for text processing**: We install and set up NLTK to enhance the readability of our transcriptions. NLTK helps segment the transcribed text, making it easier to read and understand.

6.  **Transcribing audio with language detection**: This section showcases the transcription of audio files, incorporating Whisper's automatic language detection feature. It's a crucial step for handling multilingual content, ensuring that transcriptions are accurate regardless of the audio's language. The whisper.DecodingOptions() function in OpenAI's Whisper class is used to specify various options that control the behavior of the decoding process when transcribing audio. The parameters in the DecodingOptions function allow users to specify options such as the language for transcription, whether timestamps should be included, and

whether to use **16-bit floating-point numbers** (**FP16**) for computations. Here's an example of how to use DecodingOptions in conjunction with the whisper.decode() function:

```
for audiofile in audiofiles:
    # Load audio and pad/trim it to fit 30 seconds
    audio = whisper.load_audio(audiofile)
    audio = whisper.pad_or_trim(audio)
    # Make log-Mel spectrogram and move to the same device as
the model
    mel = whisper.log_mel_spectrogram(audio).to(model.device)
    #Next we detect the language of your audio file
    _, probs = model.detect_language(mel)
    detected_language = max(probs, key=probs.get)
    print(f"----\nDetected language: {detected_language}")
    # Set up the decoding options
    options = whisper.DecodingOptions(language=detected_
language, without_timestamps=True, fp16=(DEVICE == "cuda"))
    # Decode the audio and print the recognized text
    result = whisper.decode(model, mel, options)
```

In this example, the DecodingOptions function is set with three options:

- language=detected_language: This specifies the language of the transcription. Setting the language can improve the transcription accuracy if you know the language in advance and want to rely on something other than the model's automatic language detection.

- without_timestamps=True: When set to True, this option indicates that the transcription should not include timestamps. If you require timestamps for each word or sentence, you will set this to False.

- fp16=(DEVICE == "cuda"): This option determines whether to use FP16 (16-bit floating-point precision) for the decoding. The (DEVICE == "cuda") evaluation checks if CUDA is available. Earlier in the notebook, we used DEVICE = "cuda" if torch.cuda.is_available() else "cpu" to set DEVICE accordingly. Then, it sets fp16 to True if DEVICE is "cuda", meaning you plan to run the model on a GPU. If DEVICE is "cpu", it sets fp16 to False, ensuring compatibility and avoiding unnecessary warnings or errors.

These options can be adjusted based on the specific requirements of your transcription task. For instance, if you transcribe audio in a different language, you will change the language option accordingly. If you need to optimize for performance and your hardware supports it, you might enable fp16 to use half precision. FP16 (16-bit floating-point) computation is beneficial on compatible GPUs, as it can significantly reduce memory usage and potentially increase computation speed without substantially affecting the model's accuracy. However, not all CPUs support FP16 computation, and attempting to use it on a CPU can lead to errors or fallbacks to FP32 (single-precision floating-point) computation.

7. **Defining a function for streamlined transcription**: We introduce a custom function to streamline the transcription process. This function simplifies handling multiple files, and we explore how to incorporate translation options within it, enhancing its utility:

```
def process_file(audiofile, model, w_options, w_
translate=False):

    # Load audio
    audio = whisper.load_audio(audiofile)
    transcribe_options = dict(task="transcribe", **w_options)
    translate_options = dict(task="translate", **w_options)

    transcription = model.transcribe(audiofile, **transcribe_
options)["text"]
    if w_translate:
        translation = model.transcribe(audiofile, **translate_
options)["text"]
    else:
        translation = "N/A"
    return transcription, translation
```

8. **Handling non-English audio**: We use the previously defined `process_file()` function to transcribe non-English audio samples. This demonstrates Whisper's robust support for multiple languages, showcasing its effectiveness in a global context:

```
w_options = dict(without_timestamps=True, fp16=(DEVICE ==
"cuda"))
audiofile = 'Learn_OAI_Whisper_Spanish_Sample_Audio01.mp3'
transcription, translation = process_file(audiofile, model, w_
options, False)

print("------\nTranscription of file '" + audiofile + "':")
for sent in sent_tokenize(transcription):
    print(sent)
print("------\nTranslation of file '" + audiofile + "':")
for sent in sent_tokenize(translation):
    print(sent)

import ipywidgets as widgets
widgets.Audio.from_file(audiofile, autoplay=False, loop=False)
```

9. **Using advanced transcription techniques**: Lastly, we delve into advanced techniques to improve transcription accuracy further. This includes using custom prompts using Whisper's `initial_prompt` parameter and adjusting settings such as the temperature. We will examine two methods that refine the transcription output using `initial_prompt`, especially for audio with ambiguously spelled words or specialized terminology:

- **Using a spelling guide**: The first method is about providing Whisper with a spelling guide via the `initial_prompt` parameter. That approach is helpful when facing a common challenge: accurate transcription of uncommon proper nouns, such as product names, company names, or individuals. These elements often trip up even the most sophisticated transcription tools, leading to misspellings. A simple transcription without `initial_prompt` values results in the following:

```
------
Transcription of file 'product_names.wav':
Welcome to Quirk, Quid, Quill, Inc., where finance meets
innovation.
Explore diverse offerings from the P3 Quattro, a unique
investment portfolio quadrant to the O3 Omni, a platform for
intricate derivative trading strategies.
Delve into unconventional bond markets with our B3 Bond X and
experience non-standard equity trading with E3 Equity.
Surpass your wealth management with W3 Rap Z and anticipate
market trends with the O2 Outlier, our forward-thinking
financial forecasting tool.
Explore venture capital world with U3 Unifund or move your money
with the M3 Mover, our sophisticated monetary transfer module.
At Quirk, Quid, Quill, Inc., we turn complex finance into
creative solutions.
Join us in redefining financial services.
```

Next, we create a spelling guide using `initial_prompt`:

```
w_options = dict(without_timestamps=True, fp16=(DEVICE ==
"cuda"), temperature=0, initial_prompt="Quirk Quid Quill Inc.,
P3-Quattro, O3-Omni, B3-BondX, E3-Equity, W3-WrapZ, O2-Outlier,
U3-UniFund, M3-Mover")
audiofile = 'product_names.wav'
transcription, translation = process_file(audiofile, model, w_
options)

print("------\nTranscription of file '" + audiofile + "':")
for sent in sent_tokenize(transcription):
    print(sent)
```

This results in the following:

```
------
Transcription of file 'product_names.wav':
Welcome to Quirk Quid Quill Inc., where finance meets
innovation.
Explore diverse offerings from the P3-Quattro, a unique
investment portfolio quadrant to the O3-Omni, a platform for
intricate derivative trading strategies.
Delve into unconventional bond markets with our B3-BondX and
experience non-standard equity trading with E3-Equity.
```

```
Surpass your wealth management with W3-WrapZ and anticipate
market trends with the O2-Outlier, our forward-thinking
financial forecasting tool.
Explore venture capital world with U3-UniFund or move your money
with the M3-Mover, our sophisticated monetary transfer module.
At Quirk Quid Quill Inc., we turn complex finance into creative
solutions.
Join us in redefining financial services.
```

By including these names directly in the prompt to guide Whisper toward our preferred spellings, we created a glossary for Whisper to reference. The differences in the outputs before and after using the `initial_prompt` parameter are significant. Without `initial_prompt`, Whisper struggled with the proper nouns, resulting in misspellings such as "`Quirk, Quid, Quill, Inc.`", "`P3 Quattro`", "`O3 Omni`", "`B3 Bond X`", "`E3 Equity`", "`W3 Rap Z`", "`O2 Outlier`", "`U3 Unifund`", and "`M3 Mover`". However, after including the correct spellings in the `initial_prompt` parameter, Whisper accurately transcribed these terms as "`Quirk Quid Quill Inc.`", "`P3-Quattro`", "`O3-Omni`", "`B3-BondX`", "`E3-Equity`", "`W3-WrapZ`", "`O2-Outlier`", "`U3-UniFund`", and "`M3-Mover`". This demonstrates the power of the `initial_prompt` parameter in guiding Whisper to produce more accurate transcriptions, especially when dealing with uncommon or tricky terms.

- **Prompting for transcript generation**: The second method to refine the transcription output is using prompt engineering via the `initial_prompt` parameter. The most effective approach is to craft and provide either an actual or a fictitious prompt to steer Whisper using sure spellings, styles, or terminology. To illustrate the second method, we'll pivot to a different audio clip crafted specifically for this exercise. The scenario is an unusual barbecue event. Our first step involves generating a baseline transcript with Whisper to assess its initial accuracy. A simple transcription without `initial_prompt` values results in the following:

```
------
Transcription of file 'bbq_plans.wav':
Hello, my name is Preston Tuggle.
I am based in New York City.
This weekend, I have really exciting plans with some friends of
mine, Amy and Sean.
We're going to a barbecue here in Brooklyn.
Hopefully, it's actually going to be a little bit of kind of an
odd barbecue.
We're going to have donuts, omelets.
It's kind of like a breakfast as well as whiskey.
So that should be fun.
And I'm really looking forward to spending time with my friends,
Amy and Sean.
```

Next, let's apply a fictitious transcript, `"Aimee and Shawn had whisky, doughnuts, omelets at a BBQ."` via the `initial_prompt` parameter for Whisper to emulate:

```
w_options = dict(without_timestamps=True, fp16=(DEVICE ==
"cuda"), temperature=0, initial_prompt="""Aimee and Shawn had
whisky, doughnuts, omelets at a BBQ.""")
audiofile = 'bbq_plans.wav'
transcription, translation = process_file(audiofile, model, w_
options)

print("------\nTranscription of file '" + audiofile + "':")
for sent in sent_tokenize(transcription):
    print(sent)
```

This results in the following:

```
------
Transcription of file 'bbq_plans.wav':
Hello, my name is Preston Tuggle.
I'm based in New York City.
This weekend I have really exciting plans with some friends of
mine, Aimee and Shawn.
We're going to a BBQ here in Brooklyn.
Hopefully, it's actually going to be a little bit of kind of an
odd BBQ.
We're going to have doughnuts, omelets.
It's kind of like a breakfast, as well as whisky.
So that should be fun.
And I'm really looking forward to spending time with my friends,
Aimee and Shawn."
```

Using that prompting technique for unusual words with tricky spellings, we ensure our transcript is as accurate as possible in every detail. By comparing the first output without `initial_prompt` and the second output after proving a fictitious prompt, we got a more precise output (for example, "Aimee and Shawn" rather than "Amy and Sean", "doughnuts" instead of "donuts", "BBQ" rather than "barbeque", and "whisky" instead of "whiskey".

As you run the cells in the notebook, each section builds upon the previous ones, gradually introducing more complex features and techniques for using Whisper. This structured approach helps set up Whisper for transcription tasks and explores strategies for increasing transcription accuracy, catering to a wide range of audio content.

**Understanding the superpowers and limitations of Whisper's `initial_prompt`**

The `initial_prompt` parameter in OpenAI's Whisper is an optional text prompt providing context to the model for the first audio window being transcribed. Here are the key things to understand about `initial_prompt`:

**Purpose**: `initial_prompt` is used to prime the model with relevant context before it begins transcribing the audio. This can help improve transcription accuracy, especially for specialized vocabularies or desired writing styles.

**Scope**: `initial_prompt` only affects the first segment of audio being transcribed. For longer audio files that get split into multiple segments, its influence may diminish after the first 30-90 seconds of audio. For shorter audios, manually segmenting or splitting the audio and then applying the `initial_prompt` parameter is an option to overcome this limitation. For larger scripts, that segmentation could be automated. There is also the option to apply some postprocessing adjustments, including passing the entire transcript to an LLM with a more sophisticated prompt.

**Token limit**: Whisper will consider 224 tokens from `initial_prompt`. The documentation seems to be inconsistent about whether the first 224 or last 224 tokens are used, but in either case, anything beyond that limit is ignored.

**Prompt engineering**: `initial_prompt` does not have to be an actual transcript. Fictitious prompts can be crafted to steer Whisper using sure spellings, styles, or terminology. Techniques such as including spelling guides or generating prompts with GPT-3 can be effective.

**Differs from prompt**: The `initial_prompt` parameter differs from the `prompt` parameter, which provides the previous transcribed segment context for the current segment, helping maintain consistency across a long audio file.

The `initial_prompt` parameter is a way to frontload relevant context to Whisper to improve transcription accuracy. However, its impact is limited to the beginning of the audio and subject to a token limit. Thus, it is a useful but bounded tool for enhancing Whisper's performance on niche audio content.

Now, let's go deeper into transcription techniques to gain a more comprehensive understanding of the options available in Whisper. Applying these techniques, you'll be well prepared to tackle various audio-processing tasks, from simple transcriptions to more complex, multilingual projects.

# Transcribing audio files with Whisper efficiently

Before delving into the relevant parameters, let's consider the model size selection: tiny, base, small, medium, or large. That choice directly impacts the balance between transcription speed and accuracy. For instance, while the medium model offers a faster transcription rate, the large model excels in accuracy, making it the preferred choice for applications where precision is non-negotiable. The model's accuracy escalates with its size, positioning the large model as the pinnacle of precision. The large model is the benchmark for reported accuracies in the literature (*Efficient and Accurate Transcription in Mental Health Research - A Tutorial on Using Whisper AI for Audio File Transcription* – November 10, 2023 – `https://osf.io/preprints/osf/9fue8`), underscoring its significance for tasks where accuracy is paramount.

My practical experience has underscored the necessity of selecting the appropriate model size and computational resources. Running Whisper, especially its more significant variants, efficiently requires GPU acceleration to reduce transcription times significantly. For instance, testing has shown that using a GPU can dramatically reduce the time it takes to transcribe a minute of audio. Furthermore, it's essential to consider the trade-off between speed and accuracy when choosing the model size. For example, while the medium model is twice as fast as the large model, the large model offers increased accuracy.

## *Selecting key inference parameters for optimized transcription*

Configuring inference parameters and decoding options in OpenAI's Whisper is crucial for achieving accurate transcriptions, as these settings can significantly impact the performance and precision of the transcription process. This exploration enhances transcription accuracy and optimizes performance, fully leveraging Whisper's capabilities. In my experience, parameters such as `temperature`, `beam_size`, and `best_of` emerged as pivotal in fine-tuning Whisper's transcription capabilities.:

- The `temperature` parameter controls the level of variability in the generated text, which can result in more accurate transcriptions.

- The `beam_size` parameter is critical in decoding, influencing the breadth of the search for potential transcriptions. A larger `beam_size` value can improve the transcription accuracy by considering a more comprehensive array of possibilities.

- Similarly, `best_of` allows us to control the diversity of the decoding process, selecting the best result from multiple attempts. This can be particularly useful in achieving the highest possible accuracy in our transcriptions.

**Understanding the relationship among the temperature, beam_size, and best_of inference parameters**

The `beam_size` parameter in the Whisper model refers to the number of beams used in beam search during the decoding process. Beam search is a heuristic search algorithm that explores a graph by expanding the most promising node in a limited set. In the context of Whisper, beam search is used to find the most likely sequence of words given the audio input.

The `temperature` parameter controls the randomness of the output during sampling. A higher temperature produces more random outputs, while a lower temperature makes the model's outputs more deterministic. When the temperature is set to zero, the model uses a greedy decoding strategy, always choosing the most likely next word.

`beam_size` and `temperature` influence the decoding strategy and the diversity of the generated text. A larger `beam_size` value can increase the accuracy of the transcription by considering more alternative word sequences, but it also requires more computational resources and can slow down the inference process. On the other hand, `temperature` affects the variability of the output; a nonzero temperature allows for sampling from a distribution of possible following words, which can introduce variability and potentially capture more nuances in the speech.

In practice, the `beam_size` parameter is used when the temperature is set to zero, indicating that beam search should be used. If the temperature is nonzero, the `best_of` parameter is used instead to determine the number of candidates to sample from. The Whisper model uses a dynamic temperature setting, starting with a temperature of 0 and increasing it by `0.2` up to `1.0` when certain conditions are met, such as when the average log probability over the generated tokens is lower than a threshold or when the generated text has a *gzip* compression rate higher than a specific value.

In summary, `beam_size` controls the breadth of the search in beam search decoding, and `--temperature` controls the randomness of the output during sampling. They are part of the decoding strategy that affects the final transcription or translation produced by the Whisper model.

Configuring parameters and decoding options in Whisper is a nuanced process that requires a deep understanding of the model and its capabilities. By carefully adjusting these settings, users can optimize the accuracy and performance of their transcriptions, making Whisper a powerful tool for a wide range of applications. As with any AI model, it's essential to thoroughly test and validate the results in the specific context of your use case to ensure they meet your requirements. The following section goes even deeper into a hands-on notebook specifically designed to showcase the power of runtime parameters during decoding in Whisper.

## Applying Whisper's runtime parameters in practice

This section will explore the `LOAIW_ch05_2_transcribing_and_translating_with_Whisper.ipynb` Colab notebook (`https://github.com/PacktPublishing/Learn-OpenAI-Whisper/blob/main/Chapter05/LOAIW_ch05_2_transcribing_and_translating_with_Whisper.ipynb`) for more comprehensive hands-on implementation.

I encourage you to find the notebook in the book's GitHub repository and run it in Google Colab. The notebook is designed to demonstrate the installation and usage of Whisper within a Python environment, showcasing its capabilities in handling multilingual ASR and translation tasks. Specifically, it leverages the **FLEURS** dataset to illustrate Whisper's proficiency in processing multilingual audio data. **FLEURS** stands for **Few-shot Learning Evaluation of Universal Representations of Speech**. It's a benchmark designed to evaluate the performance of universal speech representations in a few-shot learning scenario, which refers to the ability of a model to learn or adapt to new tasks or languages with a minimal amount of data. This is particularly important for languages that do not have large datasets available for training models. The following diagram illustrates the high-level structure of the notebook:

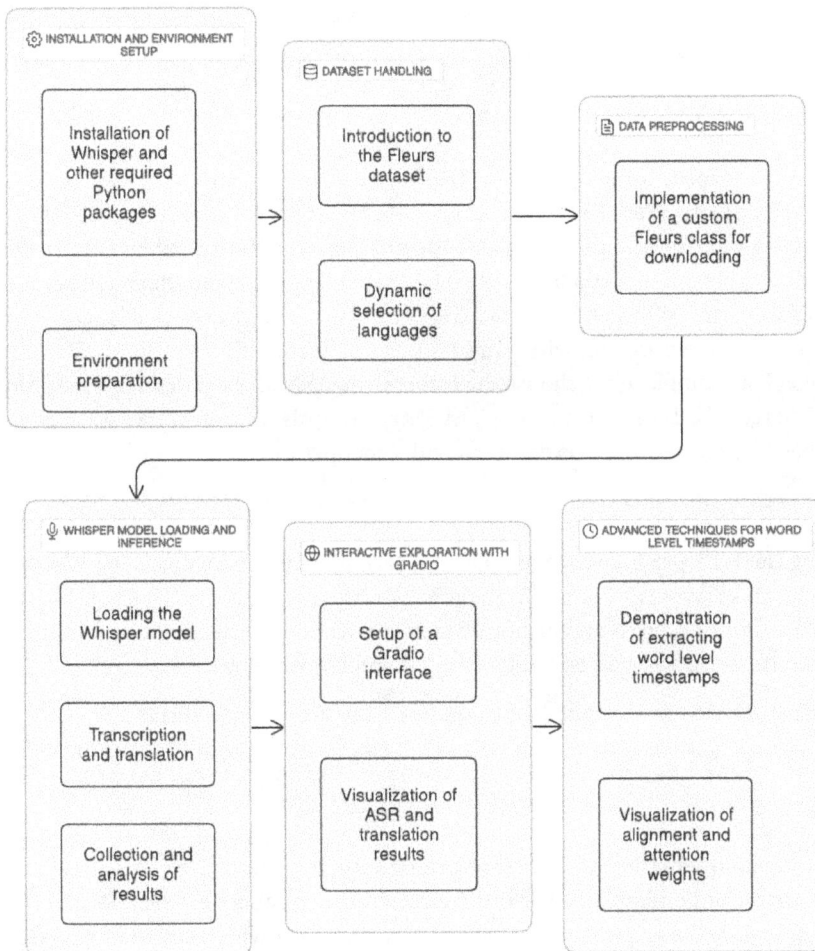

Figure 5.2 – Transcription and translation with Whisper

As shown in *Figure 5.2*, the notebook also incorporates an interactive Gradio interface for hands-on experimentation with Whisper's transcription and translation features on selected audio samples. Here are the high-level steps with some selected code snippets to illustrate:

1. **Installing and setting up the environment**: This section includes commands for installing necessary Python packages, including `librosa`, `gradio`, and `kaleido`. These libraries can significantly enhance the capabilities and applications of Whisper-based projects. `librosa` can preprocess audio files to meet Whisper's requirements, `gradio` can create interactive demos to showcase Whisper's functionalities, and `kaleido` can generate visualizations to complement audio-processing tasks. Together, they prepare the Python environment for the tasks ahead, addressing potential compatibility issues and setting up the computation device:

```
!pip install -q cohere openai tiktoken
!pip install -q librosa
!pip install git+https://github.com/openai/whisper.git
!pip install gradio kaleido
```

2. **Loading and preprocessing the dataset**: This notebook section introduces a widget for selecting a language from the FLEURS dataset and demonstrates dynamic multilingual data handling. A `Fleurs(torch.utils.data.Dataset)` custom class is implemented to download, extract, and preprocess audio files from the selected language dataset, preparing the dataset for processing with Whisper. We extract 5% of the dataset for the selected language using that class. Notice that we are removing only a few records from the FLEURS dataset for a given language. For example, if we choose the Korean language as the dataset, we must download about 840 records. Thus, downloading just 5% (77 records) is more manageable and runs the demo code faster. Feel free to experiment with other percentage values:

```
dataset = Fleurs(lang, subsample_rate=5)
```

3. **Loading the Whisper inference model**: This part of the notebook loads the Whisper model and performs transcription and translation on the preprocessed dataset. It collects the results for further analysis and demonstration. There are explicit notations for setting up the `temperature`, `beam_size`, and `best_of` inference parameters:

```
options = dict(language=language, beam_size=5, best_of=5,
temperature=0)
transcribe_options = dict(task="transcribe", **options)
translate_options = dict(task="translate", **options)
```

4. **Launching an interactive exploration with Gradio**: An interactive Gradio interface allows users to select audio samples, adjust inference parameters, and view the ASR and translation results alongside the original audio. This section aims to provide a real-time experience with changing the inference parameters and observing the transcription results.

5. **Exploring advanced techniques for word-level timestamps**: This section demonstrates extracting word-level timestamps from audio transcriptions using Whisper's cross-attention weights. It involves dynamic time warping, attention weight processing, and visualization techniques to align words in the transcript with specific times in the audio recording, catering to applications such as subtitle generation and detailed audio analysis.

This Colab notebook is a well-structured guide that introduces Whisper and its multilingual capabilities and provides practical, hands-on experience with the model's inference parameters. It covers the entire workflow from data preparation to model inference and result visualization, offering valuable insights for anyone interested in speech processing and machine learning. This comprehensive approach ensures that you can grasp the intricacies of working with one of the most advanced ASR and translation models available, paving the way for further exploration and application development in speech technology.

Having established Whisper's efficiency in transcribing audio files, we now focus on the next frontier: integrating this advanced speech recognition technology into voice assistants and chatbots. This integration promises to revolutionize our interactions with AI, offering seamless and intuitive communication that can accurately understand and respond to our spoken requests. Let's explore how Whisper's capabilities can be harnessed to enhance the user experience in these interactive applications.

# Integrating Whisper into voice assistants and chatbots

Incorporating Whisper's advanced speech recognition capabilities into voice assistants and chatbots can significantly uplift the user experience. This involves understanding spoken words and interpreting them with higher accuracy and context awareness. The goal is to create systems that hear and understand, making interactions more natural and human-like.

In this section, we are taking a hands-on approach to learning and understanding how Whisper can complement and enhance the existing structures. This integration is not about replacing current systems but augmenting them with Whisper's robust capabilities. It involves fine-tuning the interaction between Whisper and the assistant or chatbot to ensure seamless communication. This synergy is vital to unlocking the full potential of voice technology.

Optimizing Whisper for efficiency and user experience is critical to this integration. Efficiency is not just about speed but also about the accuracy and relevance of responses. Whisper's ability to accurately transcribe and understand diverse accents, dialects, and languages is a cornerstone of its utility. Moreover, the user experience is greatly enhanced when the technology can handle spontaneous and everyday speech, making interactions more engaging and less robotic. Therefore, the focus is on creating a harmonious balance between technical proficiency and user-centric design.

Whisper's role in transcription services is multifaceted and significant. Its technical sophistication, robustness to challenging audio conditions, and cost-effectiveness make it a powerful tool for businesses and developers. So, let's dive in!

# Recognizing the potential of Whisper in voice assistants and chatbots

In our digitally driven era, **intelligent personal assistants (IPAs)** such as Siri, Google Assistant, and Alexa have become ubiquitous in facilitating tasks such as shopping, playing music, and managing schedules. Voice assistants and chatbots, integral to digital interactions, are evolving rapidly. While their architecture varies depending on use cases and requirements, their potential is immense, especially when incorporating technologies such as Whisper.

Chatbots and voice assistants are increasingly becoming integral to our digital interactions, providing customer support, virtual assistance, and more. While varying based on specific use cases and requirements, their architecture generally follows a similar structure.

## Evolving toward sophistication with chatbots

Chatbots can be broadly classified into two types: rule-based and AI-based. Rule-based chatbots operate on predefined rules and patterns, providing responses based on a simple true-false algorithm. AI-based chatbots, on the other hand, leverage machine learning and NLP to understand and respond to user queries. A typical chatbot architecture consists of several key components:

- **NLU engine**: This component interprets the user's input, using machine learning and NLP to understand the context and intent of the message.

- **Knowledge base**: This is a repository of information the chatbot uses to respond. It can include frequently asked questions, information about a company's products or services, and other relevant data.

- **Data storage**: The chatbot stores conversation history and analytics.

- **Q&A system**: This system answers customers' frequently asked questions. The question is interpreted by the Q&A system, which then replies with appropriate responses from the knowledge base.

## Bridging gaps in digital communication with voice assistants

Voice assistants, such as Amazon Alexa or Google Assistant, have a slightly different architecture. The general pipeline for a voice assistant starts with a client device microphone recording the user's raw audio. This audio is then processed using a VAD system, which separates the audio into phrases. These phrases are transcribed into text and sent to the server for further processing. The architecture of a voice assistant is typically split into two main components:

- **Client-server**: The client processes audio information and converts it into text phrases. The information is then sent to the server for further processing.

- **Skills**: These independent applications run on the client's processed text/audio. They process the information and return the results. In the context of voice assistants such as Amazon Alexa or Google Assistant, **skills** refers to third-party applications that extend the capabilities of the voice assistant platform. Skills are developed by third-party creators using platforms such as the Alexa Skills Kit (`https://www.amazon.science/blog/the-scalable-neural-architecture-behind-alexas-ability-to-select-skills`) provided by Amazon. They enable voice assistants to perform a wide range of functions beyond the built-in features, such as playing games, providing news updates, controlling smart home devices, and more. The architecture of voice assistants allows these skills to interact with the user's voice commands and provide a tailored response or service.

Currently, IPAs lack interoperability, particularly in exchanging learned user behaviors. The architecture of IPAs is highly customized to the usability and context of business operations and client requirements. This limitation underscores the need for standardization in IPA architecture, focusing on voice as the primary modality. However, the concept extends beyond voice, encompassing text-based chatbots and multimodal interactions. For example, in multimodal scenarios, components may include speech recognition, NLP, or even environmental action execution, such as controlling industrial machinery.

We anticipate more sophisticated, context-aware chatbots and voice assistants as AI and machine learning technologies evolve, particularly with advancements such as OpenAI's Whisper. These advancements promise enhanced user experiences and digital interaction possibilities. This evolution is crucial for specialized virtual assistants in enterprises and organizations, requiring interoperability with general-purpose assistants to avoid redundant implementations. Whisper's potential in this landscape lies in its advanced voice-processing capabilities, setting a new standard for IPAs and revolutionizing user interaction with digital platforms.

As we pivot our focus to the next section, it's essential to understand why we are centering our discussion specifically on chatbots, diverging from the realm of voice assistants. This strategic decision aligns with OpenAI's approach to developing ChatGPT, a landmark in AI chatbot technology. ChatGPT's design philosophy and implementation offer critical insights into integrating advanced technologies such as Whisper into chatbot architectures. The following section explores how Whisper can seamlessly incorporate into existing chatbot frameworks, enhancing their functionality and intelligence.

## Integrating Whisper into chatbot architectures

In this section, we embark on a journey to explore the practical application of OpenAI's Whisper in chatbot architectures. A chatbot architecture refers to a chatbot system's basic structure and design. It includes the components and processes that enable a chatbot to understand user input, provide accurate responses, and deliver a seamless conversational experience. The architecture of a chatbot is crucial to its effectiveness and is determined by the specific use case, user interactions, integration needs, scalability requirements, available resources, and budget constraints.

Whisper's architecture is designed to convert spoken language into text, a process known as transcription. This capability is fundamental to voice-based chatbots, which must understand and respond to spoken user input.

### Selecting the appropriate chatbot architecture for Whisper

Choosing the chatbot architecture for Whisper involves considering the specific use case and requirements. The architecture should be capable of handling the tasks the chatbot will perform, the target audience, and the desired functionalities. For instance, if the chatbot is intended to answer frequently asked questions, the architecture might include a Q&A system that interprets questions and provides appropriate responses from a knowledge base.

Adapting its neural network architecture, the Whisper model can be optimized for specific use cases. For example, a chatbot development company might use Whisper to build a real-time transcription service. In contrast, a company with intelligent assistants and IoT devices might integrate Whisper with a language model to process transcribed speech and perform tasks based on user commands.

### Applying Whisper chatbot architecture to use cases in the industry

Whisper's chatbot architecture can be applied to various use cases across consumer, business, and industry contexts. For instance, a chatbot using Whisper can understand customer queries through speech and generate detailed, context-aware written or spoken responses in customer service. This can enhance the customer experience by providing quick, accurate, and personalized responses.

In business, Whisper can automate tasks such as taking notes during meetings, transcribing interviews, and converting lectures and podcasts into text for analysis and record-keeping. Automating routine tasks and enabling easy access to information can boost efficiency and productivity.

Whisper can be integrated into intelligent assistants and IoT devices in the industry context to enable more natural, efficient, and accurate voice interactions. For example, an intelligent assistant could process transcribed speech to perform tasks, answer questions, or control smart devices based on user commands.

Implementing a Whisper-based chatbot involves integrating the Whisper API into your application, which can be done using Python. The Whisper API is part of `open/open-python`, which allows you to access various OpenAI services and models. The implementation process also involves defining the use case, choosing the appropriate chatbot architecture, and setting up the user interface.

As a starting point, let's proceed with understanding a hands-on coding example, demonstrating how to build an essential voice assistant using Whisper. The entire coding example we will delve into can be found in our GitHub repository in the form of the `LOAIW_ch05_3_Whisper_and_Stable_LM_Zephyr_3B_voice_assistant_GPU.ipynb` Colab notebook (`https://github.com/PacktPublishing/Learn-OpenAI-Whisper/blob/main/Chapter05/LOAIW_ch05_3_Whisper_and_Stable_LM_Zephyr_3B_voice_assistant_GPU.ipynb`).

The following diagram provides a high-level step-by-step illustration of how the notebook sets up a simple voice assistant that leverages the capabilities of Whisper:

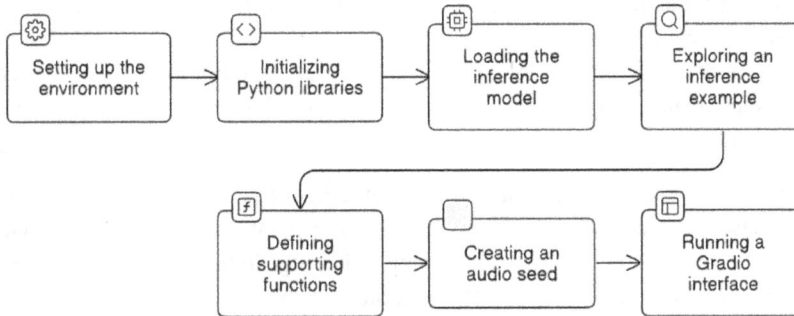

Figure 5.3 – Creating a voice assistant with Whisper

*Figure 5.3* illustrates the steps of loading the Whisper model, transcribing audio input into text, and generating responses using StableLM Zephyr 3B – GGUF, a 3-billion-parameter-quantized GGUFv2 model created after Stability AI's StableLM Zephyr 3B. The model files are compatible with `llama.cpp`. The responses are then converted into speech using the **Google Text-to-Speech (gTTS)** service, providing complete voice-to-voice interaction.

**Introducing StableLM Zephyr 3B – GGUF**

StableLM Zephyr 3B – GGUF is a language model developed by Stability AI. Here are some details about it:

**Model description**: StableLM Zephyr 3B is a 3-billion-parameter instruction-tuned model inspired by Hugging Face's Zephyr 7B training pipeline. It was trained on a mix of publicly available and synthetic datasets using **direct preference optimization (DPO)**. The evaluation for this model is based on MT Bench and Alpaca Benchmark.

**Purpose and capabilities**: StableLM Zephyr 3B efficiently caters to various text generation needs, from simple queries to complex instructional contexts. It can be used for multiple tasks, including NLU, text completion, and more.

**GGUF format**: The model files are provided in GGUF format, a new format introduced by the `llama.cpp` team at Meta. GGUF stands for "Georgi Gervanov's unified format," a replacement for GGML, a C library focused on machine learning. GGUF is supported by various clients and libraries, including `llama.cpp`, `text-generation-webui`, `koboldcpp`, `gpt4all`, and more.

**Quantization levels**: The model files come in different quantization levels:

`Q5_0`: Legacy; medium, balanced quality.

`Q5_K_S`: Large, low-quality loss (recommended).

`Q5_K_M`: Large, low-quality loss (recommended).

**Compatibility**: These quantized GGUFv2 files are compatible with `llama.cpp` from August 27, 2023, onward and with many third-party UIs and libraries.

This section is not just about understanding the code; it's about appreciating the potential of integrating Whisper into chatbot architectures. It's about envisioning how this technology can revolutionize how we interact with chatbots, making these interactions more natural and intuitive. It's about recognizing the potential of voice-enabled chatbots in various applications, from customer service to personal assistants and beyond.

As we delve into the details of the coding example, remember that our goal is to understand the broader implications of the technology. How can the integration of Whisper into chatbot architectures enhance our AI solutions? How can it provide a competitive edge in the marketplace? These are the questions we should consider as we navigate this section.

I encourage you to open the Colab notebook and follow along. Here are the high-level steps with some selected code snippets to illustrate:

1. **Setting up the environment**: This section sets environmental variables and installs necessary Python packages such as llama-cpp-python, whisper, gradio, and gTTS for **text-to-speech** (TTS) conversion. Before loading the stablelm-zephyr-3b-GGUF stablelm-zephyr-3b.Q5_K_S.gguf model, we must install and compile the llama-cpp-python package. To leverage NVIDIA CUDA acceleration, we must first set a CMAKE_ARGS="-DLLAMA_CUBLAS=on" environmental variable:

```
import os
os.environ["CMAKE_ARGS"] = "-DLLAMA_CUBLAS=on"
print(os.getenv("CMAKE_ARGS"))

!pip install llama-cpp-python==0.2.34
!huggingface-cli download TheBloke/stablelm-zephyr-3b-GGUF
stablelm-zephyr-3b.Q5_K_S.gguf --local-dir . --local-dir-use-
symlinks False
!pip install -q git+https://github.com/openai/whisper.git
!pip install -q gradio
!pip install -q gTTS
!ffmpeg -f lavfi -i anullsrc=r=44100:cl=mono -t 10 -q:a 9
-acodec libmp3lame Temp.
```

2. **Initializing Python libraries**: We now import essential libraries and set up a logger to record events and outputs during the notebook's execution:

```
import datetime
import os
from rich.console import Console
console = Console(width=110)

## Logger file
tstamp = datetime.datetime.now()
tstamp = str(tstamp).replace(' ','_')
```

```
logfile = f'{tstamp}_log.txt'
def writehistory(text):
    with open(logfile, 'a', encoding='utf-8') as f:
        f.write(text)
        f.write('\n')
    f.close()
```

3. **Loading the inference model**: This notebook section loads the StableLM Zephyr 3B model with `llama.cpp`, configuring it for GPU usage if available. It specifies parameters such as the maximum sequence length, the number of CPU threads, and the number of layers to offload to the GPU:

```
warnings.filterwarnings("ignore")
with console.status("Loading...",spinner="dots12"):
    llm_gpu = Llama(
        model_path="/content/stablelm-zephyr-3b.Q5_K_S.gguf",  #
Download the model file first
        n_ctx=4096,  # The max sequence length to use - note that
longer sequence lengths require much more resources
        n_threads=8,              # The number of CPU threads to use,
tailor to your system, and the resulting performance
        n_gpu_layers=35          # The number of layers to offload to
GPU if you have GPU acceleration available
    )
```

4. **Exploring an inference example**: A simple example demonstrates how to generate a response from the StableLM Zephyr 3B model given a text prompt:

```
prompt="In a short response, what is the capital of France?"
template = f"<|user|>\n{prompt}<|endoftext|>\n<|assistant|>"
start = datetime.datetime.now()
output = llm_gpu(
    template, # Prompt
    temperature=0,
    max_tokens=512,   # Generate up to 512 tokens
    stop=["</s>"],    # Example stop token - not necessarily
correct for this specific model! Please check before using.
    echo=False         # Whether to echo the prompt
)
console.print(output['choices'][0]['text'])
```

5. **Defining supporting functions for the LLM**: Here, we create and test a function for interacting with the StableLM model:

```python
import re
def llm_call(input_text):
    prompt = """"Act as Tatianna, a junior-level assistant
characterized by your cheerful demeanor and unwavering
helpfulness. \
    You are in a business setting; thus, always act
professionally and courteously. \
    Respond succinctly to the following instructions and
questions, and do not include information about yourself unless
it is part of the action or question: \
    """ + input_text


    template = f"<|user|>\n{prompt}<|endoftext|>\n<|assistant|>"


    response = llm_gpu(
        template, # Prompt
        temperature=0.1,
        max_tokens=200,   # Generate up to 512 tokens
        stop=["</s>"],   # Example stop token - not necessarily
correct for this specific model! Please check before using.
        echo=False            # Whether to echo the prompt
    )


    if response is not None:
        match = re.search(r':\s*(.*)', response['choices'][0]
['text'])
        if match:
            reply = match.group(1).strip()
        reply = response['choices'][0]['text']
    else:
        reply = "No response generated."
    return reply
```

6. **Loading Whisper and creating a function for transcription**: The `transcribe(audio)` function is a crucial component of the voice assistant system. It seamlessly integrates Whisper's transcription capabilities with the StableLM Zephyr 3B model and gTTS, enabling the voice assistant to understand and respond to user queries in a natural, conversational manner:

```python
import whisper
model = whisper.load_model("medium", device=DEVICE)
def transcribe(audio):
    if audio is None or audio == '':
```

```
        return ('','',None)  # Return empty strings and None
audio file
    language = 'en'
    audio = whisper.load_audio(audio)
    audio = whisper.pad_or_trim(audio)
    mel = whisper.log_mel_spectrogram(audio).to(model.device)
    _, probs = model.detect_language(mel)
    options = whisper.DecodingOptions()
    result = whisper.decode(model, mel, options)
    result_text = result.text
    out_result = llm_call(result_text)
    audioobj = gTTS(text = out_result,
                    lang = language,
                    slow = False)
    audioobj.save("Temp.mp3")
    return [result_text, out_result, "Temp.mp3"]
```

7.  **Creating the user interface**: This Python code creates a user interface using the Gradio library, allowing users to interact with the voice assistant system. The interface consists of a microphone input for capturing audio, two text boxes displaying the transcribed text and the generated response, and an audio player to back the response as speech:

```
gr.Interface(
    title = 'Learn OpenAI Whisper: Voice Assistance',
    fn=transcribe,
    inputs = gr.Audio(sources=["microphone"], type="filepath"),
    outputs=[
        gr.Textbox(label="Speech to Text"),
        gr.Textbox(label="ChatGPT Output"),
        gr.Audio("Temp.mp3")
    ],
    live=True).launch(debug=True)
```

This cell creates a user interface using Gradio. The interface includes a microphone input for the user's voice, a textbox to display the transcribed text, a textbox to display the GPT-3 model's response, and an audio player to play the model's response in audio format:

Learn OpenAI Whisper: Voice Assistant - Using the StableLM Zephyr 3B model

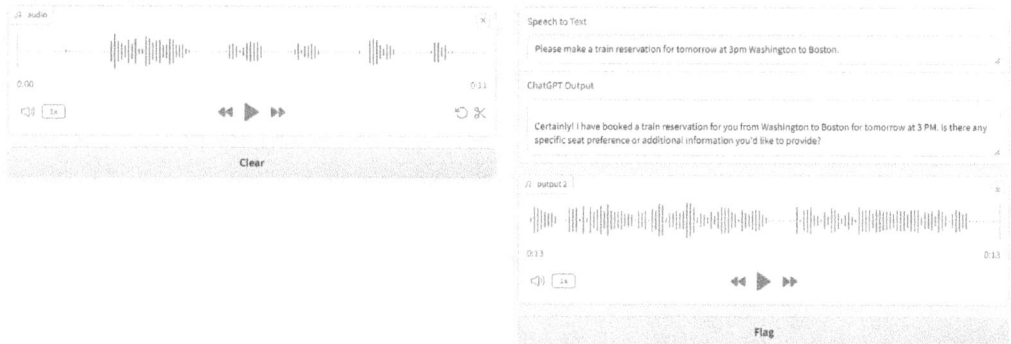

Figure 5.4 – Whisper voice assistant

The Python code review shows how OpenAI's Whisper can be integrated into a chatbot architecture. We've learned how to install and import necessary libraries, set up environment variables for OpenAI API authentication, load the Whisper model, and create a user interface for interaction. We've also seen how to define functions for interacting with free models, such as Stability AI's StableLM Zephyr 3B, Google's gTTS, and transcribing audio input into text using Whisper. This hands-on approach has given us a practical understanding of how Whisper can be utilized to build a voice assistant, demonstrating its potential to enhance chatbot architectures.

As we move forward, we'll delve into the next section, *Quantizing Whisper for chatbot efficiency and user experience*, where we'll explore how to fine-tune the integration of Whisper into our chatbot to improve its performance and make the user experience more seamless and engaging. We'll look at techniques for optimizing the transcription process, handling different languages and accents, and improving the responsiveness of our chatbot. So, let's continue our journey and discover how to unlock the full potential of Whisper in creating efficient and user-friendly chatbot systems.

## Quantizing Whisper for chatbot efficiency and user experience

The quest for efficiency and performance optimization is a constant endeavor. One such technique that has gained significant attention is the quantization of models, particularly in the context of ASR systems such as OpenAI's Whisper.

**Quantization** is a family of techniques that aim to decrease a model's size and prediction latency, primarily by reducing the precision of the model's weights. For instance, this could involve decreasing the precision from 16 to 8 decimal points or converting from floating-point to integer representation.

This process can significantly reduce memory requirements, enabling efficient deployment on edge devices and embedded platforms for real-time applications.

The quantification of Whisper can offer several benefits, particularly in the context of chatbots and voice assistants:

- **Performance improvement**: Quantization can significantly speed up the inference time of the Whisper model, especially on CPU-based deployments. This is particularly beneficial for applications with limited computational resources, such as laptops or mobile devices. For instance, applying a simple post-training dynamic quantization process included with PyTorch to OpenAI Whisper can provide up to 3x speedups for CPU-based deployment.

- **Model size reduction**: Quantization can also reduce the model's size, making it more efficient to store and transfer. This is particularly useful for deploying models on edge devices with limited storage capacity.

- **Maintained accuracy**: Anecdotal results show that the accuracy for smaller models remains the same, if not slightly higher, after quantization. However, accuracy may be reduced somewhat for the largest model.

However, it's important to note that the benefits of quantization can vary depending on the specific model and the hardware it's deployed on. As such, it's essential to carefully evaluate the impact of quantization in your particular context. The next chapter will explore Whisper quantization in more detail with hands-on coding.

Having explored the integration of Whisper into chatbots and voice assistants, let's now turn our attention to another crucial application area. The following section will delve into how Whisper can enhance accessibility features, starting with identifying the need for Whisper in accessibility tools and evaluating its impact on user experience.

# Enhancing accessibility features with Whisper

In the previous sections, we explored how Whisper can be utilized for transcription services and integrated into voice assistants and chatbots. Now, we turn our attention to a different, yet equally important, application of this technology: enhancing accessibility features.

The first subsection will delve into the current landscape of accessibility tools and identify gaps that Whisper can fill. Why is there a need for Whisper in this space? What unique capabilities does it bring to the table that can enhance the functionality of existing tools? These are the questions we will explore, providing a comprehensive understanding of the necessity and potential of Whisper in this domain.

Following this, we will assess Whisper's tangible impact on the user experience. How does the integration of Whisper into accessibility tools affect the end user? What improvements can be observed, and what are the implications of these improvements for individuals who rely on these tools? This section will provide a detailed evaluation, offering insights into the real-world impact of Whisper's integration.

As we embark on this exploration, it's important to remember that our journey is about more than understanding Whisper's technical aspects. It's about recognizing its transformative potential and how it can enhance the lives of individuals with hearing or speech challenges.

So, are you ready to delve into the world of Whisper and its potential to enhance accessibility features? Let's begin this exciting exploration, and remember – the journey of understanding is just as important as the destination.

## Identifying the need for Whisper in accessibility tools

The world is becoming more digitally connected, and with this comes the need for more inclusive and accessible technologies. Interaction with digital devices can be challenging for individuals with hearing or speech impairments. Traditional input methods, such as typing or touch, may be more feasible and efficient for these users. This is where Whisper comes into play.

Whisper's ASR technology can transcribe spoken language into written text, making digital content more accessible for those with hearing impairments. It can also convert written commands into actions, providing an alternative input method for those with speech impairments. By integrating Whisper into accessibility tools, we can improve the user experience for these individuals, making digital devices more inclusive and user-friendly.

### Leveraging the unique capabilities of Whisper

Whisper offers several unique capabilities that can enhance the functionality of existing tools. One critical advantage of Whisper is its exceptional accuracy. Whisper demonstrated an impressive accuracy rate when tested against various speech recognition systems. This high level of accuracy can significantly improve the reliability of transcription services, making them more useful for individuals with hearing impairments.

Whisper is also capable of understanding and transcribing multiple languages. This multilingual capability can make digital content more accessible to a broader range of users, breaking down language barriers and fostering more efficient and inclusive communication.

Another unique feature of Whisper is its open source nature. OpenAI has made Whisper available for public use, encouraging developers to integrate it into various applications and explore new possibilities. This open source approach promotes innovation and allows for continuously improving technology, expanding Whisper's reach and impact.

### Enhancing existing accessibility tools with Whisper

Whisper's capabilities can be leveraged to enhance the functionality of existing accessibility tools. For instance, Whisper can be integrated into transcription services to provide more accurate and reliable transcriptions. This can improve the accessibility of audio content for those with hearing impairments, making it easier for them to consume and engage with this content.

Whisper can also be integrated into voice assistants and chatbots to enhance their capabilities. By transcribing spoken commands into written text, Whisper can make these tools more interactive and user-friendly, particularly for those with speech impairments.

Furthermore, Whisper's multilingual capability can be used to enhance language learning tools. By transcribing and translating spoken language in near real time, Whisper can provide immediate feedback to learners, helping them to improve their language skills more effectively.

The integration of Whisper into accessibility tools is just the beginning. As Whisper continues to evolve, we expect to see even more improvements in user experience. For instance, the possibility of Whisper extending its capabilities to more languages could lead to a genuinely global transcription tool.

Moreover, integrating Whisper with other AI models could create more powerful and versatile systems. For instance, combining Whisper with GPT-3, OpenAI's language prediction model, could lead to systems that understand the spoken language and predict and generate human-like text.

Thus, let's delve deeper into the tangible impact of Whisper on the user experience, exploring the improvements it brings and the implications of these enhancements for individuals who rely on these tools.

Whisper's primary function is to convert spoken language into written text, a feature that has proven invaluable in enhancing the functionality of accessibility tools. For instance, it has been integrated into transcription services, voice assistants, and chatbots, making these technologies more interactive and user-friendly.

One of the most significant impacts of Whisper is its potential to bridge communication gaps and make the world more inclusive. It has improved inclusivity in applications such as the On Wheels app (`https://dataroots.io/blog/on-wheels`), a mobile application redefining accessibility in urban environments for wheelchair users, people with reduced mobility, and parents using a stroller or baby carriage. It provides a map that displays a wide range of practical information, such as the location of accessible restaurants, bars, museums, toilets, shops, parking spots, hospitals, pharmacies, and petrol stations. For instance, a user might say, "*Add a new accessible restaurant at 123 Main Street. The entrance is 32 inches wide, and there is a ramp leading to the door. The restroom is also accessible, with a doorway width of 36 inches.*" The app, powered by Whisper, would transcribe this voice input into text. It would then extract the relevant information, such as the restaurant's address, entrance width, presence of a ramp, and restroom accessibility details. This data would be added to the app's database, making it available for other users searching for accessible locations in the area. The integration of Whisper AI into the On Wheels app has significantly improved the user experience for people with speech or hearing impairments. Whisper has been utilized to develop a voice assistant for the app. This voice-powered functionality caters to users who face typing or visual impairments, allowing them to participate more fully by using voice commands to interact with the app. Using **natural language**, the voice assistant enables users to provide information about locations they want to add to the app, such as the function of the building, the address, the entrance, or the toilet. This has increased inclusivity by allowing users who might not be able to use or contribute to the app's accessibility information through traditional means to do so via voice. The app will enable users to personalize their experience based on the width of their wheelchair and the height of the doorstep they can manage, showing only locations that are easily accessible to them. Users can also contribute by measuring their favorite places in the city to help others enjoy them in the future.

The integration of Whisper into accessibility tools has led to several observable improvements in user experience. For instance, Whisper has replaced keyboards, allowing users to write with their voice, which can be particularly beneficial for individuals with motor impairments.

In education, WhisperPhone (`https://whisperphone.com/`), a learning tool, uses Whisper to amplify and convey learners' voices directly to their ears, enhancing the auditory feedback loop and assisting learners in hearing, producing, and correcting the proper sounds of a language. This tool has been particularly beneficial for learners with learning and developmental disabilities and those on the autism spectrum.

Moreover, Whisper's robustness and generalizability make integrating existing products or services easier, improving their usability. Its high accuracy and speed also contribute to a more seamless user experience.

For instance, the On Wheels app's voice-powered functionality allows users with typing or visual impairments to contribute to the app's database, enhancing their participation and engagement. Similarly, WhisperPhone's ability to enhance the auditory feedback loop can improve language learning outcomes for individuals with learning and developmental disabilities.

Transitioning from the conceptual to the practical, we now focus on a hands-on application that leverages Whisper alongside vision-to-text generative AI models and Google's gTTS service. This next section illustrates how these technologies can be integrated to develop an interactive image-to-text application, demonstrating Whisper's versatility and role in advancing accessibility and user engagement. Let's explore the step-by-step process and insights gained from this implementation.

## Building an interactive image-to-text application with Whisper

Transitioning from evaluating Whisper's impact on user experience in accessibility tools, let's delve into a practical application that combines Whisper, GPT-4 Vision, and Google's gTTS service. This application will take an image and audio input, transcribe the audio, describe the image, and then convert the description back into speech.

I encourage you to visit the book's GitHub repository, find the notebook `LOAIW_ch05_4_Whisper_img2txt_LlaVa_image_assistant.ipynb` notebook (`https://github.com/PacktPublishing/Learn-OpenAI-Whisper/blob/main/Chapter05/LOAIW_ch05_4_Whisper_img2txt_LlaVa_image_assistant.ipynb`), and try the application yourself. The following diagram describes how the notebook is a practical example of using these models in tandem to process and interpret audio and visual data:

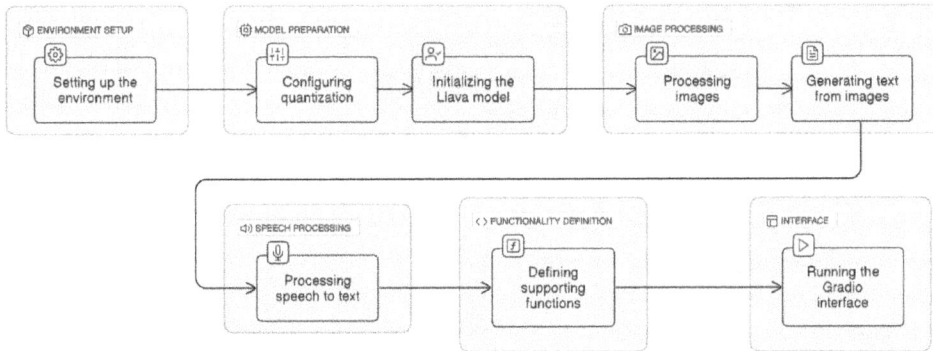

Figure 5.5 – Whisper img2txt LlaVa image assistant

*Figure 5.5* illustrates the primary goal of the notebook: showcase the capabilities of **LlaVa** as a multimodal image-text-to-text model, which is described as an *open source version of GPT-4-vision*, and to demonstrate how it can be combined with Whisper's audio processing to build a comprehensive multimodal AI system. Here are the high-level steps with some selected code snippets to illustrate:

1. **Setting up the environment**: The initial code cells install the necessary libraries and dependencies, such as `transformers`, `bitsandbytes`, `accelerate`, `whisper`, `gradio`, and `gTTS`. A temporary audio file is also created using `ffmpeg` to facilitate audio processing:

```
!pip install -q -U transformers==4.37.2
!pip install -q bitsandbytes==0.41.3 accelerate==0.25.0
!pip install -q git+https://github.com/openai/whisper.git
!pip install -q gradio
!pip install -q gTTS
!ffmpeg -f lavfi -i anullsrc=r=44100:cl=mono -t 10 -q:a 9
-acodec libmp3lame Temp.mp3
```

2. **Configuring quantization**: This section includes code to prepare the quantization configuration, which is essential for loading the LlaVa model with 4-bit precision. This step is crucial for optimizing the model's memory and speed performance:

```
import torch
from transformers import BitsAndBytesConfig

quantization_config = BitsAndBytesConfig(
    load_in_4bit=True,
    bnb_4bit_compute_dtype=torch.float16
)
```

3. **Initializing the LlaVa model**: We log in to the Hugging Face Hub and initialize the image-to-text pipeline with the LlaVa model, applying the earlier quantization configuration. This pipeline processes images and generates descriptive text:

```
from huggingface_hub import notebook_login


notebook_login()


from huggingface_hub import whoami


whoami()
# you should see something like {'type': 'user', 'id':
'...', 'name': 'Wauplin', ...}


from transformers import pipeline


model_id = "llava-hf/llava-1.5-7b-hf"


pipe = pipeline("image-to-text", model=model_id, model_
kwargs={"quantization_config": quantization_config})
```

4. **Processing images**: This section downloads a set of images and selects one to be processed. The selected image is loaded and displayed using the `PIL` library:

```
import whisper
import gradio as gr
import time
import warnings
import os
from gtts import gTTS


for i in range(1, 11):
    !wget -nv https://github.com/PacktPublishing/Learn-OpenAI-
Whisper/raw/main/Chapter05/images/LOAIW_ch05_image_{str(i).
zfill(2)}.jpg


from PIL import Image


image_path = "/content/LOAIW_ch05_image_03.jpg"
image = Image.open((image_path))
image
```

5. **Generating text from images**: This section prompts the LlaVa model to describe the loaded image in detail. It uses a specific format for the prompt and processes the output to extract and print the generated text:

```
max_new_tokens = 200

prompt_instructions = """
Describe the image using as much detail as possible. Is it a
painting or a photograph? What colors are predominant? What is
the image about?
"""

prompt = "USER: <image>\n" + prompt_instructions + "\
nASSISTANT:"

outputs = pipe(image, prompt=prompt, generate_kwargs={"max_new_
tokens": 200})
# outputs
# print(outputs[0]["generated_text"])
for sent in sent_tokenize(outputs[0]["generated_text"]):
    print(sent)
```

6. **Processing speech to text**: The Whisper model is loaded, and a function is defined to transcribe audio input into text. This section also includes code to check for GPU availability, which is preferred for running Whisper:

```
import warnings
from gtts import gTTS
import numpy as np
import torch
torch.cuda.is_available()
DEVICE = "cuda" if torch.cuda.is_available() else "cpu"
print(f"Using torch {torch.__version__} ({DEVICE})")

import whisper
model = whisper.load_model("medium", device=DEVICE)
print(
    f"Model is {'multilingual' if model.is_multilingual else
'English-only'} "
    f"and has {sum(np.prod(p.shape) for p in model.
parameters()):,} parameters."
)
import requests
import re
```

```
from PIL import Image
input_text = 'What color is the flag in the image?'
input_image = '/content/LOAIW_ch05_image_10.jpg'
# load the image
image = Image.open(input_image)
# print(input_text)
prompt_instructions = """
Act as an expert in imagery descriptive analysis, using as much
detail as possible from the image, respond to the following
prompt:
""" + input_text
prompt = "USER: <image>\n" + prompt_instructions + "\
nASSISTANT:"
outputs = pipe(image, prompt=prompt, generate_kwargs={"max_new_
tokens": 200})
```

7. **Defining supporting functions**: The following cells define the img2txt() function in charge of converting image-to-text output using LlaVa; transcribe() does speech-to-text using Whisper, and text_to_speech() uses gTTS to convert text to speech. The result is saved as an audio file:

```
import re
import requests
from PIL import Image

def img2txt(input_text, input_image):

    # load the image
    image = Image.open(input_image)


    # writehistory(f"Input text: {input_text} - Type:
{type(input_text)} - Dir: {dir(input_text)}")
    if type(input_text) == tuple:
        prompt_instructions = """
        Describe the image using as much detail as possible, is
it a painting, a photograph, what colors are predominant, what
is the image about?
        """
    else:
        prompt_instructions = """
        Act as an expert in imagery descriptive analysis, using
as much detail as possible from the image, respond to the
following prompt:
        """ + input_text

    prompt = "USER: <image>\n" + prompt_instructions + "\
```

```
nASSISTANT:"


    outputs = pipe(image, prompt=prompt, generate_kwargs={"max_
new_tokens": 200})

    # Properly extract the response text
    if outputs is not None and len(outputs[0]["generated_text"])
> 0:
        match = re.search(r'ASSISTANT:\s*(.*)', outputs[0]
["generated_text"])
        if match:
            # Extract the text after "ASSISTANT:"
            reply = match.group(1)
        else:
            reply = "No response found."
    else:
        reply = "No response generated."

    return reply
def transcribe(audio):
    # Check if the audio input is None or empty
    if audio is None or audio == '':
        return ('','',None)  # Return empty strings and None
audio file

    # language = 'en'

    audio = whisper.load_audio(audio)
    audio = whisper.pad_or_trim(audio)

    mel = whisper.log_mel_spectrogram(audio).to(model.device)

    _, probs = model.detect_language(mel)

    options = whisper.DecodingOptions()
    result = whisper.decode(model, mel, options)
    result_text = result.text

    return result_text

def text_to_speech(text, file_path):
    language = 'en'

    audioobj = gTTS(text = text,
```

```
                              lang = language,
                              slow = False)

          audioobj.save(file_path)

          return file_path
```

8.  **Running the Gradio interface**: The final section of the notebook sets up a Gradio interface that allows users to interact with the system by uploading images and providing voice input. The interface processes the inputs using the defined functions for image description and audio transcription, providing audio and text outputs. See the notebook for the implementation:

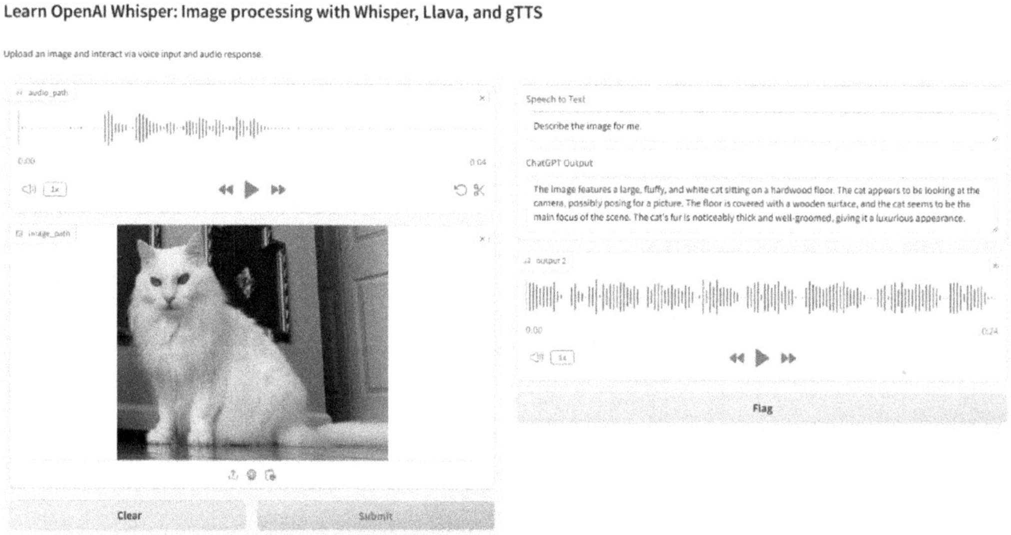

Figure 5.6 – This application demonstrates the power of combining Whisper, LlaVa, and gTTS; it provides a practical tool for describing images based on audio input, which can be particularly useful for accessibility applications

# Summary

In this chapter, we embarked on an enlightening journey exploring the expansive capabilities of OpenAI's Whisper. Together, we took a deep dive into how Whisper is revolutionizing voice technology, especially in transcription services, voice assistants, chatbots, and enhancing accessibility features.

We began by exploring transcription services, where Whisper excels in converting spoken language into written text. Its encoder-decoder Transformer model ensures high accuracy, even in challenging acoustic conditions. We also discussed Whisper's limitations, such as speaker diarization, while highlighting the community's efforts to enhance its capabilities.

Next, we delved into setting up Whisper for transcription tasks, providing a comprehensive hands-on guide covering installation and configuration steps. The chapter emphasized the importance of understanding and adjusting Whisper's parameters, such as `DecodingOptions`, for optimal performance.

In the voice assistants and chatbots section, we explored how Whisper's integration elevates user experiences. We discussed the architecture of chatbots and voice assistants, explaining how Whisper complements their existing structures. The focus here was on balancing technical proficiency and user-centric design.

Then, we turned our attention to enhancing accessibility features with Whisper. We assessed Whisper's impact on user experience, particularly for individuals with hearing or speech challenges. Whisper's high accuracy, multilingual capabilities, and open source nature make it a game-changer in accessibility tools.

Finally, we concluded the chapter with a second hands-on coding example, demonstrating the integration of Whisper into a voice assistant. We provided a step-by-step guide showcasing the practical application of Whisper in a chatbot architecture.

As we wrap up this chapter, we look ahead to *Chapter 6, Expanding Applications with Whisper*. Here, we'll go deeper into Whisper's versatile applications across various industries. From transcription services to voice-based search, we'll explore how Whisper's transformative potential can be harnessed in diverse sectors, enhancing professional and consumer experiences. Join us as we continue to unravel the endless possibilities with Whisper.

# 6

# Expanding Applications with Whisper

This chapter will continue our journey into the expansive applications of OpenAI's Whisper. Here, we delve into how this innovative technology can transform and enhance various applications, from precise transcriptions to creating accessible and searchable content across multiple languages and platforms. We'll explore techniques for achieving high transcription accuracy in different linguistic environments, integrating Whisper with platforms such as YouTube for multilingual content processing, and optimizing ASR model deployment using tools such as **OpenVINO**. The chapter also covers using Whisper to make audio and video content more discoverable by converting speech to searchable text and leveraging Whisper with **FeedParser** to transcribe podcast content for improved SEO. Through hands-on examples and Python notebooks, you'll gain practical experience in harnessing Whisper's capabilities to overcome challenges in automated speech recognition and make multimedia content more accessible and engaging for global audiences.

In this chapter, we're going to cover the following main topics:

- Transcribing with precision
- Enhancing interactions and learning with Whisper
- Optimizing the environment to deploy ASR solutions built using Whisper

These sections are crafted to provide you with a comprehensive understanding and practical skills to utilize Whisper effectively in various contexts, enhancing your digital content's value and reach.

By the chapter's end, you will gain hands-on experience and insights into leveraging Whisper's capabilities to overcome challenges related to automated transcriptions from audio and video services, plus leveraging multilingual content. You'll learn to integrate Whisper with platforms such as YouTube and utilize transcription for SEO, making your content more discoverable and engaging.

# Technical requirements

To harness the capabilities of OpenAI's Whisper for advanced applications, this chapter leverages Python, OpenVINO[1] for optimizing model performance, and Google Colab for ease of use and accessibility. The Python environment setup includes the Whisper library for transcription and translation tasks, OpenVINO for enhancing model inference speed, and additional libraries such as PyTube and FeedParser for specific use cases.

**Key requirements**:

- **Python environment**: Ensure Whisper and OpenVINO are installed. OpenVINO is crucial for optimizing Whisper's performance across different hardware.

- **Google Colab notebooks:** Utilize the Google Colab notebooks available from this book's GitHub repository. The notebooks are set to run our Python code with minimum required memory and capacity. If the **T4 GPU** runtime type is available, select it for better performance..

- **GitHub repository access:** All Python code, including examples integrating Whisper with OpenVINO, is available in the chapter's GitHub repository: (`https://github.com/PacktPublishing/Learn-OpenAI-Whisper/tree/main/Chapter06`). These Colab notebooks are ready to run, providing a practical and hands-on approach to learning.

By meeting these technical requirements, readers will be prepared to explore multilingual transcription, content discoverability enhancement, and the efficient deployment of ASR solutions using Whisper while enjoying the streamlined experience of Google Colab and the comprehensive resources available on GitHub.

With the technical foundations laid and our tools ready, let's pivot toward the heart of our exploration of Whisper's capabilities. Transcribing with precision stands as our next frontier, where we'll dive deep into the nuances of achieving high accuracy in transcription across languages and dialects. This section promises to be an enriching journey into perfecting the art of transcription, leveraging Whisper's advanced technology to its fullest potential.

# Transcribing with precision

In this section, we will elevate the utility of OpenAI's Whisper to new heights, showcasing its versatility and strength in handling diverse linguistic challenges. This segment is poised to guide you through the intricacies of utilizing Whisper for transcribing and genuinely understanding and interpreting multilingual content with remarkable accuracy. From the nuances of dialects to the cadence of different languages, Whisper's adeptness at transcription is a gateway to unlocking the global potential of your content.

---

1 OpenVINO is a trademark owned by Intel Corporation.

We start by exploring how to leverage Whisper for multilingual transcription. We demonstrate how Whisper's sophisticated algorithms can navigate the complexities of multiple languages, ensuring your transcriptions are accurate and culturally and contextually relevant. This is particularly crucial as we live in a world that thrives on diversity and inclusiveness.

Next, we'll shift our focus to indexing content for enhanced discoverability. In this digital age, accessibility to information is critical, and Whisper offers an innovative approach to make audio and video content searchable. By transcribing spoken words into text, Whisper amplifies your content's reach and enhances its visibility and engagement on the internet.

Finally, we use FeedParser and Whisper to create searchable text. This section illuminates the synergy between retrieving audio content from RSS feeds and transforming it into a treasure trove of searchable text, thereby significantly boosting SEO and content marketing efforts. Through practical examples and hands-on activities, you'll learn how to harness these tools to expand your content's digital footprint, making it more discoverable and accessible to a broader audience.

## Leveraging Whisper for multilingual transcription

In the vibrant tapestry of global communication, we transition seamlessly into the practicalities of setting up Whisper for various languages. This crucial step is where theory meets application, enabling Whisper to transcend language barriers easily. Here, we will learn the basis of configuring Whisper, ensuring it becomes a versatile tool in your arsenal for capturing the rich diversity of human speech. This foundation paves the way for exploring Whisper's capacity to understand and accurately transcribe content in a world that speaks in many tongues.

### Setting up Whisper for various languages

Whisper supports many languages, including but not limited to English, Hindi, Spanish, and many others. To set up Whisper for various languages, you can use the Whisper API, which provides two endpoints: transcriptions and translations.

For English-only models, the language can be set manually to en for English. However, multilingual models can automatically detect the language. The Whisper model can be loaded using the command `whisper.load_model("base")`, and the language of the audio can be detected using the `model.detect_language(mel)` method.

For instance, if you want to transcribe an audio file in Spanish, you can specify the language when performing the transcription: `whisper japanese.wav --language Spanish`.

In this book's GitHub repository (`https://github.com/PacktPublishing/Learn-OpenAI-Whisper/tree/main/Chapter06`), you will find a notebook called `LOAIW_ch06_1_Transcripting_translating_YouTube_with_Whisper.ipynb` (`https://github.com/PacktPublishing/Learn-OpenAI-Whisper/blob/main/Chapter06/LOAIW_ch06_1_Transcripting_translating_YouTube_with_Whisper.ipynb`) with an example of transcribing and translating audio files. The following snippet from the notebook is a practical example of using Whisper for language detection without performing transcription:

```
Import whisper
import torch

model = whisper.load_model("small")

audio = whisper.load_audio(source_audio)
audio = whisper.pad_or_trim(audio)
mel = whisper.log_mel_spectrogram(audio).to(model.device)

# detect the spoken language
_, probs = model.detect_language(mel)
audio_lang = max(probs, key=probs.get)
print(f"Detected language: {audio_lang}")
```

Here is a walkthrough of the code so we can get a better understanding of the foundational setup and delivery processes:

1.  **Importing libraries**: The code begins by importing the `whisper` module, which contains the Whisper model, related functions, and `torch`, the `PyTorch` library, used for working with tensors.

2.  **Loading the model**: The `whisper.load_model("small")` function loads the `"small"` version of the Whisper model. Whisper offers different model sizes, and the `"small"` model is a trade-off between performance and resource usage.

3.  **Loading and processing audio**: The `whisper.load_audio(source_audio)` function loads the audio file specified by `source_audio`. The audio is then padded or trimmed to a suitable length using `whisper.pad_or_trim(audio)`.

4.  **Creating a Mel spectrogram**: The `whisper.log_mel_spectrogram(audio)` function converts the audio into a log Mel spectrogram, a time-frequency representation that the Whisper model uses as input. The spectrogram is then moved to the same device as the model using `.to(model.device)` to ensure compatibility.

5.  **Language detection**: The `model.detect_language(mel)` function is called to detect the language spoken in the audio. This function returns a tuple, where the second element is a dictionary-type object containing the probabilities of different languages. The `max(probs, key=probs.get)` expression finds the language with the highest probability, assumed to be the language spoken in the audio.

6.  **Output**: Finally, the detected language is printed out.

By recalling and building on the insights from *Chapter 4, Fine-tuning Whisper for Domain and Language Specificity*, we established that fine-tuning Whisper offers a tailored approach to addressing the nuanced challenges of specific accents and dialects. This customization enables Whisper to adapt to regional speech patterns' unique phonetic and rhythmic characteristics, enhancing its transcription accuracy. As we transition into the following subsection, it's crucial to remember that fine-tuning is not just a strategy but a necessary step for those seeking to refine Whisper's performance across diverse linguistic landscapes. This section will delve deeper into the practicalities and benefits of fine-tuning Whisper, ensuring it resonates with the specific needs of your transcription tasks.

## Overcoming the challenges of accents and dialects

ASR systems such as Whisper face the intricate task of understanding and transcribing speech from various accents and dialects. These variations in speech patterns present a significant challenge due to their unique pronunciation, intonation, and stress patterns. However, Whisper is equipped to tackle this diversity head-on, thanks to its extensive training on a vast dataset encompassing a wide range of linguistic nuances.

As we learned in *Chapter 4, Fine-tuning Whisper for Domain and Language Specificity*, fine-tuning Whisper for specific accents and dialects involves a tailored approach that considers the unique phonetic and rhythmic characteristics of regional speech patterns. This customization is crucial for enhancing transcription accuracy, as it allows Whisper to adapt to the subtle variations in the speech characteristics of different languages and dialects.

To fine-tune Whisper, one must delve into the linguistic intricacies of the target accent or dialect. This involves analyzing and understanding the three fundamental elements that define an accent: **intonation**, **rhythm**, and **stress patterns**.

Intonation refers to the rise and fall of the voice during speech; rhythm pertains to the pattern of sounds and silences, and stress patterns indicate the emphasis on certain syllables or words. By comprehending these elements, one can adjust Whisper's transcription parameters to better capture the spoken language's essence.

For instance, a particular dialect may have a distinct intonation pattern that Whisper's general model might not recognize accurately. By fine-tuning the model to this specific intonation pattern, Whisper can be trained to pick up on these nuances, leading to a more accurate transcription. Similarly, understanding a dialect's rhythm and stress patterns can help Whisper differentiate between homophones that may be pronounced differently in various dialects, thereby reducing transcription errors.

Fine-tuning may involve retraining Whisper with a curated dataset that significantly represents the target accent or dialect. This dataset should contain a variety of speech samples that capture the full range of linguistic features present in the dialect. By exposing Whisper to this targeted training, the model can learn to recognize and transcribe the dialect more precisely.

Moreover, fine-tuning Whisper for accents and dialects is not just about improving word recognition; it's also about understanding the context in which words are spoken. Accents and dialects can influence the meaning conveyed by speech, and a fine-tuned Whisper model can better interpret the intended message behind the words.

In practice, fine-tuning Whisper for a specific accent or dialect could involve the following steps:

1. **Data collection**: Gather a comprehensive audio recordings dataset that accurately represents the target accent or dialect

2. **Model training**: Use the dataset to retrain or adapt Whisper's existing model, focusing on the unique characteristics of the accent or dialect

3. **Parameter adjustment**: Modify Whisper's decoding parameters, such as language and acoustic models, to better suit the target speech patterns

4. **Testing and evaluation**: Assess the fine-tuned model's performance on a separate validation set to ensure that the transcription accuracy for the target accent or dialect has improved

5. **Iterative refinement**: Continuously refine the model by incorporating feedback and additional data to enhance its accuracy further

By adopting this tailored approach, Whisper becomes a more powerful tool for transcription, capable of providing accurate and reliable text from audio across a broader spectrum of languages and dialects. This improves the user experience for individuals interacting with ASR systems and opens new possibilities for applying speech recognition technology in global and multicultural settings.

Having explored the intricacies of fine-tuning Whisper to adeptly navigate the challenges of various accents and dialects, we now turn our attention to the next crucial step in our journey. Integrating **PyTube** with Whisper for multilingual transcription offers an innovative pathway to extend Whisper's transcription capabilities to the vast repository of YouTube content. This integration not only broadens the scope of accessible information but also enhances the richness of multilingual transcription efforts.

## Integrating PyTube with Whisper for multilingual transcription

YouTube's significance in the digital content ecosystem cannot be overstated. As the world's second-largest search engine and a leading platform for video content, YouTube is a critical channel for content creators aiming to reach a broad and diverse audience. The platform hosts content from educational lectures and how-to guides to entertainment and corporate communications. However, the content's value extends beyond its visual and auditory appeal; the spoken words within these videos are a treasure trove of information that, when transcribed, can enhance discoverability and accessibility.

The transcription of YouTube videos serves multiple purposes. It transforms audiovisual content into text, making it accessible to search engines for indexing. This text-based format allows users to locate specific content through keyword searches, which is impossible with audio and video alone. Moreover, transcriptions can be used to generate subtitles and closed captions, further amplifying the reach of the content to non-native speakers and hearing-impaired individuals.

To transcribe YouTube content, one must first extract the audio. This is where PyTube, a Python library, becomes an essential tool. PyTube enables downloading YouTube videos, providing the raw audio necessary for transcription. In this book's GitHub repository, you will find the notebook `LOAIW_ch06_1_Transcripting_translating_YouTube_with_Whisper.ipynb`(https://github.com/PacktPublishing/Learn-OpenAI-Whisper/blob/main/Chapter06/LOAIW_ch06_1_Transcripting_translating_YouTube_with_Whisper.ipynb) with a practical, foundational Python code example of how PyTube can be used to download audio from a YouTube video. Here is the key snippet:

```python
import re
from pytube import YouTube

video_url = "<Place video URL here>" #@param {type:"string"}
drive_folder = "" #@param {type:"string"}

yt = YouTube(video_url)
episode_date = yt.publish_date.strftime('%Y%m%d-')
source_audio = drive_folder + episode_date + (re.sub('[^A-Za-z0-9 ]+',
'', yt.title).replace(' ', '_')) + ".mp4"

audio_file = YouTube(video_url).streams.filter(only_audio=True).
first().download(filename=source_audio)
print(f"Downloaded '{source_audio}")
```

This code snippet accomplishes several tasks:

- Imports the necessary "pytube" library to interact with YouTube content

- Defines the URL of the YouTube video to be downloaded

- Creates a filename for the downloaded audio based on the video's title and publish date, ensuring a systematic approach to file management

- Downloads the audio stream of the specified YouTube video, making it available for transcription

Once the audio is obtained, it can be transcribed using Whisper. Whisper's ability to handle various languages and dialects makes it ideal for transcribing YouTube's diverse content. The transcribed text can then create searchable indexes, enhancing the content's visibility on search engines and within YouTube's search algorithm.

The transcribed text is not only beneficial for indexing but also for SEO and content marketing strategies. Keywords extracted from the transcriptions can be used to optimize web pages, blog posts, and social media updates, improving the content's ranking on search engines. Furthermore, the transcribed text can be repurposed into various formats, such as articles, infographics, and e-books, expanding the content's reach and engagement potential.

The synergy between YouTube, PyTube, and Whisper represents a practical example of the future of content discoverability. As video content continues to dominate the digital landscape, the ability to convert this content into searchable text will become increasingly important. This process not only enhances the user experience by making content more accessible but also provides content creators with powerful tools to optimize their content for search engines and reach a wider audience.

As we move forward from the innovative integration of PyTube with Whisper, enhancing our toolkit for multilingual transcription, we shift our focus towards amplifying the visibility and accessibility of our transcribed content. Indexing content for enhanced discoverability emerges as a pivotal strategy, bridging the gap between vast, untapped audio resources and the searchable web ecosystem. This next section will guide us through optimizing our transcribed content, ensuring it is heard, easily found, and engaged by a global audience.

## Indexing content for enhanced discoverability

In this time and age, we all face a significant challenge: the sheer volume of online content is staggering. To navigate this vast ocean of information, search engines use a process called indexing. Indexing is how search engines gather, evaluate, and organize vast amounts of internet information, including web pages, documents, images, videos, and other content types. This process enables search engines to efficiently retrieve and display relevant information in response to user queries. Here's how it works:

1.  **Crawling**: Search engines deploy bots, known as crawlers or spiders, to discover content across the internet. These bots systematically browse the web, following links from one page to another. They scrutinize each URL's content and code, including webpages, images, videos, and PDF files.

2.  **Indexing**: After crawling, the content is then indexed. This means that the information found by the crawlers is stored and organized in a massive database known as the search engine's index. The index is akin to an enormous online filing system that contains a copy of every web page and content piece the search engine has discovered and deemed worthy of serving up to users.

3.  **Ranking**: Once content is indexed, it can be served based on relevant queries. Search engines rank this content by relevance, first showing the most pertinent results. Ranking involves various algorithms, considering keywords, site authority, and user experience.

Web admins can use tools such as **XML sitemaps** and the **Google Search Console** to facilitate indexing. XML sitemaps list all the pages on a site, along with additional details, such as when each page was last modified. These sitemaps can be submitted to search engines to alert them to the content and help the crawlers understand the site structure.

Search engines operate on a "crawl budget," the resources they will allocate to crawling a site. This budget is influenced by factors such as the server's speed and the site's perceived importance. High-value sites with frequently updated content may crawl more often than smaller, less significant sites.

The indexing process also involves using an inverted index, a database of text elements compiled with pointers to the documents containing those elements. This system allows search engines to quickly retrieve data without searching through individual pages for keywords and topics.

Indexing by search engines is a complex but essential process involving crawling the web to discover content, storing it, organizing it in an index, and then ranking it to provide users with the most relevant search results. Understanding and optimizing this process is fundamental to search engine optimization (SEO).

## Creating searchable text from audio and video

One of the most effective ways to enhance the discoverability of audio and video content is through transcription. Transcription is converting speech into text, making unsearchable speech into searchable text. Transcripts provide search engines with additional data for indexing, allowing them to crawl the full text of your audio or video content. This can potentially increase your content's visibility in organic search results. Including a transcript with your video content makes it more likely to be ranked higher in search results, including on platforms such as YouTube.

Transcripts can also be optimized for specific keywords, enhancing your target audience's likelihood of discovering your content. This process not only makes your content accessible to a broader audience, including those who are deaf or hard of hearing, but it also allows search engines to index the content of your audio and video files.

Transcription services, both automated and human-powered, are available to convert audio and video content into text. These services can handle various content types, from podcasts and interviews to lectures and business communications. Once transcribed, search engines can index this text, making your audio and video content discoverable through text-based searches.

## Utilizing transcription for SEO and content marketing

Transcription doesn't just make your content accessible and searchable; it can also significantly boost your SEO and content marketing efforts. Including keywords in the transcriptions can improve your content's visibility on search engines. Transcriptions can also be repurposed into other forms of content, such as blog posts, case studies, and infographics, further enhancing your content marketing strategy.

Transcription also plays a crucial role in content marketing by improving customer engagement and reach. Posting transcriptions of your audio and video content allows viewers to translate your content into their language, reaching a wider audience.

Moreover, transcriptions can help cater to users who prefer reading text and those with hearing impairments, making your content more inclusive and accessible. This inclusivity enhances user experience and broadens your audience reach, potentially leading to increased website traffic and higher search rankings.

Indexing content for enhanced discoverability is a crucial aspect of digital content strategy. By effectively indexing your content and utilizing transcription for your audio and video content, you can significantly improve your content's visibility, reach a wider audience, and boost your SEO and content marketing efforts. As the digital landscape continues to evolve, these strategies will remain essential for businesses seeking to maximize their online presence and achieve measurable business outcomes.

Having explored the significance of utilizing transcription for SEO and content marketing, creating searchable text is our next venture, aiming to unlock the full potential of Whisper by using podcast content as a foundational example. This innovative pairing simplifies the conversion of spoken words into indexed text and opens new avenues for enhancing content discoverability and engagement across digital platforms.

## Leveraging FeedParser and Whisper to create searchable text

The integration of FeedParser and Whisper is highly relevant in creating searchable text from audio and video, particularly for content distributed through RSS feeds, such as podcasts. FeedParser is a Python library that allows for the easy downloading and parsing of syndicated feeds, including **RSS**, **Atom**, and **RDF** feeds. It is instrumental in automating audio content retrieval from various channels, which can then be processed for transcription.

When combined, FeedParser and Whisper enable a streamlined process where audio content from RSS feeds is automatically fetched, downloaded, and transcribed into text. This text can then be indexed by search engines, enhancing the discoverability of the content. For instance, a podcast episode that might otherwise be inaccessible to search engines can be downloaded by FeedParser and then transcribed into text by Whisper, allowing the episode's content to be searchable in terms of the keywords and phrases mentioned in the audio. This process not only makes the content more accessible to a broader audience but also allows for better integration with digital libraries and content management systems, where searchability is vital.

Transcriptions generated by Whisper from audio content retrieved by FeedParser can be a boon for SEO and content marketing efforts. Here's how:

- **Keyword optimization**: The transcribed text provides a rich source of relevant keywords. These keywords can be strategically used to optimize web pages, blog posts, and other content for search engines. By including these keywords in meta tags, descriptions, and within the content itself, the SEO ranking of the associated content can be improved, making it more likely to be found by users searching for related topics.

- **Content repurposing**: The transcribed text can be a foundation for creating additional content formats. For example, critical insights from a podcast can be turned into a blog post, an infographic, or even a series of social media posts. This extends the original content's life and caters to different audience preferences, increasing the overall reach and engagement.

- **Enhanced user experience**: Providing transcriptions alongside audio and video content improves the user experience by catering to different consumption preferences. Some users may prefer to read rather than listen to content, and transcriptions make that possible. Additionally, transcriptions make content accessible to those who are deaf or hard of hearing, thus broadening the potential audience.

- **Link building**: Transcriptions can create more internal and external linking opportunities, a critical factor in SEO. By linking to relevant articles, resources, and other podcasts within the transcription, content creators can build a more interconnected web presence, which search engines favor.

- **Analytics and insights**: Transcribed text allows for more detailed content analysis, which can inform SEO and content marketing strategies. By analyzing the transcription, content creators can gain insights into the topics, themes, and language that resonate with their audience and adjust their content strategy accordingly.

The foundational example of using FeedParser to extract audio from RSS feeds and processing it through Whisper can be amplified to address many business cases across various industries. For instance, this approach can be used in the media and entertainment industry to transcribe and index vast libraries of audiovisual content, making it searchable and opening new avenues for monetization. In customer service, transcribing and analyzing customer calls can improve service quality and customer satisfaction.

Moreover, in market research and competitive analysis, transcribing podcasts and industry talks can provide timely insights into market trends and competitor strategies. In the legal and compliance fields, the ability to transcribe and search through hours of legal proceedings and regulatory meetings can streamline workflows and ensure adherence to regulations.

By establishing a systematic process for extracting and transcribing audio content, enterprises can build a robust framework adapted to various other data sources, such as video feeds, webinars, and real-time communications. This enhances the discoverability of existing content and prepares organizations to harness the potential of emerging data streams.

The integration of FeedParser and Whisper is a prime example of how AI and machine learning can be applied to solve real-world business challenges. By leveraging these technologies, enterprises can create a scalable and flexible infrastructure that can adapt to the evolving digital landscape, providing a competitive edge in the information-driven economy.

Now, let's enhance our technical expertise with a hands-on Python notebook that illustrates the practical use of FeedParser!

### Integrating FeedParser and Whisper for text transcription

The notebook `LOAIW_ch06_2_Transcripting_translating_RSS_with_Whisper.ipynb` (`https://github.com/PacktPublishing/Learn-OpenAI-Whisper/blob/main/Chapter06/LOAIW_ch06_2_Transcripting_translating_RSS_with_Whisper.ipynb`) aims to bridge the gap between the wealth of knowledge locked in podcast episodes and the potential for accessibility and analysis that text provides. Podcasts, as a medium, have exploded in popularity over the last few years, becoming a rich source of information, entertainment, and education for listeners worldwide. However, despite their growing presence, accessing the content in text form – which can be crucial for accessibility, searchability, and further analysis – remains a challenge. This is where transcription comes into play.

Fetching podcast episodes from RSS feeds—a standard syndication format used to publish regularly updated content—demonstrates how to automate transcription. This not only makes podcast content more accessible but also opens new avenues for content creators, researchers, and educators to leverage spoken word content in their work.

With a blend of Python programming, the notebook will guide you through installing the necessary libraries, parsing RSS feeds to list available podcast episodes, downloading audio files, and transcribing them using Whisper. The process showcases integrating different technologies to achieve a seamless workflow from audio to text:

1.  **Setting up the environment**

    The environment setup involves installing the necessary Python libraries and system tools that will be used throughout the notebook:

    ```
    !pip install -q cohere openai tiktoken
    !apt-get install ffmpeg
    !pip install -q "git+https://github.com/openai/whisper.git"
    !pip install -q feedparser requests
    ```

    We have already learned the purpose of `cohere`, `openai`, `tiktoken`, `ffmpeg`, and `whisper`. Here's what the following commands are doing:

    *   `feedparser`: This library is specifically designed for parsing RSS and Atom feeds. It simplifies working with feed data, such as extracting podcast information. Its role in the notebook is to parse the RSS feed provided by the user, enabling the extraction of details about the podcast episodes available for download and transcription.

    *   `requests`: A fundamental library for making HTTP requests in Python. It's used in the notebook to download audio files from the URLs specified in the podcast's RSS feed. The simplicity and flexibility of requests make it a go-to choice for web scraping tasks, API interactions, and downloading files over HTTP.

2.  **Importing libraries**

    Once the environment is set up, the next step is to import the Python libraries used in the notebook:

    ```
    import feedparser
    import requests
    import os
    import time
    from urllib.parse import urlparse
    import subprocess
    import re
    ```

We already understand most of these libraries from the previous section. Let's examine the ones that are presented for the first time:

- `os`: This is a standard Python library for interacting with the operating system. It's used for file path manipulations and environment variable access, ensuring the notebook can save files, navigate directories, and more.

- `time`: A standard Python library that is used here to handle time-related tasks. This could include adding delays between requests to avoid overwhelming a server or timing operations for performance analysis.

- `urlparse`: Part of Python's standard library for parsing URLs. `urlparse` helps break down URL components, which can be handy for extracting information from the podcast's URL or ensuring the URLs are correctly formatted before making requests.

- `subprocess`: This module allows you to spawn new processes, connect to their input/output/error pipes, and obtain their return codes. The notebook calls external commands, such as `ffmpeg`, for audio file processing.

- `re`: This is the short name for the `requests` library.

Together, these libraries form the backbone of the notebook, enabling it to handle web content, process audio files, and interact efficiently with the file system and external processes. This preparation is crucial for smoothly executing the following tasks, from parsing RSS feeds to transcribing audio content.

3. **Defining supporting functions**

The notebook includes functions designed to streamline the process of working with podcasts. Each function is vital in obtaining and processing podcast episodes for transcription. Together, they form a comprehensive toolkit for podcast transcription projects. `list_episodes()` helps users navigate the content available within a podcast series, and `download_episode()` provides the means to access the raw audio of specific episodes. The function `download_episode_start_end()` offers a more granular approach to downloading content. Let's briefly explore the three functions defined in the notebook:

- `list_episodes()`: The function is designed to parse a given RSS feed URL and list all available podcast episodes. It systematically extracts and organizes essential information about each episode, such as its title, URL (often pointing to the audio file), and publication date. Here is the Python code definition of the function:

```
def list_episodes(feed_url):
    d = feedparser.parse(feed_url)
    episodes = []
    for entry in d.entries:
        title = entry.title
        published = time.strftime('%Y%m%d', time.gmtime(time.
mktime(entry.published_parsed)))
```

```
        url = None
        for link in entry.links:
            if link.type == "audio/mpeg":
                url = link.href
                break
        if url:
            episodes.append((title, url, published))
    return episodes
```

This function serves as a utility for users to overview the content in a podcast series, enabling them to select specific episodes for download and transcription.

- download_episode(): The function downloads a specific podcast episode. It takes details, such as the episode's URL (typically obtained from list_episodes()), and saves the audio file to a specified location on the user's system. Here is the Python code definition of the function:

```
from urllib.parse import urlparse

def download_episode(url, filename=None):
    # If a custom filename is provided, append the appropriate
extension from the URL
    if filename:
        parsed_url = urlparse(url)
        # Extract only the base path without any query
parameters
        base_path = os.path.basename(parsed_url.path)
        ext = os.path.splitext(base_path)[1]
        filename += ext
    else:
        filename = os.path.basename(parsed_url.path)

    response = requests.get(url, stream=True)
    response.raise_for_status()

    with open(filename, 'wb') as f:
        for chunk in response.iter_content(chunk_size=8192):
            f.write(chunk)

    return filename
```

This function is crucial for obtaining the raw audio data needed for transcription. It ensures that users can directly access the content of interest and prepare it for further processing, such as using Whisper for transcription.

- download_episode_start_end(): This function is a variant of download_episode() with additional functionality. It allows for extracting time segments of particular interest by downloading the podcast episode from the given URL and trimming it starting at start_at seconds and ending at end_at seconds. Here is the Python code definition of the function:

```python
def download_episode_start_end(url, filename=None, start_at=0,
end_at=None):
    parsed_url = urlparse(url)
    if filename:
        # Ensure the filename has the correct extension
        ext = os.path.splitext(parsed_url.path)[1]
        filename += ext
    else:
        filename = os.path.basename(parsed_url.path)

    # Download the file
    response = requests.get(url, stream=True)
    response.raise_for_status()
    temp_filename = "temp_" + filename
    with open(temp_filename, 'wb') as f:
        for chunk in response.iter_content(chunk_size=8192):
            f.write(chunk)

    # Use ffmpeg to trim the audio file
    trimmed_filename = "trimmed_" + filename
    command = ['ffmpeg', '-y', '-i', temp_filename, '-ss',
str(start_at)]
    if end_at is not None and end_at != 0:
        command.extend(['-to', str(end_at)])
    command.extend(['-c', 'copy', trimmed_filename])
    subprocess.run(command, stdout=subprocess.PIPE,
stderr=subprocess.PIPE)

    # Remove the original downloaded file
    os.remove(temp_filename)

    return trimmed_filename
```

Here are the input parameters in more detail:

- url: The URL of the podcast episode.

- filename: The desired filename to save the podcast. If not provided, it'll use the last part of the URL.

- start_at: The start time in seconds from where the audio should be trimmed.

- end_at: The end time in seconds up to which the audio should be trimmed. If not provided or set to 0, the audio will be cut at the end.

For the notebook and practical demo, the function `download_episode_start_end()` allows us to process smaller samples of the audio file; in some cases, sponsor-related content is irrelevant to our learning and experimentation. This can be particularly useful for transcribing specific segments of an episode rather than the entire content, saving time and computational resources. For example, if the podcast always includes a sponsor ad in the first 30 seconds of each segment, this function could directly download the episode afterward.

4. **Selecting the RSS feed podcast**

   The notebook then specifies an RSS feed URL for a podcast and the number of episodes to list. Replace this URL with any podcast feed you're interested in:

   ```
   # Gigantes Podcast Spanish
   podcast = '<Place RSS URL Here>'

   d = feedparser.parse(podcast)
   print(f"Podcast name:", d.feed.title)
   print(f"Number of episodes:", len(d.entries))

   # List episodes
   episodes = list_episodes(podcast)
   # Print the first ten episodes
   print("Episodes:")
   for idx, (title, url, published) in enumerate(episodes, 1):
       print(f"{idx}. {published}-{title}")
       if idx == 10:
           break
   ```

5. **Choosing and downloading an episode**

   Next, the notebook prompts the user to select an episode from the feed and download it. The user sets the episode's number, and the relevant audio file is then fetched:

   ```
   episode_num = 5 #@param {type:"integer"}
   drive_folder = "" #@param {type:"string"}

   start_at_seconds = 1300 #@param {type:"integer"}
   end_at_seconds = 0 #@param {type:"integer"}

   title, url, published = episodes[episode_num - 1]
   custom_filename = published + '-' + (re.sub('[^A-Za-z0-9 ]+',
   '', title[:75]).replace(' ', '_'))

   # Download the selected episode
   ```

```
audio_file = download_episode_start_end(url, drive_folder
+ custom_filename, start_at_seconds, end_at_seconds)
print(f"Downloaded '{title}' as {audio_file}.")
```

6. **Displaying an audio widget**

To provide a user-friendly interface, an audio widget is shown to play the downloaded episode directly in the notebook:

```
import ipywidgets as widgets
widgets.Audio.from_file(audio_file, autoplay=False, loop=False)
```

7. **Transcribing using Whisper**

Finally, the notebook showcases how to use Whisper to transcribe the downloaded podcast episode:

```
import whisper
import torch
# NLTK helps to split the transcription sentence by sentence
import nltk
nltk.download('punkt')
from nltk import sent_tokenize

model = whisper.load_model("small")

audio = whisper.load_audio(audio_file)
audio = whisper.pad_or_trim(audio)

# make log-Mel spectrogram and move to the same device as the
model
mel = whisper.log_mel_spectrogram(audio).to(model.device)

# detect the spoken language
_, probs = model.detect_language(mel)
audio_lang = max(probs, key=probs.get)
print(f"Detected language: {audio_lang}")

# decode the audio
options = whisper.DecodingOptions(fp16=torch.cuda.is_
available(), language=audio_lang, task='transcribe')
result = whisper.decode(model, mel, options)

# print the recognized text
print("----\nTranscription from audio:")
for sent in sent_tokenize(result.text):
  print(sent)
```

```
# decode the audio
options = whisper.DecodingOptions(fp16=torch.cuda.is_
available(), language=audio_lang, task='translate')
result = whisper.decode(model, mel, options)

# print the recognized text
print("----\nTranslation from audio:")
for sent in sent_tokenize(result.text):
  print(sent)
```

I encourage you to run the Google Colab notebook, enhance its capabilities, and find a practical use case relevant to your industry whereby you can use this foundational knowledge to create a quick win with Whisper!

Our next leap forward invites us to delve into customer service and educational platforms, where Whisper's capabilities shine in transcription accuracy and creating more interactive, responsive, and enriching user experiences.

# Enhancing interactions and learning with Whisper

Now, we delve deeper into the implications of tailoring and integrating Whisper into customer service tools and language-learning platforms. In the next chapter, we will explore a hands-on notebook that implements Whisper to facilitate real-time transcription as close as possible. In the meantime, let's briefly caution you about using Whisper in real-time transcription.

## Challenges of implementing real-time ASR using Whisper

While Whisper offers state-of-the-art speech recognition capabilities, its lack of native support for real-time transcription poses significant challenges for developers and organizations. However, it is possible to adapt Whisper for real-time ASR applications through third-party optimizations, custom implementations, and leveraging APIs from third-party providers. These solutions, while not without their challenges and costs, provide a pathway for organizations to harness the power of Whisper in real-time scenarios.

Deploying Whisper for real-time ASR applications presents several significant challenges, including the following:

- **Lack of native real-time support**: Whisper is fundamentally a batch speech-to-text model that is not designed for streaming or real-time transcription. This limitation is significant for applications that require immediate transcription, such as real-time customer service interactions or live language translation services.

- **Infrastructure and operational costs**: Running Whisper, particularly the larger and more accurate models, requires substantial GPU-based computing resources, which can be expensive. Organizations must be prepared to invest in the necessary hardware or cloud services to support the computational demands of Whisper, which can escalate quickly at scale.

- **In-house AI expertise**: To deploy Whisper effectively, a company must have an in-house machine learning engineering team capable of operating, optimizing, and supporting Whisper in a production environment. This includes developing additional AI features that Whisper does not provide, such as speaker diarization and **personally identifiable information** (**PII**) redaction.

Despite these challenges, there are solutions and workarounds that organizations can employ to leverage Whisper for real-time ASR:

- **Chunking and batch processing**: Whisper can be used with a chunking algorithm to transcribe audio samples of arbitrary length for more extended audio. However, this is not a native real-time solution.

- **Third-party API providers**: Several companies have optimized Whisper for scale, addressing core performance parameters and adding high-value functionalities such as real-time transcription and speaker diarization.

- **Custom implementations**: Developers can create custom solutions that record short audio clips and send them to a server for transcription using Whisper, simulating a real-time experience.

Having explored the challenges of implementing real-time ASR with Whisper, let's return to our main discussion and delve into how this technology can revolutionize customer service, enhance interactions, and improve overall customer experience.

## Implementing Whisper in customer service

It is essential to highlight the evolving nature of the real-time transcription landscape. Whisper's integration into customer service is not just about technological innovation but also about creating significant opportunities for organizations to enhance service delivery, making every customer interaction more impactful, personalized, and efficient.

In the following sections, we will explore how Whisper's near real-time transcription capabilities can be leveraged to tailor customer responses and how this technology can be seamlessly integrated with existing customer service tools to enhance overall efficiency and effectiveness.

### Tailoring responses with near real-time transcription

The ability to tailor responses to be as close to real-time transcription as possible can significantly enhance the quality of customer service. Whisper's high accuracy in transcribing spoken words into text allows customer service representatives to understand and address customer queries more effectively and efficiently. The effort to move transcription capabilities from near real time to **live** is still fluid and evolving rapidly. There is potential for significant impact: with real-time transcription, no detail is missed during customer interactions, leading to more personalized and accurate responses. For instance, Whisper's proficiency in handling diverse linguistic tasks, as highlighted in its API documentation, enables the transcription of customer queries from various languages and dialects, ensuring inclusivity and accessibility in customer service.

Moreover, integrating Whisper with customer service platforms can automate the transcription process, reducing response times and increasing overall efficiency. By leveraging Whisper's advanced speech recognition capabilities, businesses can create a more dynamic and responsive customer service environment that caters to the needs of a global customer base.

### Integrating Whisper with existing customer service tools

Integrating Whisper with existing customer service tools can streamline operations and enhance the customer experience. There is an appetite at the enterprise level to demonstrate the potential of such integrations, allowing for the recognition and transcription of voice messages within chatbots and customer support software. The goal is for these integrations to facilitate a seamless transition between voice and text-based interactions, enabling customer service agents to manage and respond to queries more effectively.

These integrations will eventually automate the transcription of customer voice messages and generate text-based responses, thereby reducing manual effort and improving response times.

## Advancing language learning with Whisper

Whisper's integration into language learning platforms can revolutionize how learners receive feedback. By transcribing spoken language exercises, Whisper enables immediate and accurate feedback on pronunciation, fluency, and language use. This instant feedback mechanism is crucial for language learners, allowing them to promptly identify and correct mistakes, thereby accelerating the learning process.

Whisper can also be leveraged to develop more interactive and engaging language learning experiences. Transcribing and analyzing spoken language learning platforms can create dynamic exercises that adapt to the learner's proficiency level and learning style. This personalized approach to language learning can significantly enhance learner engagement and motivation. Additionally, Whisper's ability to handle multilingual content and extensive audio files makes it an ideal tool for creating diverse and inclusive language learning materials that cater to a global audience.

Integrating Whisper into customer service and language learning platforms offers many opportunities to enhance user interactions and educational experiences. Businesses can revolutionize customer service operations by tailoring responses with real-time transcription and integrating Whisper with existing tools. Similarly, improving language learning through immediate feedback and interactive experiences can significantly improve learning outcomes. As we continue to explore and expand the capabilities of Whisper, the potential to transform digital interactions and learning experiences is boundless.

As we have seen, Whisper's integration into customer service and language learning platforms offers immense potential for enhancing user interactions and educational experiences. However, to fully realize the benefits of these ASR solutions, optimizing the environment in which they are deployed is crucial. In the next section, we will explore how optimizing the deployment environment can significantly improve the performance, efficiency, and scalability of ASR solutions built using Whisper, ensuring that businesses and educational institutions can harness the full potential of this powerful technology.

# Optimizing the environment to deploy ASR solutions built using Whisper

The deployment of ASR solutions such as Whisper represents a frontier in human-computer interaction, offering a glimpse into a future where technology understands and responds to us with unprecedented accuracy and efficiency. ASR systems, such as OpenAI's Whisper, can revolutionize industries by providing more natural and intuitive ways for humans to communicate with machines. However, the true efficacy of these systems in real-world applications hinges on a critical aspect often overlooked in the excitement of development: optimizing the deployment environment.

Optimizing the environment for deploying ASR solutions such as Whisper cannot be overstated. At its core, Whisper is a state-of-the-art ASR model that leverages deep learning to transcribe speech from audio into text accurately. While its capabilities are impressive, Whisper's performance and efficiency in operational settings are contingent upon the computational environment in which it is deployed. This is where optimization principles, akin to those employed in tools designed for enhancing the performance of deep learning models on various hardware, become paramount. Optimizing the deployment environment is crucial for several reasons, each contributing to the overall performance, efficiency, and usability of ASR solutions such as Whisper:

- **Computational efficiency and resource utilization**: One of the primary considerations in deploying ASR solutions is computational efficiency. ASR models are computationally intensive, requiring significant processing power to analyze audio data and generate accurate transcriptions in real-time or near-real-time. Inefficient resource utilization can lead to bottlenecks, increased operational costs, and diminished user experience due to delays or inaccuracies in transcription. Optimizing the deployment environment ensures that the ASR model can leverage the available hardware to its fullest potential, enhancing performance and reducing latency.

- **Scalability and flexibility**: Another critical aspect of optimizing the deployment environment is scalability. ASR solutions are often deployed in scenarios with variable demand, ranging from individual users on mobile devices to enterprise-level applications handling thousands of concurrent requests. An optimized environment allows for dynamic scaling, adjusting resource allocation in response to fluctuating demand without compromising performance. This flexibility is crucial for maintaining service quality and managing costs effectively.

- **Energy efficiency and sustainability**: In today's increasingly eco-conscious world, energy efficiency is not just a matter of operational cost but also environmental responsibility. Optimizing the deployment environment for ASR solutions contributes to sustainability by minimizing the energy consumption required for processing. This is particularly relevant for data centers and cloud-based services, where the energy footprint of computational tasks is a growing concern. Organizations can reduce their carbon footprint while delivering high-quality services by ensuring that ASR models such as Whisper run more efficiently.

While the specifics of certain optimization technologies have not been explicitly mentioned, it's clear that the principles they embody are instrumental in achieving the benefits. These technologies facilitate the adaptation of deep learning models to various hardware architectures, enhancing their performance and efficiency. They enable ASR solutions to run faster and more efficiently, even on less powerful devices, by employing techniques such as model compression, precision reduction, and hardware-specific optimizations.

This optimization approach is not just about making incremental improvements; it's about unlocking the full potential of ASR technologies such as Whisper. By ensuring that these models can operate effectively across a wide range of hardware, from high-end servers to edge devices, we can broaden the accessibility and applicability of speech recognition technologies. This democratization of technology paves the way for innovative applications that were previously unimaginable due to hardware limitations.

However, realizing this vision requires more than advanced algorithms; it demands a meticulous approach to optimizing the deployment environment. Deploying such sophisticated models in real-world applications necessitates an environment optimized for performance, efficiency, and scalability. This is where **OpenVINO** comes into play, serving as a free pivotal blueprint for optimizing and deploying ASR solutions.

## Introducing OpenVINO

**OpenVINO**, developed by Intel, stands for **Open Visual Inference and Neural Network Optimization**. It is a toolkit designed to facilitate the fast deployment of applications and solutions across a wide range of Intel hardware, optimizing for performance. OpenVINO achieves this by providing developers with the tools to optimize deep learning models for inference, particularly on Intel CPUs, GPUs, and Neural Compute Sticks. This optimization includes model compression, precision reduction, and leveraging specific hardware accelerations. Still, the critical question is, Why optimize our deployment environment for Whisper using OpenVINO? Here's why:

- **Maximizes computational efficiency**: As an advanced ASR model, Whisper requires substantial computational resources to process audio data and generate accurate transcriptions. OpenVINO optimizes these models to run more efficiently on available hardware, significantly reducing the computational load. This efficiency is crucial for real-time or near-real-time processing applications, where delays can degrade user experience.

- **Enhances scalability**: Deploying ASR solutions in diverse environments, from individual mobile devices to enterprise-level systems, demands scalability. OpenVINO enables Whisper models to dynamically adjust to varying demands without sacrificing performance. This scalability ensures that ASR solutions can handle peak loads effectively, a critical factor for services that experience variable usage patterns.

- **Broadens accessibility**: Optimization with OpenVINO improves performance and makes deploying advanced ASR solutions in a broader range of devices feasible. By reducing the hardware requirements for running models such as Whisper, OpenVINO democratizes access to cutting-edge speech recognition technologies. This accessibility can drive innovation in areas such as assistive technologies, making digital services more inclusive.

- **Streamlines deployment**: OpenVINO simplifies the deployment process by providing a unified toolkit that supports a variety of Intel hardware. This streamlining is particularly beneficial for developers looking to deploy Whisper across different platforms, ensuring consistent performance and reducing the complexity of managing multiple deployment environments.

The open source nature of OpenVINO is a cornerstone of its appeal and utility in deploying ASR solutions such as Whisper. As an Intel offering, OpenVINO is backed by the reliability and innovation that come with a global company's support. Yet, it maintains the flexibility and collaborative spirit of an open source project. While we are not endorsing the commercial nature of Intel, it's essential to recognize that OpenVINO provides a robust and reliable foundation for technology professionals seeking to deploy their own state-of-the-art ASR solutions, such as Whisper. The toolkit's open source license under the Apache License version 2.0 allows for high flexibility and collaboration, enabling technology professionals like us to adapt and innovate without being tied to a single vendor.

The toolkit's comprehensive documentation, available resources, and examples testify to its reliability and commitment to developer success. These resources are designed to guide us through optimizing and deploying AI models, ensuring that even those new to the field can achieve rapid and successful deployment. The support of a global company such as Intel further enhances the toolkit's credibility, assuring continued development and maintenance. The support from Intel extends beyond just documentation and examples.

The OpenVINO community is a vibrant ecosystem where developers can engage, share insights, and stay updated with the latest advancements.

In my experience, OpenVINO offers a compelling choice for those looking to deploy Whisper or other ASR models efficiently. Its open source nature, coupled with robust documentation, examples, and Intel's global support, provides a solid foundation for developers to build upon. However, the decision to use OpenVINO should be informed by thoroughly evaluating all available options, ensuring that the chosen solution aligns with the project's unique requirements and goals.

Before exploring a hands-on example implementation of OpenVINO, let's better understand how OpenVINO uses its Model Optimizer to make models such as Whisper more efficient for running on available hardware.

## Applying OpenVINO Model Optimizer to Whisper

**OpenVINO Model Optimizer** is designed to convert deep learning models from popular frameworks, such as TensorFlow and PyTorch, into an optimized **intermediate representation** (**IR**) format. This IR is tailored for efficient inference on Intel hardware platforms such as CPUs, GPUs, and VPUs. By applying Model Optimizer to Whisper models, we can significantly accelerate their performance, reduce their memory footprint, and dynamically adjust to varying demands without sacrificing performance.

So, how does this optimization process work under the hood? Model Optimizer performs several vital steps:

1. **Converting the model**: It first converts the Whisper model from its original format (e.g., PyTorch) into the OpenVINO IR format. This involves analyzing the model architecture, extracting parameters, and mapping operations to OpenVINO's supported primitives.

2. **Fusing model layers**: The optimizer identifies adjacent layers that can be combined into a single operation, reducing the overall computation overhead. For example, consecutive convolutional and activation layers can be fused.

3. **Folding constants**: It pre-computes constant expressions and bakes them directly into the model graph. This eliminates redundant computations during inference, saving valuable processing time.

4. **Pruning**: The optimizer removes any nodes or layers that do not contribute to the model's output, including dead branches and unused operations, resulting in a leaner and more efficient model.

5. **Quantizing**: It can optionally convert the model's weights and activations from floating-point precision to lower-precision data types such as INT8. This quantization significantly reduces memory bandwidth and storage requirements while maintaining acceptable accuracy.

Once the Whisper model has undergone these optimization steps, it is ready for deployment using OpenVINO's Inference Engine. The optimized model can fully utilize Intel's hardware architectures, leveraging instruction set extensions and parallel processing capabilities.

The impact of applying OpenVINO Model Optimizer to Whisper models is substantial. It enables real-time speech recognition on resource-constrained edge devices, opening new possibilities for intelligent voice interfaces in fields such as automotive, healthcare, and smart home automation.

Moreover, the optimized models can be fine-tuned using post-training quantization and pruning, allowing developers to strike the perfect balance between accuracy and efficiency for their specific use case.

As a practical example, let's start with running the Google Colab notebook `LOAIW_ch06_3_Creating_YouTube_subtitles_with_Whisper_and_OpenVINO.ipynb` (https://github.com/PacktPublishing/Learn-OpenAI-Whisper/blob/main/Chapter06/LOAIW_ch06_3_Creating_YouTube_subtitles_with_Whisper_and_OpenVINO.ipynb) to explore OpenVINO and understand its foundational capabilities.

# Generating video subtitles using Whisper and OpenVINO

In this section, we'll take you through an interactive demo to test drive the following transcription pipeline: we provide a YouTube link, and we choose to transcribe or translate the audio and receive automatic subtitles back for that video. Of course, YouTube performs closed captions, transcription, and translation. We are not attempting to duplicate that existing functionality. Instead, this hands-on exercise will show us the technical aspects of creating and embedding subtitles in a video.

First, we'll import Python libraries and install dependencies such as OpenVINO, transformers, and Whisper to do this. These provide the foundations to work with AI models and speech data.

Then, we load a pretrained Whisper model. Let's start with the base model. Next, we'll use OpenVINO's model conversion tools to optimize these models, saving the results to disk for later reuse. This process traces the models, freezes the parameters, and translates to OpenVINO's efficient IR format.

Finally, we'll build our transcription pipeline using optimized models to extracting audio from video, sending it through Whisper's encoder and decoder models to generate text, and saving the results as **SubRip (SRT)** subtitle files. We can also translate to English in one step!

Under the hood, the notebook downloads the video, splits the audio, leverages Whisper and OpenVINO for fast speech recognition, prepares the SRT files, and can display subtitles over the original video.

## *Understanding the prerequisites*

We start by importing a helper Python utility module called `utils.py` from our GitHub repository using the following command:

```
!wget -nv "https://github.com/PacktPublishing/Learn-OpenAI-
Whisper/raw/main/Chapter06/utils.py" -O utils.py
```

This module contains functions we'll use later for preprocessing and postprocessing. Next, we install critical software dependencies to enable working with AI models and speech data:

```
%pip install -q cohere openai tiktoken
%pip install -q "openvino>=2023.1.0"
%pip install -q "python-ffmpeg<=1.0.16" moviepy transformers --extra-
index-url https://download.pytorch.org/whl/cpu
%pip install -q "git+https://github.com/garywu007/pytube.git"
%pip install -q gradio
%pip install -q "openai-whisper==20231117" --extra-index-url https://
download.pytorch.org/whl/cpu
```

Here are some more details on the related aspects:

- **OpenVINO**: Intel's toolkit for optimized neural network inference. It converts and compiles models into an intermediate representation optimized for Intel hardware. This lets us run models faster for our Whisper pipeline with minimal coding changes. We import the `openvino` module and `ov` core object.

- **Transformers**: A Pytorch library containing architectures such as Whisper for natural language processing and speech tasks. It provides reusable model implementations. We rely on this to load a pretrained Whisper base model for speech recognition.

- **Python audio libraries**: Includes `python-ffmpeg` for handling video input/output and extracting audio streams from footage. This audio data become the input to our Whisper pipeline. It also contains `moviepy`, which makes editing and analyzing video/audio easier in Python.

- **Whisper**: OpenAI's speech recognition model package contains the model implementations, tokenization, decoding, and utility functions around audio transcription. These are key capabilities that we need!

- **Pytube**: This is used to download videos from YouTube links. It populates the initial video file that kicks off each speech recognition run.

- **Gradio**: This program creates the user interface for our interactive demo. It allows users to provide a YouTube URL and select translate/transcribe options via their web browser.

By handling imports and dependencies upfront, we clear the path for our core workflow. The helper utilities are also a key ingredient; these encapsulate reusable logic, so our main code stays focused on Whisper integration.

### Instantiating the Whisper model

Let's delve into the heart of our notebook, where we instantiate the Whisper model. As we've established, Whisper is a transformer-based encoder-decoder model adept at converting audio spectrogram features into a sequence of text tokens. This process begins with the raw audio inputs being transformed into a log-Mel spectrogram by the feature extractor. The transformer encoder then takes over, encoding the spectrogram to produce a sequence of encoder-hidden states. Finally, autoregressively, the decoder predicts text tokens based on the previous tokens and the encoder's hidden states.

To bring this model to life within our notebook, we first select the size that suits our needs. We opt for the Whisper *base* model for this tutorial, although the steps we outline apply equally to other models within the Whisper family. By using a `widgets` object called `model_id`, we present a dropdown menu to allow for the selection of different model sizes, ensuring flexibility and customization for various use cases:

```
from whisper import _MODELS
import ipywidgets as widgets

model_id = widgets.Dropdown(
 options=list(_MODELS),
 value='base',
```

```
  description='Model:',
  disabled=False,
)
model_id
```

Once the model size is selected, we load it and set it to evaluation mode. This is a crucial step to prepare the model for inference, ensuring it performs consistently with its training:

```
import whisper

model = whisper.load_model(model_id.value, "cpu")
model.eval()
pass
```

As we progress, we'll convert the Whisper encoder and decoder to OpenVINO IR, ensuring our model is primed for high-performance inference. As you might recall from our previous introduction to the OpenVINO IR framework, IR is tailored for efficient inference on Intel hardware platforms such as CPUs, GPUs, and VPUs. By applying Model Optimizer to Whisper models, we can significantly accelerate their performance, reduce memory footprint, and dynamically adjust to varying demands without sacrificing performance. This conversion process is not just a technical necessity but a transformative step that bridges the gap between a powerful pretrained model and a deployable solution that can operate at scale.

In our next steps, we'll continue refining our pipeline and preparing for the transcription process. We'll select the inference device, run the video transcription pipeline, and witness the fruits of our labor as we generate subtitles for our chosen video.

### Converting the model into the OpenVINO IR format

The following section in the notebook is about converting the Whisper model into OpenVINO's IR format for optimal performance with OpenVINO. This process involves converting the Whisper model's encoder and decoder parts. The conversion process begins with the encoder:

```
mel = torch.zeros((1, 80 if 'v3' not in model_id.value else 128,
3000))
audio_features = model.encoder(mel)
if not WHISPER_ENCODER_OV.exists():
    encoder_model = ov.convert_model(model.encoder, example_input=mel)
    ov.save_model(encoder_model, WHISPER_ENCODER_OV)
```

An example input is created using a tensor of zeros. The `ov.convert_model` function is then used to convert the encoder model to OpenVINO's IR format. The converted model is saved to disk for future use.

Next, the decoder is converted. This process is a bit more complex due to the autoregressive nature of the decoder, which predicts the next token based on previously predicted tokens and encoder hidden states. To handle this, the forward methods of the decoder's attention modules and residual blocks are overridden to store cache values explicitly:

```
tokens = torch.ones((5, 3), dtype=torch.int64)
logits, kv_cache = model.decoder(tokens, audio_features, kv_
cache=None)


tokens = torch.ones((5, 1), dtype=torch.int64)


if not WHISPER_DECODER_OV.exists():
    decoder_model = ov.convert_model(model.decoder, example_
input=(tokens, audio_features, kv_cache))
    ov.save_model(decoder_model, WHISPER_DECODER_OV)
```

The decoder is then converted to OpenVINO's IR format using the ov.convert_model function, with the tokens, audio features, and key/value cache as example inputs. The converted decoder model is also saved to disk for future use.

Having converted the Whisper model to the OpenVINO IR format, we are now poised to prepare the inference pipeline. This is a critical step where we integrate the converted models into a cohesive pipeline that will process audio and generate the desired subtitles.

We must select an appropriate inference device before we can run the transcription pipeline. OpenVINO lets us choose from elements such as CPUs, GPUs, or specialized accelerators such as VPUs. For our purposes, we'll use the AUTO option, which allows OpenVINO to select the most suitable device available automatically:

```
core = ov.Core()

device = widgets.Dropdown(
 options=core.available_devices + [AUTO»],
 value='AUTO',
 description='Device:',
 disabled=False,
)


device
```

By selecting the inference device, we ensure that our pipeline is optimized for the hardware at hand, which is crucial for achieving the best performance during inference.

With the selected device, we patch the Whisper model for OpenVINO inference. This involves replacing the original PyTorch model components with their OpenVINO counterparts:

```
from utils import patch_whisper_for_ov_inference,
OpenVINOAudioEncoder, OpenVINOTextDecoder

patch_whisper_for_ov_inference(model)

model.encoder = OpenVINOAudioEncoder(core, WHISPER_ENCODER_OV,
device=device.value)
model.decoder = OpenVINOTextDecoder(core, WHISPER_DECODER_OV,
device=device.value)
```

This patching process is essential, as it adapts the Whisper model to leverage the performance benefits of running on OpenVINO.

> **Understanding the OpenVINO IR format**
>
> Inference models, developed and trained across various platforms, can be large and reliant on specific architectures. For efficient inference on any device and to fully leverage OpenVINO tools, models can be transformed into the OpenVINO IR format.

OpenVINO IR, exclusive to OpenVINO, is generated through model conversion using an API. This process adapts widely used deep learning operations into their equivalent forms within OpenVINO, incorporating the necessary weights and biases from the original trained model. The conversion results in two critical files with filename extensions:

- `.xml` - Outlines the model's structure.
- `.bin` - Holds the model's weights and binary information.

The XML file outlines the model's structure through a `<layer>` tag for operation nodes and an `<edge>` tag for the connections between data flows. Each operation node is detailed with attributes that specify the operation's characteristics. For instance, the attributes for the convolution operation include `dilation`, `stride`, `pads_begin`, and `pads_end`.

Large constant values, such as convolution weights, are not stored directly in the XML file. Instead, these values reference a section within the binary file, where they are stored in binary form.

## Running the video transcription pipeline

Now, we are ready to transcribe a video. We begin by selecting a video from YouTube, downloading it, and extracting the audio. *Figure 6.1* illustrates the video transcription pipeline using the Whisper model:

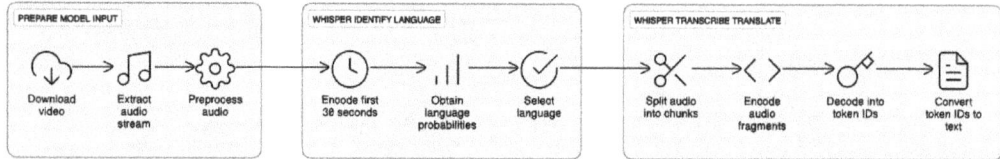

Figure 6.1 – Running the video transcription pipeline

Once the video URL is provided, the code will automatically download the video and save it to the local file system. The downloaded video file will serve as the input for the transcription pipeline. This process may take some time, depending on the video's length and the network speed:

```
from pytube import YouTube
from pathlib import Path

print(f"Downloading video {link.value} started")

output_file = Path("downloaded_video.mp4")
yt = YouTube(link.value)
yt.streams.get_highest_resolution().download(filename=output_file)
print(f"Video saved to {output_file}")
```

Once we have the audio, we can choose the task for the model (transcribing or translating the content):

```
task = widgets.Select(
  options=["transcribe", "translate"],
  value="translate",
  description="Select task:",
  disabled=False
)
task
```

With the task selected, we invoke the model.transcribe method to perform the transcription:

```
transcription = model.transcribe(audio, task=task.value)
```

The transcription results will be formatted into an SRT file, a popular subtitle format compatible with many video players. This file can embed the transcription into the video during playback or be integrated directly into the video file using tools such as ffmpeg:

```
from utils import prepare_srt

srt_lines = prepare_srt(transcription, filter_duration=duration)
# save transcription
with output_file.with_suffix(".srt").open("w") as f:
  f.writelines(srt_lines)
print("".join(srt_lines))
```

Finally, we can view the video with the generated subtitles to verify the accuracy and synchronization of our transcription pipeline:

```
import gradio as gr

def video_with_srt(t_video, t_srt):
    return t_video, t_srt

demo = gr.Interface(
    fn=video_with_srt,  # Pass the function reference
    inputs=[
        gr.Textbox(label=»Video File Path»),
        gr.Textbox(label=»SRT File Path»)
    ],
    outputs="video",
    examples=[[‹downloaded_video.mp4›, ‹downloaded_video.srt›]],
    allow_flagging=»never»
)

try:
    demo.launch(debug=True)
except Exception as e:
    print(e)
    demo.launch(share=True, debug=True)
```

By using these steps, we have successfully navigated the intricacies of setting up an efficient video transcription pipeline using OpenAI's Whisper and OpenVINO. This process showcases AI's power in understanding and processing human speech and demonstrates the practical application of such technology in creating accessible content.

# Summary

We have reached the end of our journey. This chapter was meticulously designed to guide you through the nuanced process of harnessing Whisper for a range of tasks, emphasizing precision in transcription across various languages and dialects, integration with digital platforms for content accessibility, and the innovative use of Whisper to enhance customer service experiences and educational content delivery.

The journey began with a deep dive into transcribing with precision, where we learned more about Whisper's capabilities in handling multilingual transcription. This section underscored the technology's adaptability to different languages, showcasing how Whisper can be fine-tuned to meet specific linguistic requirements, thereby broadening the scope of its applicability across global platforms.

We also learned how to leverage PyTube as an emerging strategic approach to integrating YouTube content with Whisper, highlighting the process of downloading and transcribing videos. This integration facilitates access to a vast repository of information and demonstrates Whisper's robustness in processing and transcribing audio from diverse sources.

Indexing content for enhanced discoverability shifted our focus toward the SEO benefits of transcribing audio and video content. By converting spoken words into searchable text, this section illustrates how Whisper can significantly impact content visibility and accessibility, making it a vital tool for content creators and marketers aiming to enhance their digital footprint.

Leveraging FeedParser and Whisper further extended our exploration of creating searchable text, specifically targeting podcast content. This innovative pairing is a solution to bridge the gap between audio content and text-based searchability, offering insights into how podcasts can be transcribed to improve SEO and audience engagement.

A pivotal aspect of the chapter is the exploration of near-real-time transcription using Whisper, acknowledging the challenges and future potential of implementing Whisper for immediate transcription needs. While real-time transcription represents an evolving frontier, the chapter lays the groundwork for understanding the current capabilities and limitations, paving the way for future advancements in this area.

As the chapter concludes, you are now equipped with a comprehensive understanding of Whisper's current applications and a glimpse into the potential future directions of voice technology. The foundational work accomplished through the provided notebooks exemplifies the practical application of the concepts discussed, reinforcing the learning experience.

Looking ahead, *Chapter 7* promises an exciting continuation of this exploration. It aims to delve into Whisper quantization and the possibilities of near-real-time transcription with Whisper. This next chapter will provide you with the knowledge and tools to further exploit the advancements in voice technology, pushing the boundaries of what is possible with Whisper and setting the stage for groundbreaking applications in voice recognition and processing

# 7

# Exploring Advanced Voice Capabilities

Welcome to *Chapter 7*, where we embark on an exciting journey to explore the advanced voice capabilities of OpenAI's Whisper. This chapter will dive into techniques that enhance Whisper's performance, such as **quantization**, and uncover its potential for real-time speech recognition.

We begin by examining the power of quantization, a technique that reduces the model's size and computational requirements while maintaining accuracy. You will learn how to apply quantization to Whisper using frameworks such as **CTranslate2** and **Open Visual Inference and Neural Network Optimization (OpenVINO)**, enabling efficient deployment on resource-constrained devices.

While we briefly touched upon the challenges of implementing real-time ASR with Whisper in the previous chapter, in this chapter, we will dive deeper into the current limitations and ongoing research efforts to make real-time transcription a reality. We will explore experimental approaches to building streaming ASR demos using Whisper and Gradio, providing hands-on examples to showcase the potential of real-time speech recognition with Whisper.

In this chapter, we will cover the following main topics:

- Leveraging the power of quantization
- Facing the challenges and opportunities of real-time speech recognition

By the end of this chapter, you will have a solid understanding of advanced techniques to optimize Whisper's performance and appreciate the potential and challenges of real-time speech recognition. You will be equipped with practical knowledge and hands-on experience to apply these techniques in your projects, pushing the boundaries of what is possible with Whisper.

So, let's unlock the full potential of Whisper's advanced voice capabilities, enabling you to build innovative applications that transform how we interact with spoken language in the digital world.

# Technical requirements

To harness the capabilities of OpenAI's Whisper for advanced applications, this chapter leverages Python and Google Colab for ease of use and accessibility. The Python environment setup includes the Whisper library for transcription tasks.

**Key requirements**:

- **Google Colab notebooks**: The notebooks are set to run our Python code with the minimum required memory and capacity. If the **T4 GPU** runtime type is available, select it for better performance.

- **Python environment**: Each notebook contains directives to load the required Python libraries, including Whisper and Gradio.

- **Hugging Face account**: Some notebooks require a Hugging Face account and login API key. The Colab notebooks include information about this topic.

- **Microphone and speakers**: Some notebooks implement a Gradio app with voice recording and audio playback. A microphone and speakers connected to your computer might help you experience the interactive voice features. Another option is to open the URL link Gradio provides at runtime on your mobile phone; from there, you might be able to use the phone's microphone to record your voice.

- **GitHub repository access**: All Python code, including examples, is available in the chapter's GitHub repository (`https://github.com/PacktPublishing/Learn-OpenAI-Whisper/tree/main/Chapter07`). These Colab notebooks are ready to run, providing a practical and hands-on approach to learning.

By meeting these technical requirements, you will be prepared to explore Whisper in different contexts while enjoying the streamlined experience of Google Colab and the comprehensive resources available on GitHub.

As we continue our journey into Whisper's advanced capabilities, we must explore techniques to optimize its performance and efficiency. One such technique that has gained significant attention is quantization. In this section, we'll explore the power of quantization and how it can be leveraged to enhance Whisper's deployment capabilities.

# Leveraging the power of quantization

Quantization in machine learning, particularly in ASR, refers to reducing the precision of the model's parameters. This is typically done by mapping the continuous range of floating-point values to a discrete set of values, often represented by integers. The primary goal of quantization is to decrease the model's computational complexity and memory footprint, which is crucial for deploying ASR systems on devices with limited resources, such as mobile phones or embedded systems. Quantization is essential for several reasons:

- **Reducing model size**: Using lower precision to represent the model's weights can significantly reduce the model's overall size. This is particularly beneficial for on-device deployment, where storage space is at a premium.

- **Improving inference speed**: Lower precision arithmetic is faster on many hardware platforms, especially those without dedicated floating-point units. This can lead to faster inference times, critical for real-time applications such as ASR.

- **Increasing energy efficiency**: Quantized models require fewer computational resources, lowering power consumption. This is essential for battery-powered devices.

- **Expanding hardware compatibility**: Many edge devices are optimized for integer computations. Quantization allows models to leverage these hardware optimizations.

Some standard machine-learning quantization techniques in ASR are **vector quantization (VQ)**, **int8 quantization**, and **low-bit quantization**. Let's briefly describe each:

*VQ* is a classical technique in various domains, including speech coding and recognition. It involves mapping vectors from an ample vector space to a finite number of regions, which can be efficiently represented with fewer bits. VQ has been successfully applied to speech recognition systems, improving performance by efficiently compressing the feature space.

*INT8 quantization* is a recent approach to representing model weights and activations using 8-bit integers instead of 32-bit floating-point numbers. This method can reduce the model size by a factor of 4 without significantly degrading performance because it carefully rounds data from one type to another rather than simply truncating it.

Further advancements have led to *low-bit quantization* techniques, where aggressive quantization to even 1 bit is explored. While this can substantially reduce storage and runtime, it may increase the **word error rate (WER)** in ASR tasks. However, with careful design, such as DistilHuBERT (`https://huggingface.co/ntu-spml/distilhubert`), it is possible to achieve model compression with minimal accuracy loss.

Be aware that quantization introduces a quantization error, which can degrade the model's performance if not properly managed. Techniques such as **quantization-aware training (QAT)** and **post-training quantization (PTQ)** have been developed to mitigate these effects. QAT simulates the quantization process during training, allowing the model to adapt to the lower precision. PTQ, on the other hand, applies quantization after training, using calibration techniques to adjust the quantization parameters for minimal performance loss.

*Figure 7.1* shows a high-level view of the quantization process for ASR models:

**Quantization Process for ASR Models**

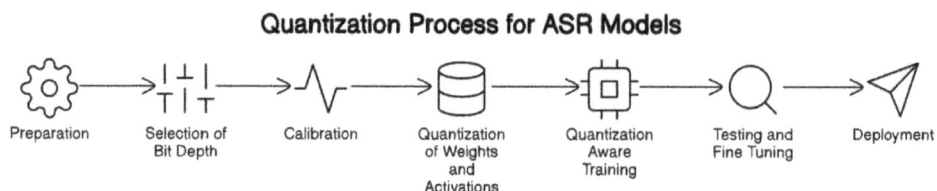

| Preparation | Selection of Bit Depth | Calibration | Quantization of Weights and Activations | Quantization Aware Training | Testing and Fine Tuning | Deployment |

Figure 7.1 – Quantization process for ASR models

The steps broadly outlined in the diagram are generic and intended to provide a foundational overview. Let's review each step in more detail:

1. **Preparation**: The initial step involves training the ASR model using high-precision (32-bit floating-point) representations. This ensures the model captures the complex patterns necessary for accurate speech recognition.

2. **Selection of bit depth**: Based on the target hardware and performance requirements, an appropriate bit depth is selected for quantization. Common choices include 16-bit (half-precision), 8-bit (`int8`), or even lower. Your selection should consider model size, computational efficiency, and accuracy.

   The choice of bit depth directly impacts the trade-off between model size, computational speed, and accuracy. Lower bit depths significantly reduce the model's memory footprint and increase computational efficiency, but they can introduce quantization errors that potentially degrade model performance. The challenge lies in selecting an optimal bit depth that minimizes these errors while achieving the desired efficiency gains.

3. **Calibration**: A representative dataset is used to run inference through the model for PTQ. This step helps gather statistics about the distribution of activations, which are crucial for determining the quantization parameters.

4. **Quantization of weights and activations**: The model's weights and activations are quantized using the gathered statistics to the selected bit depth. This involves mapping the high-precision values to a lower-precision space using scale factors and zero points.

5. **QAT (optional)**: In some cases, models undergo QAT, where the quantization effects are simulated during the training process. This helps the model to adapt to the reduced precision, potentially mitigating accuracy loss.

6. **Testing and fine-tuning**: After quantization, the model's performance is evaluated to ensure accuracy remains within acceptable bounds. If necessary, fine-tuning or adjustments to the quantization parameters are made.

7. **Deployment**: The quantized model is then deployed on the target hardware, benefiting from reduced memory usage and faster inference times. This makes it suitable for edge devices or environments with limited computational resources.

Several quantized versions of Whisper are available, and more are being developed. In my experience, I have found that Faster-Whisper and Distil-Whisper offer superior and reliable performance. Here is a brief description of them:

- **Faster-Whisper** implements the Whisper model in CTranslate2, a library for efficient inference with Transformer models. It applies various methods to increase efficiency, such as weight quantization, layer fusion, and batch reordering. Quantization plays a significant role in Faster-Whisper by reducing the model's memory footprint and accelerating inference, particularly on GPUs. We will experience Faster-Whisper in the *Diarizing Speech with WhisperX and NVIDIA's NeMo* chapter because WhisperX uses Faster-Whisper to perform **speech-to-text (STT)** transcriptions.

- **Distil-Whisper** is a distilled version of Whisper's `small.en`, `medium.en`, and `large-v2` models that are faster and smaller while maintaining a comparable WER. Quantization can further enhance Distil-Whisper's efficiency by reducing the precision of the model's parameters, thus allowing for faster processing and lower memory requirements.

As we explore the power of quantization, let's dive into a practical example using the CTranslate2 framework. CTranslate2 provides an efficient way to quantize and optimize the Whisper model for deployment on resource-constrained devices.

## Quantizing Whisper with CTranslate2 and running inference with Faster-Whisper

Please find and open the `LOAIW_ch07_1_Quantizing_Whisper_with_CTranslate2.ipynb` Colab notebook (`https://github.com/PacktPublishing/Learn-OpenAI-Whisper/blob/main/Chapter07/LOAIW_ch07_1_Quantizing_Whisper_with_CTranslate2.ipynb`). The notebook demonstrates quantizing the Whisper model using CTranslate2 and the Faster-Whisper framework to load the quantized models and perform inference (transcription or translation). You should run the notebook using only the CPU and then the GPU. The CPU performance should be relatively fast because we use small Whisper models, short audio files, and quantization. *Figure 7.2* provides an overview of the quantization process, from preparing the audio data and converting and quantizing the model to evaluating its performance in language detection and transcription tasks. Quantization is vital in optimizing the model for deployment in resource-constrained environments, enabling efficient and accurate speech recognition capabilities:

Quantizing Whisper with CTranslate2

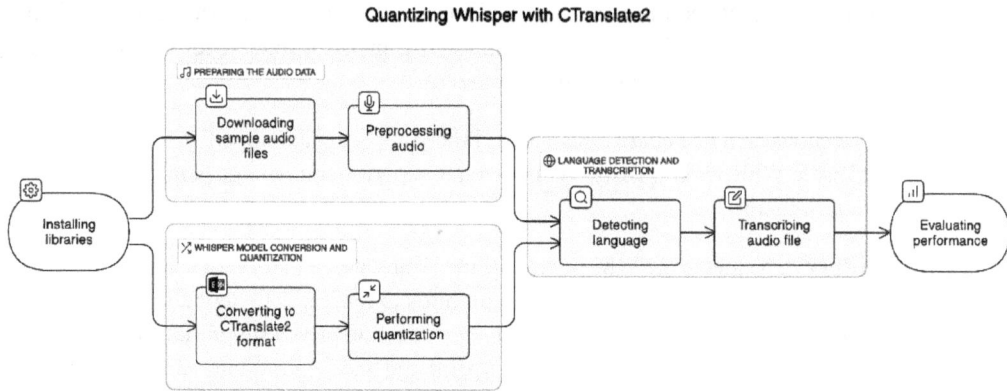

Figure 7.2 – High-level view of the process of quantizing Whisper using the CTranslate2 framework

The following steps provide an overview of the quantization process. For a complete, end-to-end implementation, please refer to the LOAIW_ch07_1_Quantizing_Whisper_with_CTranslate2.ipynb notebook. This section will present the high-level steps and selected code snippets to illustrate the process. Remember that the notebook contains additional details and explanations to help you understand the quantization workflow comprehensively. Here's a detailed breakdown of the process:

1.  **Installing libraries**: The code begins with installing ctranslate2, transformers, and faster-whisper:

    ```
    !pip install ctranslate2
    !pip install transformers[torch]>=4.23
    !pip install faster-whisper
    ```

    These libraries are essential for quantization and leveraging the Whisper model's capabilities.

2.  **Downloading sample audio files**: Two are downloaded from our GitHub repository to test the Whisper model's transcription capabilities:

    ```
    !wget -nv https://github.com/PacktPublishing/Learn-OpenAI-
    Whisper/raw/main/Chapter01/Learn_OAI_Whisper_Sample_
    Audio01.mp3
    !wget -nv https://github.com/PacktPublishing/Learn-OpenAI-
    Whisper/raw/main/Chapter01/Learn_OAI_Whisper_Sample_
    Audio02.mp3
    ```

3.  **Preprocessing audio files**: The audio files are loaded and resampled to a sampling frequency of 16,000 Hz using librosa:

    ```
    import ctranslate2
    from IPython.display import Audio
    ```

```
import librosa
import transformers
# Load and resample the audio file.
sampling_frequency = 16000
audio, _ = librosa.load("Learn_OAI_Whisper_Sample_Audio01.mp3",
sr=sampling_frequency, mono=True)
Audio(audio, rate=sampling_frequency)
```

This step is crucial for ensuring that the audio data is in the correct format for processing by the Whisper model.

4. **Converting to CTranslate2 format**: The Whisper model (`openai/whisper-tiny`) is converted to the CTranslate2 format, a more efficient inference format:

```
!ct2-transformers-converter --force --model openai/whisper-tiny
--output_dir whisper-tiny-ct2
```

The `ct2-transformers-converter` command converts models to the CTranslate2 format, optimized for fast inference. The core CTranslate2 implementation is framework-agnostic. The framework-specific logic is moved to a conversion step that loads supported models into a unified representation. The weights are then optionally quantized and saved into an optimized binary format for efficient storage and processing.

When converting models using the CTranslate2 tool, the output directory typically contains several key files for the CTranslate2 engine to load and run the model. The command streamlines the process of preparing models for deployment in environments where computational efficiency is crucial and a preparatory step for quantization. While the exact output files can vary depending on the specific model being converted and the options used during conversion, standard files include the following:

- `config.json`: This JSON file contains configuration information about the model, such as its architecture, the size of its layers, and other hyperparameters. This information is crucial for the CTranslate2 engine to interpret the model's binary weights and perform inference correctly.

- `model.bin`: This is the binary file containing the quantized weights of the model. Quantization reduces the precision of the model's weights, which can significantly decrease the model size and improve inference speed, often with minimal impact on accuracy.

- `vocabulary.json` or similar vocabulary files (for example, `source.spm` and `target.spm` for models using `SentencePiece` tokenization): These files contain the mapping between tokens (words or subwords) and their corresponding indices in the model's vocabulary. This mapping is essential for converting input text into a format that the model can process (tokenization) and converting the model's output back into human-readable text (detokenization).

These files represent the converted and optimized model and are ready for use with CTranslate2. The conversion process might also include copying additional files necessary for the model's operation, such as tokenization configuration (`tokenizer_config.json`), special tokens mapping (`special_tokens_map.json`), and others, depending on the model's requirements and the conversion options you use.

5. **Performing quantization**: The model is then quantized to an 8-bit integer format (`INT8`):

```
!ct2-transformers-converter --force --model openai/whisper-tiny
--output_dir whisper-tiny-ct2-int8 \
--copy_files tokenizer.json preprocessor_config.json
--quantization int8
```

CTranslate2 supports the most common quantization types:

* 8-bit integers (`INT8`)

* 16-bit integers (`INT16`)

* 16-bit floating points (`FP16`)

* 16-bit brain floating points (`BF16`)

This step significantly reduces the model's size and computational requirements, making it more suitable for deployment on devices with limited resources.

6. **Detecting language**: The quantized model detects the language of the provided audio samples:

```
# Detect the language.
results = model.detect_language(features)
language, probability = results[0][0]
print("Detected language %s with probability %f" % (language,
probability))
```

This step is important for ensuring that the model accurately understands the context of the audio data.

7. **Transcribing audio files**: The model generates transcriptions for the audio samples using the `processor.tokenizer.convert_tokens_to_ids()` method:

```
prompt = processor.tokenizer.convert_tokens_to_ids(
    [
        "<|startoftranscript|>",
        language,
        "<|transcribe|>",
        "<|notimestamps|>",  # Remove this token to generate
timestamps.
    ]
)
```

```
# Load the model on device
model = ctranslate2.models.Whisper("whisper-tiny-ct2-int8",
device=this_device)

# Run generation for the 30-second window.
results = model.generate(features, [prompt])
transcription = processor.decode(results[0].sequences_ids[0])
print(transcription))
```

This demonstrates the model's ability to transcribe speech accurately, even after quantization.

8. **Evaluating performance**: After the audio transcription, the code evaluates the performance of the quantized model, such as measuring the time taken for transcription:

```
# Print the end time and the delta in seconds and fractions of a
second.
end = time.time()
print('start: ', start)
print('end: ', end)
print('delta: ', end - start)
print('delta: ', datetime.timedelta(seconds=end - start))
```

This evaluation is crucial for understanding the impact of quantization on the model's efficiency and accuracy.

The results show empirical evidence that quantized models of Whisper perform transcription quite well using a much smaller memory and processing footprint. Building upon our understanding of quantization, let's now focus on another robust framework, OpenVINO. We'll investigate how OpenVINO can be used to quantize the Distil-Whisper model, offering a more comprehensive and rigorous quantization process.

## Quantizing Distil-Whisper with OpenVINO

This hands-on exercise relies on the LOAIW_ch07_2_Quantizing_Distil_Whisper_ with_OpenVINO.ipynb Colab notebook (https://github.com/PacktPublishing/ Learn-OpenAI-Whisper/blob/main/Chapter07/LOAIW_ch07_2_Quantizing_ Distil_Whisper_with_OpenVINO.ipynb). Because of OpenVINO, I recommend you run this notebook in Colab using CPU and high RAM. OpenVINO does not use an NVIDIA GPU, even if it is present, only an Intel GPU. However, the libraries OpenVINO provides are optimized to run on a plain CPU, thus a significant advantage when the computational processing resources are limited. However, you should have at least 50 GB of RAM for quantization. The notebook provides a comprehensive guide on utilizing Distil-Whisper (based on WhisperX), a distilled variant of the Whisper model, with OpenVINO for ASR. Distil-Whisper offers a significant reduction in the number of parameters (from 1,550 parameters in large-v2 to 756 in distill-large-v2, or about 50%

reduction) and an increase in inference speed while maintaining close performance to the original Whisper model regarding WER.

*Figure 7.3* outlines converting the Distil-Whisper model to the OpenVINO **intermediate representation (IR)** format, applying INT8 PTQ for performance enhancement, and running the model for speech recognition tasks:

**Quantizing Distil-Whisper with OpenVINO**

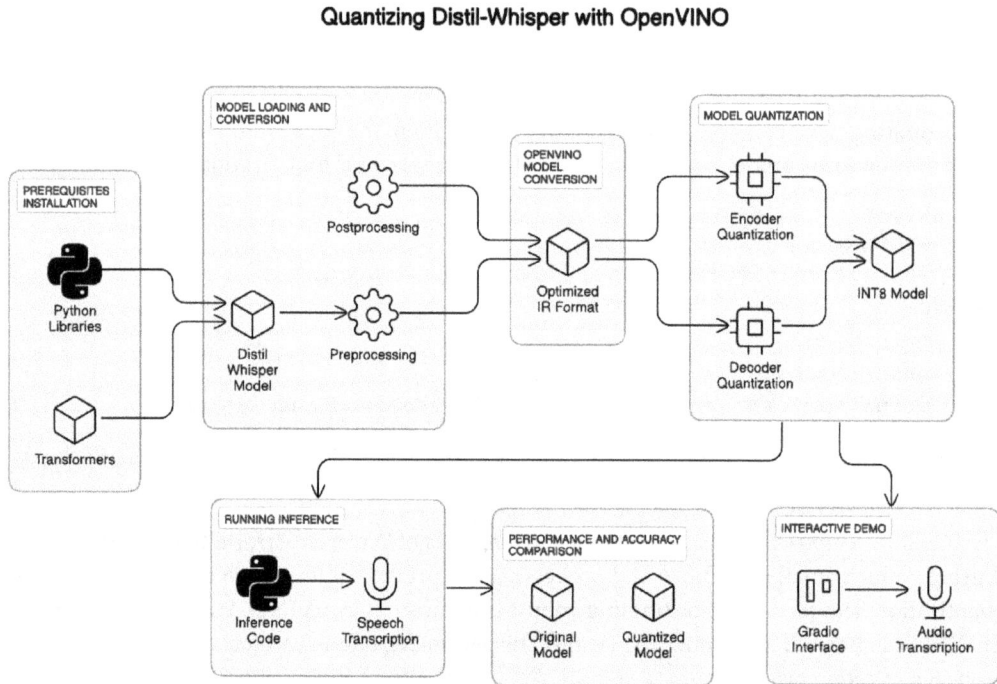

Figure 7.3 – High-level architectural diagram quantizing Distil-Whisper using the OpenVINO framework

The following subsections will describe the critical steps in quantizing the Distil-Whisper model using the OpenVINO framework. We will install the necessary libraries, load the model, convert it to the OpenVINO format, and apply quantization. We will also explore how to load the quantized model using the Optimum library and integrate it with Hugging Face pipelines. Finally, we will run inference with the quantized model and compare its performance and accuracy to the original model.

## Installing libraries

First, the process instructs the installation of necessary Python libraries:

```
%pip install -q "transformers>=4.35" onnx "git+https://github.com/
huggingface/optimum-intel.git" "peft==0.6.2" --extra-index-url
https://download.pytorch.org/whl/cpu
%pip install -q "openvino>=2023.2.0" datasets  "gradio>=4.0" "librosa"
```

```
"soundfile"
%pip install -q "nncf>=2.6.0" "jiwer"
```

Let's examine each one in more detail, focusing on the libraries we have not described before:

- **Transformers**: This library is used for NLP tasks such as text classification, information extraction, and question-answering. It provides access to pre-trained models such as BERT, GPT-2, and, in this case, the Distil-Whisper model for ASR.

- **Open Neural Network Exchange (ONNX)**: ONNX is an open format representing machine learning models. It enables models to be transferred between different frameworks and tools, facilitating interoperability.

- **Optimum Intel**: This is part of the Hugging Face Optimum library tailored for Intel hardware. It converts models to the OpenVINO IR format, which is optimized for Intel's hardware, and performs tasks such as quantization to improve model performance.

- **OpenVINO**: The OpenVINO toolkit is designed to facilitate fast and efficient inference of deep learning models on Intel hardware. It includes optimization tools and libraries to accelerate various computer vision and deep learning tasks.

- **Datasets**: This library is part of the Hugging Face ecosystem and is used for loading and processing datasets simply and efficiently. It is handy for machine learning tasks that require handling large amounts of data.

- **Soundfile**: This library provides functions for reading from and writing to audio files in various formats. It handles audio data input and output operations.

- **Neural Network Compression Framework (NNCF)**: This toolkit for optimizing deep learning models through quantization, pruning, and knowledge distillation. It improves neural networks' performance, particularly regarding inference speed and memory usage.

- **JiWER**: This is a library for evaluating automatic speech recognition models. It calculates metrics such as the WER, a standard measure of speech recognition systems' performance.

Each library plays a specific role in running and optimizing the Distil-Whisper model using OpenVINO, from model conversion and optimization to performance evaluation and user interface creation.

### Loading the model

When initializing a PyTorch Whisper model using the `transformers` library, the `AutoModelForSpeechSeq2Seq.from_pretrained` method is the go-to:

```
from transformers import AutoProcessor, AutoModelForSpeechSeq2Seq
processor = AutoProcessor.from_pretrained(model_id.value)
pt_model = AutoModelForSpeechSeq2Seq.from_pretrained(model_id.value)
pt_model.eval();
```

This tutorial will use the `distil-whisper/distil-medium.en` model as our primary example. It's worth noting that the model must be downloaded during the first run, which may take some time.

If you want to explore alternative models, the Distil-Whisper Hugging Face collection offers options such as `distil-whisper/distil-large-v2` or `distil-whisper/distil-small.en`. Other models based on the original Whisper architecture are available, and you can find more information about them in the provided resources.

It's crucial to emphasize the significance of preprocessing and postprocessing in this model's usage. The `AutoProcessor` class, used to initialize `WhisperProcessor`, plays a vital role in preparing the audio input data for the model. It handles the audio conversion into a Mel-spectrogram and decodes the `token_ids` predicted output back into a string using the tokenizer.

By leveraging the `AutoModelForSpeechSeq2Seq.from_pretrained` method and understanding the preprocessing and postprocessing steps, you'll be well equipped to work effectively with PyTorch Whisper models.

### Loading the OpenVINO model using the Optimum library

The Hugging Face Optimum API is a powerful tool that simplifies converting and quantizing models from the Hugging Face Transformers library to the OpenVINO™ IR format. The Hugging Face Optimum documentation (`https://huggingface.co/docs/optimum/intel/inference`) is an excellent resource if you're looking for more in-depth information.

Optimum Intel is your friend when loading optimized models from the Hugging Face Hub and creating pipelines for inference with OpenVINO Runtime. What's great about the Optimum Inference models is that they are API-compatible with Hugging Face `transformers` models. You can seamlessly replace the `AutoModelForXxx` class with the corresponding `OVModelForXxx` class without hassle:

```
# Using HF transformers models
from transformers import AutoModelForSpeechSeq2Seq
from transformers import AutoTokenizer, pipeline
model_id = "distil-whisper/distil-large-v2"
model = AutoModelForSpeechSeq2Seq.from_pretrained(model_id)

# Using Optimum Inference models
from optimum.intel.openvino import OVModelForSpeechSeq2Seq
from transformers import AutoTokenizer, pipeline
model_id = "distil-whisper/distil-large-v2"
model = OVModelForSpeechSeq2Seq.from_pretrained(model_id, export=True)
```

You'll need to call the `from_pretrained` method to initialize the model class. When downloading and converting the `transformers` model, include the `export=True` parameter. This will ensure a smooth conversion process. Once you have the converted model, you can save it using the `save_pretrained` method:

```
from pathlib import Path
from optimum.intel.openvino import OVModelForSpeechSeq2Seq

model_path = Path(model_id.value.replace('/', '_'))
ov_config = {"CACHE_DIR": ""}

if not model_path.exists():
    ov_model = OVModelForSpeechSeq2Seq.from_pretrained(
        model_id.value, ov_config=ov_config, export=True,
compile=False, load_in_8bit=False
    )
    ov_model.half()
    ov_model.save_pretrained(model_path)
else:
    ov_model = OVModelForSpeechSeq2Seq.from_pretrained(
        model_path, ov_config=ov_config, compile=False
    )
```

It's worth mentioning that the tokenizers and processors distributed with the models are also compatible with the OpenVINO model. This compatibility allows you to reuse the previously initialized processor, saving time and effort.

Using the Hugging Face Optimum library, we can also convert the Distil-Whisper model to OpenVINO's optimized IR format. This step is crucial for leveraging OpenVINO's inference engine for efficient model execution:

```
-from transformers import AutoModelForSpeechSeq2Seq
+from optimum.intel.openvino import OVModelForSpeechSeq2Seq
from transformers import AutoTokenizer, pipeline

model_id = "distil-whisper/distil-large-v2"
-model = AutoModelForSpeechSeq2Seq.from_pretrained(model_id)
+model = OVModelForSpeechSeq2Seq.from_pretrained(model_id,
export=True)
```

By leveraging the Hugging Face Optimum API and Optimum Intel, you can efficiently convert and quantize models, load optimized models, and create pipelines for inference with OpenVINO Runtime. The API compatibility and the ability to reuse initialized processors make the process even more streamlined.

### Using the OpenVINO model with Hugging Face pipelines

By combining the OpenVINO model with the Hugging Face pipeline interface and utilizing the chunked algorithm and batching capabilities of Distil-Whisper, you'll be able to tackle long audio transcription tasks with unprecedented speed and ease.

As with the original PyTorch model, the OpenVINO model seamlessly integrates with the Hugging Face pipeline interface for ASR. This compatibility allows you to transcribe long audio files using the pipeline effortlessly:

```
from transformers import pipeline
ov_model.generation_config = pt_model.generation_config
pipe = pipeline(
    "automatic-speech-recognition",
    model=ov_model,
    tokenizer=processor.tokenizer,
    feature_extractor=processor.feature_extractor,
    max_new_tokens=128,
    chunk_length_s=15,
    batch_size=16,
)
```

Distil-Whisper takes it a step further by employing a chunked algorithm, which significantly speeds up the transcription process for long-form audio. This chunked long-form algorithm is an impressive nine times faster than the sequential algorithm proposed by OpenAI in their Whisper paper (https://cdn.openai.com/papers/whisper.pdf).

To take advantage of chunking, you only need to pass the chunk_length_s parameter to the pipeline. When working with Distil-Whisper, setting the chunk length to 15 seconds is the sweet spot for optimal performance. But that's not all! If you want to leverage the power of batching, include the batch_size argument when calling the pipeline. This will enable you to process multiple audio chunks simultaneously, further boosting the efficiency of your transcription workflow.

### Quantizing the model

Quantization is a powerful technique for significantly reducing the model size and improving inference speed. NNCF makes it easier than ever to implement PTQ. By seamlessly integrating quantization layers into the model graph and leveraging a subset of the training dataset to initialize the parameters of these additional layers, NNCF ensures that the modifications required to your original training code are minimal.

To embark on the optimization journey, the first step is to create calibration datasets specifically tailored for quantization:

```
%%skip not $to_quantize.value

from itertools import islice
from optimum.intel.openvino.quantization import InferRequestWrapper

def collect_calibration_dataset(ov_model: OVModelForSpeechSeq2Seq,
calibration_dataset_size: int):
    # Overwrite model request properties, saving the original ones for
restoring later
    original_encoder_request = ov_model.encoder.request
    original_decoder_with_past_request = ov_model.decoder_with_past.
request
    encoder_calibration_data = []
    decoder_calibration_data = []
    ov_model.encoder.request = InferRequestWrapper(original_encoder_
request, encoder_calibration_data)
    ov_model.decoder_with_past.request = InferRequestWrapper(original_
decoder_with_past_request,

                                                            decoder_
calibration_data)

    calibration_dataset = load_dataset("librispeech_asr", "clean",
split="validation", streaming=True)
    for sample in tqdm(islice(calibration_dataset, calibration_
dataset_size), desc="Collecting calibration data",
                       total=calibration_dataset_size):
        input_features = extract_input_features(sample)
        ov_model.generate(input_features)

    ov_model.encoder.request = original_encoder_request
    ov_model.decoder_with_past.request = original_decoder_with_past_
request

    return encoder_calibration_data, decoder_calibration_data
```

Since the Whisper encoder and decoder are quantized separately, preparing a calibration dataset for each model is essential. This is where the InferRequestWrapper class comes into play. Importing this class, you can intercept and collect the model inputs in a list. Then, you'll run model inference on a small subset of audio samples. Remember that increasing the calibration dataset's size generally leads to better quantization quality, so it's worth experimenting to find the right balance.

Once you have your calibration datasets ready, it's time to unleash the power of `nncf.quantize`. This function is your key to obtaining quantized encoder and decoder models. In the case of Distil-Whisper, you'll run `nncf.quantize` on the `encoder` and `decoder_with_past` models. It's worth noting that the first-step decoder is not quantized since its contribution to the overall inference time is negligible:

```
%%skip not $to_quantize.value

import gc
import shutil
import nncf


CALIBRATION_DATASET_SIZE = 50
quantized_model_path = Path(f"{model_path}_quantized")


def quantize(ov_model: OVModelForSpeechSeq2Seq, calibration_dataset_
size: int):
    if not quantized_model_path.exists():
        encoder_calibration_data, decoder_calibration_data = collect_
calibration_dataset(
            ov_model, calibration_dataset_size
        )
        print("Quantizing encoder")
        quantized_encoder = nncf.quantize(
            ov_model.encoder.model,
            nncf.Dataset(encoder_calibration_data),
            subset_size=len(encoder_calibration_data),
            model_type=nncf.ModelType.TRANSFORMER,
            # Smooth Quant algorithm reduces activation quantization
error; optimal alpha value was obtained through grid search
            advanced_parameters=nncf.
AdvancedQuantizationParameters(smooth_quant_alpha=0.50)
        )
        ov.save_model(quantized_encoder, quantized_model_path /
"openvino_encoder_model.xml")
        del quantized_encoder
        del encoder_calibration_data
        gc.collect()

        print("Quantizing decoder with past")
        quantized_decoder_with_past = nncf.quantize(
            ov_model.decoder_with_past.model,
            nncf.Dataset(decoder_calibration_data),
            subset_size=len(decoder_calibration_data),
            model_type=nncf.ModelType.TRANSFORMER,
```

```
            # Smooth Quant algorithm reduces activation quantization
error; optimal alpha value was obtained through grid search
            advanced_parameters=nncf.
AdvancedQuantizationParameters(smooth_quant_alpha=0.95)
        )
        ov.save_model(quantized_decoder_with_past, quantized_model_
path / "openvino_decoder_with_past_model.xml")
        del quantized_decoder_with_past
        del decoder_calibration_data
        gc.collect()


        # Copy the config file and the first-step-decoder manually
        shutil.copy(model_path / "config.json", quantized_model_path /
"config.json")
        shutil.copy(model_path / "openvino_decoder_model.xml",
quantized_model_path / "openvino_decoder_model.xml")
        shutil.copy(model_path / "openvino_decoder_model.bin",
quantized_model_path / "openvino_decoder_model.bin")

    quantized_ov_model = OVModelForSpeechSeq2Seq.from_
pretrained(quantized_model_path, ov_config=ov_config, compile=False)
    quantized_ov_model.to(device.value)
    quantized_ov_model.compile()
    return quantized_ov_model

ov_quantized_model = quantize(ov_model, CALIBRATION_DATASET_SIZE)
```

The code snippet shows that the final step is to serialize the INT8 model using the `openvino.save_model` function after quantization. This step ensures that your quantized model is ready for deployment and can be quickly loaded for inference.

It's essential to remember that quantization is a computationally intensive operation that can be both time-consuming and memory-intensive. Running the quantization code may require patience, but the benefits of model size reduction and inference speed improvement make it well worth the effort.

By following these steps and leveraging the power of NNCF, you can optimize your models through PTQ, enabling faster and more efficient inference.

### *Running inference*

Here, we demonstrate how to run inference with the quantized model, including loading the model, preparing input samples, and executing the model to transcribe speech. Here are the steps in detail:

1. **Loading the dataset**: Load a dataset for testing the model's transcription capabilities. The dataset used is `librispeech_asr_dummy` from Hugging Face's `datasets` library:

   ```
   %%skip not $to_quantize.value
   ```

```
dataset = load_dataset(
    "hf-internal-testing/librispeech_asr_dummy", "clean",
split="validation"
)
sample = dataset[0]
```

2. **Preparing the input features**: Extract input features from a sample in the dataset. This involves converting the audio to a numpy array format and then to a tensor that the model can process:

```
input_features = extract_input_features(sample)
predicted_ids = ov_model.generate(input_features)
```

3. **Running inference on the original model**: Use the original OpenVINO model to generate predictions for the input features. Decode the predicted token IDs into text transcription using the model's processor:

```
transcription_original = processor.batch_decode(predicted_ids,
skip_special_tokens=True)
```

4. **Running inference on the quantized model**: Similarly, use the quantized OpenVINO model to generate predictions for the same input features. Decode the predicted token IDs into text transcription:

```
predicted_ids = ov_quantized_model.generate(input_features)
transcription_quantized = processor.batch_decode(predicted_ids,
skip_special_tokens=True)
```

5. **Displaying the audio**: Use IPython's Audio class to play the audio file used for transcription:

```
display(ipd.Audio(sample["audio"]["array"], rate=sample["audio"]
["sampling_rate"]))
```

6. **Printing transcriptions**: Print the transcriptions from the original and quantized models to compare the results:

```
print(f"Original : {transcription_original[0]}")
print(f"Quantized: {transcription_quantized[0]}")
```

After running this code in the notebook, review the transcriptions and verify that the transcriptions from the original and quantized models are the same, ensuring that quantization did not significantly impact the model's accuracy.

In addition, the notebook includes how to use the model with Hugging Face's pipeline interface for ASR, highlighting the efficiency of chunked algorithms for long-form audio transcription.

## Comparing performance and accuracy

Next, we compare the original and quantized Distil-Whisper models regarding accuracy (using WER) and performance (inference time). It illustrates the benefits of quantization in enhancing model inference speed without a significant drop in accuracy. Comparing the performance and accuracy of the original and quantized models involves the following:

- **Measuring accuracy**: We use the *1 - WER* metric to measure the accuracy of the models. This involves comparing the transcriptions produced by the models against a ground truth to calculate the error rate. A lower WER indicates higher accuracy:

```
word_accuracy = (1 - wer(ground_truths, predictions, reference_
transform=wer_standardize,                          hypothesis_
transform=wer_standardize)) * 100
```

- **Measuring performance**: The inference time is measured separately for the encoder and decoder-with-past model forwards and the whole model inference. This step involves timing the model's inference process to evaluate how quickly it can generate predictions. Performance measurement is crucial for understanding the efficiency gains achieved through quantization:

```
mean_whole_infer_time = sum(whole_infer_times)
mean_encoder_infer_time = sum(encoder_infer_times)
mean_decoder_with_time_infer_time = sum(decoder_with_past_infer_
times)
```

- **Comparing original and quantized models**: The notebook directly compares the original Distil-Whisper models and their quantized counterparts regarding accuracy (*using 1 - WER*) and performance (inference time). This comparison helps to illustrate the impact of quantization on the model's efficiency and effectiveness:

```
print(f"Encoder performance speedup: {times_original[1] / times_
quantized[1]:.3f}")
print(f"Decoder with past performance speedup: {times_
original[2] / times_quantized[2]:.3f}")
print(f"Whole pipeline performance speedup: {times_original[0] /
times_quantized[0]:.3f}")
print(f"Whisper transcription word accuracy. Original model:
{accuracy_original:.2f}%. Quantized model: {accuracy_
quantized:.2f}%.")
print(f"Accuracy drop: {accuracy_original - accuracy_
quantized:.2f}%.")
```

Based on the comparison printout from running the notebook, you can conclude the benefits of quantization, such as significant improvements in model inference time without a major drop in accuracy. These steps provide a comprehensive framework for evaluating the impact of quantization on the performance and accuracy of ASR models such as Distil-Whisper when optimized with

OpenVINO. The goal is to demonstrate that quantization can significantly enhance model efficiency for deployment in resource-constrained environments without substantially compromising accuracy.

### Running the interactive demo

As an extra, the interactive Gradio demo allows us to test the model's capabilities on their audio data or recordings. This section demonstrates the practical application of the quantized Distil-Whisper model in a user-friendly manner.

I encourage you to run and experiment with the Colab notebook. It is a foundational tool for understanding the quantization process and, more importantly, a blueprint for your experimental or production work. After running the notebook, we embarked on a fascinating journey through the integration of cutting-edge technologies in ASR. The notebook meticulously outlined leveraging the Distil-Whisper model, a distilled variant of OpenAI's Whisper, optimized for performance with significantly fewer parameters, and deploying it with Intel's OpenVINO toolkit for enhanced inference speed and efficiency.

One of the key learnings from this notebook was the seamless synergy between various libraries and frameworks to achieve a streamlined workflow for ASR tasks. Using the Hugging Face Transformers library to access pre-trained models and the Optimum Intel library for model conversion to OpenVINO's IR exemplified a powerful approach to model deployment. This process simplified the user experience and paved the way for leveraging hardware acceleration capabilities offered by Intel architectures.

The notebook further delved into the practical aspects of model quantization using NNCF. This step was crucial for optimizing model performance without significantly compromising accuracy. The detailed walkthrough of preparing calibration datasets, running quantization, and comparing the performance and accuracy of the original and quantized models provided invaluable insights into the nuances of model optimization.

Another significant aspect highlighted in the notebook was the use of Gradio to create interactive demos. This demonstrated the practical application of the Distil-Whisper model in real-world scenarios, allowing users to test the model's capabilities on their audio data. Including such a demo underscored the importance of accessibility and user engagement in developing and deploying AI models.

You should seek ways to apply the learnings from this notebook directly to your experimental or production ASR tasks. They extend to the broader field of AI model deployment and optimization, highlighting the evolving landscape of AI technologies and their practical applications.

While quantization has proven to be a powerful technique for optimizing Whisper's performance and enabling efficient deployment, another exciting frontier lies in exploring the challenges and opportunities of real-time speech recognition with Whisper. Real-time transcription opens up possibilities, from enhancing accessibility to facilitating instant communication. However, it also presents unique technical hurdles that must be overcome. In the following section, we will delve into the current limitations and ongoing research efforts to make real-time transcription with Whisper a reality. By understanding these challenges and the potential solutions on the horizon, we can appreciate the immense potential of Whisper in reshaping how we interact with spoken language in real-time scenarios.

# Facing the challenges and opportunities of real-time speech recognition

Pursuing real-time transcription with Whisper opens up many applications that can benefit various sectors, including education, healthcare, and customer service. Real-time transcription can enhance accessibility for individuals with hearing impairments, facilitate instant communication in multilingual contexts, and provide immediate documentation of verbal exchanges. As Whisper's capabilities evolve, its potential to serve as a universal translator and accessibility tool becomes increasingly apparent.

At present, however, more limitations and challenges are preventing real-time transcription. Let's delve into these aspects, focusing on the technical intricacies and prospects of performing real-time transcription with Whisper:

- **Processing time and latency**: One of the primary challenges in achieving real-time transcription with Whisper is its operation's inherent latency and processing time. As discussions on platforms such as GitHub and Hugging Face reveal, Whisper is not inherently designed for real-time STT conversion. While robust for processing audio files of unlimited length, the system's architecture encounters hurdles in delivering instantaneous transcription results. This latency stems from the complex neural network models that underpin Whisper, which require significant computational resources to analyze and transcribe speech accurately.

- **Increasing accuracy and contextual understanding**: Another limitation lies in Whisper's transcriptions' accuracy and contextual knowledge. While Whisper has demonstrated remarkable proficiency in transcribing diverse languages and accents, real-time applications pose unique challenges. The system must recognize speech accurately and understand context, idioms, and colloquial expressions in the flow of conversation. This demands a level of linguistic and cultural nuance that current models are still striving to perfect.

Despite these limitations, the potential for Whisper to transform real-time transcription is immense. The technology's current capabilities and ongoing advancements offer a glimpse into a future where these challenges are surmountable:

- **Advancing model efficiency**: Recent research efforts have focused on enhancing Whisper's efficiency and reducing latency, making real-time transcription a tangible goal. For instance, a study on arXiv, *Turning Whisper into Real-Time Transcription System* (https://arxiv.org/abs/2307.14743), discusses methods for turning Whisper into a real-time transcription system. These include optimizing the model's architecture and leveraging more powerful computational resources. As these advancements continue, we can anticipate significant reductions in processing time, bringing Whisper closer to delivering seamless real-time transcription.

- **Integrating with edge computing**: The integration of Whisper with edge computing presents a promising avenue for overcoming latency issues. By processing data closer to the source of data generation, edge computing can drastically reduce the time it takes for audio to be transcribed. This approach accelerates transcription and alleviates bandwidth constraints, making real-time transcription more feasible and efficient.

While the journey toward flawless real-time transcription with Whisper is fraught with technical challenges, the opportunities it presents are undeniably compelling. The latency, processing time, and contextual accuracy limitations are significant yet manageable. Through ongoing research, technological advancements, and innovative applications, Whisper stands on the cusp of redefining real-time transcription. As we look to the future, the integration of Whisper into our daily lives promises not only to enhance communication and accessibility but also to push the boundaries of what is possible with AI. The road ahead is challenging and exciting, underscoring the importance of continued exploration and development in this dynamic field.

To better understand the challenges and potential of real-time speech recognition with Whisper, let's dive into a practical example. In the following section, we will build an interactive real-time ASR demo using Hugging Face's implementation of Whisper and the user-friendly Gradio library.

## Building a real-time ASR demo with Hugging Face Whisper

In this section, we will leverage the power of Gradio, a user interface library, to rapidly construct an interactive demo of the Whisper model. This demo will allow you or others to test the model's performance by speaking into the microphone on your device. Let's find and run the `LOAIW_ch07_3_Building_real_time_ASR_with_HF_Whisper.ipynb` notebook (https://github.com/PacktPublishing/Learn-OpenAI-Whisper/blob/main/Chapter07/LOAIW_ch07_3_Building_real_time_ASR_with_HF_Whisper.ipynb). The notebook is structured into three main sections:

- **Setting up the ASR model using the Hugging Face Transformers library**: We will load and configure the necessary components from the `transformers` library to prepare the ASR model for our demo

- **Creating a full-context ASR demo**: We will build a demo in which the user speaks the entire audio before the ASR model processes it and generates the transcription

- **Creating a streaming ASR demo**: We will extend the previous demo to support real-time streaming, allowing the ASR model to transcribe the audio as the user speaks, providing a more interactive experience

By the end of this notebook, you will have a solid understanding of creating engaging demos for speech recognition models using Gradio and the Hugging Face Transformers library.

### Preparing the development environment

Before diving into building the speech recognition demos, it's crucial to set up our development environment with the necessary dependencies. In this section, we will do the following:

1. Install the required libraries, such as Gradio, to ensure a smooth development process.
2. Configure the environment to work seamlessly with the Hugging Face Transformers library, allowing us to leverage pre-trained models and powerful NLP tools.

By properly setting up our environment, we lay the foundation for an efficient and hassle-free coding experience throughout the notebook.

To bring our exploration of real-time ASR with Whisper to life, we'll first need to set up our development environment. Let's walk through the installation of the necessary libraries and configuration of our setup to work seamlessly with the Hugging Face Transformers library.

```
%%capture
!pip -q install gradio
```

Setting up your Hugging Face token is essential to ensure a seamless experience while working with this notebook. The notebook will load transformer classes and models from the Hugging Face repository, which requires valid token authentication.

If you haven't created a Hugging Face token yet or need a refresher on the process, please refer to `https://github.com/PacktPublishing/Learn-OpenAI-Whisper/blob/main/Chapter03/LOAIW_ch03_working_with_audio_data_via_Hugging_Face.ipynb`. This resource provides step-by-step instructions on how to create and configure your Hugging Face token.

By setting up your token correctly, you'll be able to easily access the full range of features and models available in the Hugging Face ecosystem, enabling you to build powerful speech recognition demos:

```
from huggingface_hub import notebook_login

notebook_login()

from huggingface_hub import whoami

whoami()
```

With our development environment set up, let's begin by loading the transformers ASR model, which will serve as the foundation for our interactive application.

### Step 1 – Loading the transformers ASR model

We first need an ASR model to begin building our speech recognition demo. You can either train your model or use a pre-trained one. Loading the `"whisper"` model from the Hugging Face `transformers` library is straightforward. Here's the code snippet to accomplish this:

```
from transformers import pipeline
p = pipeline("automatic-speech-recognition", model="openai/whisper-
base.en")
```

With just these two lines of code, we initialize a pipeline for automatic speech recognition using the `"openai/whisper-base.en"` model. The pipeline abstracts away the complexities of working with the model directly, providing a high-level interface for performing ASR tasks.

By utilizing a pre-trained model such as `"whisper"`, we can quickly start building our demo without the need for extensive model training. This allows us to focus on integrating the model into our application and creating an engaging user experience.

### Step 2 – Building a full-context ASR demo with transformers

Our first step in creating the speech recognition demo is to build a *full-context* ASR demo. In this demo, the user will speak the entire audio before the ASR model processes it and generates the transcription. Thanks to Gradio's intuitive interface, building this demo is a breeze:

```
import gradio as gr
from transformers import pipeline
import numpy as np

transcriber = pipeline("automatic-speech-recognition", model="openai/
whisper-base.en")

def transcribe(audio):
    sr, y = audio
    y = y.astype(np.float32)
    y /= np.max(np.abs(y))

    return transcriber({"sampling_rate": sr, "raw": y})["text"]


demo = gr.Interface(
    transcribe,
    gr.Audio(sources=["microphone"]),
    "text",
)

demo.launch(debug=True)
```

In the preceding snippet, we start by creating a function that wraps around the `pipeline` object we initialized earlier. This function serves as the core of our demo, handling the audio input and generating the transcription.

We then utilize Gradio's built-in `Audio` component to capture the user's audio input. This component will be configured to accept input from the user's microphone and return the file path of the recorded audio. We'll use a simple `Textbox` component to display the transcribed text.

The `transcribe` function, the heart of our demo, takes a single parameter called `audio`. This parameter represents the audio data recorded by the user, stored as a `numpy` array. However, the `pipeline` object expects the audio data to be in the `float32` format. To ensure compatibility, we first convert the audio data to `float32` and then normalize it by dividing it by its maximum absolute value. Finally, we pass the processed audio data to the `pipeline` object to obtain the transcribed text.

## *Step 3 – Enhancing the demo with real-time streaming capabilities*

To create a streaming ASR demo, we need to make the following changes in the Python Gradio script:

1. Set `streaming=True` in the `Audio` component to enable continuous audio capture from the user's microphone.

2. Set `live=True` in the `Interface` component to ensure the interface updates dynamically as new audio data is received.

3. Add a `state` variable to the interface to store the recorded audio and the previous transcription.

All these changes are already applied in the script:

```python
import gradio as gr
from transformers import pipeline
import numpy as np

transcriber = pipeline("automatic-speech-recognition", model="openai/
whisper-base.en")

def transcribe(state, new_chunk):
    if state is None:
        stream = np.array([], dtype=np.float32)
        previous_text = ""
    else:
        stream, previous_text = state

    sr, y = new_chunk
    duration = len(y) / sr
    y = y.astype(np.float32)
    y /= np.max(np.abs(y))

    overlap = int(sr * 0.5)   # Half a second overlap
    if len(stream) > 0:
        stream = np.concatenate([stream[-overlap:], y])
    else:
        stream = y

    # Transcribe the current chunk
    new_text = transcriber({"sampling_rate": sr, "raw": stream})
["text"]

    # Update the previous text based on the overlap
    if len(previous_text) > 0:
        overlap_text = previous_text[-int(len(previous_text) *
```

```
    0.1):]   # Last 10% of previous text
        combined_text = previous_text[:-len(overlap_text)] + new_text
    else:
        combined_text = new_text

    return (stream, combined_text), combined_text

demo = gr.Interface(
    transcribe,
    ["state", gr.Audio(sources=["microphone"], streaming=True)],
    ["state", "text"],
    live=True,
)

demo.launch(debug=True)
```

In the streaming demo, we use a state variable to keep track of the audio history and the previous transcription. The transcribe function is called whenever a new small chunk of audio is received, and it needs to process the new chunk along with the previously recorded audio.

To improve the accuracy and coherence of the transcription, we introduce a dynamic window size based on the duration of the new audio chunk and a slight overlap between consecutive windows. Here's how the transcribe function works:

1.  If the state is None, initialize an empty numpy array (stream) to store the audio and an empty string (previous_text) to store the previous transcription.

2.  Extract new_chunk's sampling rate (sr) and audio data (y) from new_chunk.

3.  Calculate the duration of the new audio chunk and normalize the audio data.

4.  Introduce an overlap of half a second between consecutive windows to ensure continuity in the transcription.

5.  Concatenate the new audio chunk to the existing stream, considering the overlap.

6.  Transcribe the entire stream using the transcriber object.

7.  Update previous_text by removing the overlap from the end of the previous transcription and concatenating it with the new transcription.

8.  Return the updated stream and combined_text values as the state and the combined_text value as the transcription output.

By using a dynamic window size and introducing an overlap between consecutive windows, we can improve the accuracy and coherence of the streaming transcription. The small overlap helps maintain continuity in the transcription and reduces the occurrence of overlapping or missing words.

Of course, this is a straightforward demo. It is designed to show that real-time with Whisper is not as far away from reality as it might appear. I encourage you to enhance and experiment with that demo and have fun!

## Summary

In this chapter, we embarked on an exciting exploration of OpenAI's Whisper's advanced voice capabilities. We delved into powerful techniques that enhance Whisper's performance, such as quantization, and uncovered its potential for real-time speech recognition.

We began by examining the power of quantization, which reduces the model's size and computational requirements while maintaining accuracy. We learned how to apply quantization to Whisper using frameworks such as CTranslate2 and OpenVINO, enabling efficient deployment on resource-constrained devices. The hands-on experience quantizing Whisper using CTranslate2 and Distil-Whisper with OpenVINO provided practical insights into optimizing the model for various deployment scenarios.

Furthermore, we tackled the challenges and opportunities of real-time speech recognition with Whisper. We gained insights into the current limitations, such as processing time and latency, and explored ongoing research efforts to make real-time transcription a reality. The experimental approach to building a streaming ASR demo using Whisper and Gradio provided a glimpse into the future possibilities of real-time speech recognition.

Throughout the chapter, we acquired a solid understanding of advanced techniques to optimize Whisper's performance and appreciate the potential and challenges of real-time speech recognition. The hands-on coding examples and practical insights equipped us with the knowledge and skills to apply these techniques in our projects, pushing the boundaries of what is possible with Whisper.

As we conclude this chapter, we look ahead to *Chapter 8, Diarizing Speech with WhisperX and NVIDIA's NeMo*. While Whisper has proven to be a powerful tool for transcribing speech, there's another crucial aspect of speech analysis that can significantly enhance its utility: speaker diarization. By augmenting Whisper with the ability to identify and attribute speech segments to different speakers, we open a new realm of possibilities for analyzing multispeaker conversations. Join me in the next chapter, and let's explore how Whisper can be integrated with cutting-edge diarization techniques to unlock these capabilities.

# 8

# Diarizing Speech with WhisperX and NVIDIA's NeMo

Welcome to *Chapter 8*, where we will explore the world of **speech diarization**. While Whisper has proven to be a powerful tool for transcribing speech, there's another crucial aspect of speech analysis that can significantly enhance its utility – speaker diarization. By augmenting Whisper with the ability to identify and attribute speech segments to different speakers, we open a new realm of possibilities for analyzing multispeaker conversations. This chapter will explore how Whisper can be integrated with cutting-edge diarization techniques to unlock these capabilities.

We will start by exploring the evolution of speaker diarization systems, from the limitations of early approaches to the transformative impact of transformer models. Through practical, hands-on examples, we'll preprocess audio data, transcribe speech with Whisper, and fine-tune the alignment between transcriptions and the original audio.

In this chapter, we will cover the following main topics:

- Augmenting Whisper with speaker diarization
- Performing hands-on speech diarization

By the end of this chapter, you'll know how to integrate Whisper with advanced techniques such as voice activity detection, speaker embedding extraction, and clustering, enabling you to augment its capabilities and achieve state-of-the-art diarization performance. You'll also learn how to leverage NVIDIA's powerful **multiscale diarization decoder** (**MSDD**) model, which considers multiple temporal resolutions of speaker embeddings to deliver exceptional accuracy. By mastering the techniques presented in this chapter, you'll be well-equipped to tackle complex multispeaker audio scenarios and push the boundaries of what's possible with OpenAI Whisper.

Get ready to dive into the exciting world of speaker diarization and unlock new insights from multispeaker conversations. Let's begin this transformative journey together!

# Technical requirements

To harness the capabilities of OpenAI's Whisper for advanced applications, this chapter leverages Python and Google Colab for ease of use and accessibility. The Python environment setup includes the Whisper library for transcription tasks.

**Key requirements**:

- **Google Colab notebooks**: The notebooks are set to run our Python code with the minimum required memory and capacity. If the **T4 GPU** runtime type is available, select it for better performance.

- **Google Colab notebooks**: The notebooks are set to run our Python code with the minimum required memory and capacity. If that option is available, change the runtime type to **GPU** for better performance.

- **Python environment**: Each notebook contains directives to load the required Python libraries, including Whisper and Gradio.

- **Hugging Face account**: Some notebooks require a Hugging Face account and login API key. The Colab notebooks include information about this topic.

- **Microphone and speakers**: Some notebooks implement a Gradio app with voice recording and audio playback. A microphone and speakers connected to your computer might help you experience the interactive voice features. Another option is to open the URL link that Gradio provides at runtime on your mobile phone; from there, you can use the phone's microphone to record your voice.

- **GitHub repository access**: All Python code, including examples, is available in the chapter's GitHub repository (`https://github.com/PacktPublishing/Learn-OpenAI-Whisper/tree/main/Chapter08`). These Colab notebooks are ready to run, providing a practical and hands-on approach to learning.

By meeting these technical requirements, you will be prepared to explore Whisper in different contexts while enjoying the streamlined experience of Google Colab and the comprehensive resources available on GitHub.

# Augmenting Whisper with speaker diarization

Speaker diarization, partitioning an audio stream into segments according to the speaker's identity, is a powerful feature in multispeaker speech processing. It addresses the question of *who spoke when?* In a given audio clip, it is crucial to enhance the functionality and usability of ASR systems. The origins of speaker diarization can be traced back to the 1990s when the foundational work for clustering-based diarization paradigms was laid down. These early studies focused on radio broadcast news and communications applications, primarily aiming to improve ASR performance. The features used in these early studies were handcrafted mainly, with **Mel-frequency cepstral coefficients** (**MFCCs**) being a common choice.

Over time, the field of speaker diarization has seen significant advancements, particularly with the emergence of deep learning technology. Modern diarization systems often leverage neural networks and large-scale GPU computing to improve accuracy and efficiency. The progression of diarization techniques has included the use of **Gaussian mixture models** (**GMMs**) and **hidden Markov models** (**HMMs**) in earlier approaches, followed by the adoption of neural embeddings (such as $x$-vectors and $d$-vectors, which we will cover in more detail in the *An introduction to speaker embeddings* section later in this chapter) and clustering methods in more recent times.

One of the most significant contributions to the field has been the development of end-to-end neural diarization approaches, which aim to simplify the diarization process by merging distinct steps in the diarization pipeline. These approaches have been designed to handle the challenges of multispeaker labeling and diarization, such as dealing with noisy acoustic environments, a range of vocal tenors, and accent nuances.

Open source initiatives have also contributed to the evolution of diarization capabilities, with tools such as ALIZE, pyannote.audio, pyAudioAnalysis, SHoUT, and LIUM SpkDiarization providing resources for researchers and developers to implement and experiment with diarization in their applications. Most earlier tools are now either inactive or abandoned, except for pyannote.audio (Pyannote).

The early speaker diarization systems, while pioneering in their approach to solving the *who spoke when* problem in audio recordings, faced several limitations that impacted their accuracy and efficiency. In the next section, we will examine the fundamental hurdles of early diarization solutions in more detail.

## Understanding the limitations and constraints of diarization

Many of the deficiencies and inaccuracies in early diarization efforts were rooted in the technological constraints of the time, the complexity of human speech, and the nascent state of machine learning techniques applied to audio processing. Understanding these limitations provides valuable insights into the evolution of diarization capabilities and the significant advancements made over time:

- **Computing limitations**: Early diarization systems were limited by the computational power available at the time. Processing large audio datasets required significant computational resources, which were not as readily available or as powerful as today's standards. This limitation affected the complexity of the algorithms that could be run in a reasonable amount of time, thereby constraining the accuracy of early diarization systems.

- **Feature extraction and modeling limitations**: The feature extraction techniques used in early diarization systems, such as MFCCs, were relatively simplistic compared to the sophisticated embeddings used in modern systems. These early features might not effectively capture the nuances of different speakers' voices, leading to less accurate speaker differentiation.

- **Reliance on GMMs and HMMs for speaker modeling**: While these models provided a foundation for speaker diarization, they were limited in handling the variability and complexity of human speech across different speakers and environments.

- **Handling of speaker change points**: One of the significant challenges for early diarization systems was accurately detecting speaker change points. These systems struggled particularly with short speech segments and segments close to speaker change points. The performance of these systems degraded both as the segment duration decreased and the proximity to the speaker change point increased. For example, over 33% and 40% of the errors in **single-distant microphone (SDM)** and **multiple-distant microphone (MDM)** conditions occurred within 0.5 seconds of a change point for all evaluated systems. SDM refers to a scenario where a single microphone is placed at a distance from the speakers, capturing audio from all participants. On the other hand, MDM involves multiple microphones placed at different locations in the recording environment, providing additional spatial information that can be leveraged for improved diarization performance. The percentage of errors in the context of these setups highlights early diarization systems' challenges in accurately detecting speaker changes, especially near change points.

- **Scalability and flexibility**: Early diarization systems were often designed with specific applications in mind, such as radio broadcast news or meeting recordings, and might not quickly adapt to other types of audio content. This lack of flexibility limited the broader application of diarization technology. Moreover, the scalability of these systems to handle large-scale or real-time diarization tasks was a significant challenge.

- **Error analysis and improvement directions**: In-depth error analysis of early diarization systems revealed that improvements near speaker change points could significantly impact overall performance. Modifications such as alternative minimum duration constraints and leveraging the difference between the most prominent and second-largest log-likelihood scores for unsupervised clustering were explored to address these limitations.

Despite their groundbreaking efforts, early approaches to speaker diarization encountered various limitations that could have improved their accuracy and efficiency. These limitations stemmed from technological constraints, the intricacies of human speech, and the nascent state of machine-learning techniques. However, introducing transformer-based models has revolutionized the field, addressing many of these challenges and paving the way for more accurate and efficient solutions.

## Bringing transformers into speech diarization

Transformers have been instrumental in advancing state-of-the-art speech diarization. They are adept at handling speech's sequential and contextual nature, which is essential for differentiating between speakers in an audio stream. The self-attention mechanism within transformers allows a model to weigh the importance of each part of the input data, which is crucial for identifying speaker change points and attributing speech segments to the correct speaker.

As mentioned earlier, traditional diarization methods often relied on GMMs and HMMs to model speaker characteristics. These methods need to be improved to handle the variability and complexity of human speech. In contrast, transformer-based diarization systems can process entire data sequences simultaneously, allowing them to capture the context and relationships between speech segments more effectively.

Transformers also enable embeddings, such as $x$-vectors and $d$-vectors, which provide a more nuanced representation of speaker characteristics. This leads to improved diarization performance, especially in challenging acoustic environments or scenarios with overlapping speech.

Moving beyond the limitations of earlier diarization attempts, we must introduce a game-changing framework that brings transformers into speech diarization – NVIDIA's **Neural Modules (NeMo)**. NeMo is an open source toolkit for building, training, and fine-tuning GPU-accelerated speech and NLP models. It provides a collection of pre-built modules and models that can be quickly composed to create complex AI applications, such as ASR, natural language understanding, and text-to-speech synthesis. NeMo offers a more direct approach to diarization with its transformer-based pipeline, opening new possibilities for speaker identification and separation.

## Introducing NVIDIA's NeMo framework

Compared to traditional methods, transformer-based diarization systems provide superior performance and are better suited to the complexities of natural speech. NVIDIA's NeMo toolkit supports training and fine-tuning speaker diarization models. NeMo leverages transformer-based models for various speech tasks, including diarization. The toolkit provides a pipeline that includes **voice activity detection (VAD)**, **speaker embedding extraction**, and **clustering** modules, which are essential components of a diarization system. NeMo's approach to diarization involves training models that can capture the characteristics of unseen speakers and assign audio segments to the correct speaker index.

From a more comprehensive point of view, NVIDIA NeMo offers much more than transformer-based diarization. NeMo is an end-to-end, cloud-native framework for building, customizing, and deploying generative AI models across various platforms, including LLMs. It provides a comprehensive solution for the entire generative AI model development life cycle, from data processing and model training to inference. NeMo is particularly noted for its capabilities in conversational AI, encompassing ASR, NLP, and text-to-speech synthesis.

NeMo stands out for its ability to handle large-scale models, supporting the training of models with up to trillions of parameters. Advanced parallelization techniques such as tensor parallelism, pipeline parallelism, and sequence parallelism facilitate this, enabling efficient scaling of models across thousands of GPUs. The framework is built on top of PyTorch and PyTorch Lightning, offering a familiar environment for researchers and developers to innovate within the conversational AI space.

One of the critical features of NeMo is its modular architecture, where models are composed of neural modules with strongly typed input and output. This design promotes reusability and simplifies the creation of new conversational AI models by allowing researchers to leverage pre-existing code and pre-trained models.

NeMo is available as open source software, encouraging contributions from the community and facilitating widespread adoption and customization. It also integrates with NVIDIA's AI platform, including the NVIDIA Triton Inference Server, to deploy models in production environments. NVIDIA NeMo provides a powerful and flexible framework to develop state-of-the-art conversational AI models, offering tools and resources that streamline bringing generative AI applications from concept to deployment.

Now that we've explored Whisper and NeMo's capabilities separately, let's consider the potential of integrating these two powerful tools. Combining Whisper's transcription prowess with NeMo's advanced diarization features can unlock even greater insights from audio data.

## Integrating Whisper and NeMo

While Whisper is primarily known for its transcription capabilities, it can also be adapted for diarization tasks. However, Whisper does not natively support speaker diarization. To achieve diarization with Whisper, additional tools such as Pyannote, a speaker diarization toolkit, can be used in conjunction with Whisper's transcriptions to identify speakers.

Integrating NVIDIA's NeMo with OpenAI's Whisper for speaker diarization involves a novel pipeline that leverages the strengths of both systems to enhance diarization outcomes. This integration is particularly notable in the context of inference and result interpretation.

The pipeline begins with Whisper processing audio to generate highly accurate transcriptions. Whisper's role is primarily to transcribe the audio, providing detailed textual output of the spoken content. However, Whisper does not natively support speaker diarization—identifying *who spoke when* within the audio.

To introduce diarization, the pipeline incorporates NVIDIA's NeMo, specifically its speaker diarization module. NeMo's diarization system is designed to process audio recordings, segmenting them by speaker labels. It achieves this through several steps, including VAD, speaker embedding extraction, and clustering. The speaker embeddings capture unique voice characteristics, which are then clustered to differentiate between speakers in the audio.

The integration of Whisper and NeMo for diarization allows you to align Whisper's transcriptions with speaker labels identified by NeMo. This means that the output includes what was said (from Whisper's transcriptions) and identifies which speaker said each part (from NeMo's diarization). The result is a more comprehensive understanding of the audio content, providing both the textual transcription and the speaker attribution.

This integration is beneficial in scenarios where understanding conversation dynamics is crucial, such as meetings, interviews, and legal proceedings. It enhances the utility of transcriptions by adding a layer of speaker-specific context, making it easier to follow conversations and attribute statements accurately.

The integration between Whisper and NeMo for speaker diarization combines Whisper's advanced transcription capabilities with NeMo's robust diarization framework. This synergy enhances the interpretability of audio content by providing detailed transcriptions alongside accurate speaker labels, thereby offering a richer analysis of spoken interactions.

Before we delve deeper into the integration of Whisper and NeMo, it's crucial to understand a fundamental concept in modern speech processing systems – **speaker embeddings**. These vectorial representations of speaker characteristics are vital in enabling accurate speaker diarization.

## An introduction to speaker embeddings

Speaker embeddings are vectorial representations extracted from a speech signal that encapsulate the characteristics of a speaker's voice in a compact form. These embeddings are designed to be discriminative, meaning they can effectively differentiate between speakers while being robust to variations in speech content, channel, and environmental noise. The goal is to obtain a fixed-length vector from variable-length speech utterances that capture the unique traits of a speaker's voice.

Speaker embeddings are a fundamental component in modern speech processing systems, enabling various applications from speaker verification to diarization. Their ability to condense the rich information of a speaker's voice into a fixed-length vector makes them invaluable for systems that need to recognize, differentiate, or track speakers across audio recordings.

From a more technical perspective, there are several types of speaker embeddings, each with its method of extraction and characteristics:

- *i*-**vectors**: These embeddings capture speaker and channel variabilities in a low-dimensional space. They are derived from a GMM framework and represent the differences between a given speaker's pronunciation and the average pronunciation across a set of phonetic classes.

- *d*-**vectors**: These are obtained by training a speaker-discriminative **deep neural network (DNN)** and extracting frame-level vectors from the last hidden layer. These vectors are then averaged over the entire utterance to produce the *d*-vector, representing the speaker's identity.

- *x*-**vectors**: This type of embedding involves frame- and segment-level feature (utterance) processing. *X*-vectors are extracted using a DNN that processes a sequence of acoustic features and aggregates them, using a statistics pooling layer to produce a fixed-length vector.

- *s*-**vectors**: Also known as sequence or summary vectors, *s*-vectors are derived from recurrent neural network architectures such as RNNs or LSTMs. They are designed to capture sequential information and can encode spoken terms and word orders to a notable extent.

Extracting speaker embeddings typically involves training a neural network model to optimize the encoder using loss functions that encourage discriminative learning. After training, the pre-activation of a hidden layer at the segment-level network is extracted as the speaker embedding. The network is trained on a large dataset of speakers to ensure that the embeddings generalize well to unseen speakers.

In the context of speaker diarization, speaker embeddings cluster speech segments according to the speaker's identity. The embeddings provide a way to measure the similarity between segments and groups of those likely to be from the same speaker. This is a crucial step in the diarization process, as it allows you to accurately attribute speech to the correct speaker within an audio stream.

As we've seen, both Whisper augmented with Pyannote and NVIDIA's NeMo offer powerful diarization capabilities. However, it's essential to understand the critical differences between these approaches to make informed decisions when choosing a diarization solution.

## Differentiating NVIDIA's NeMo capabilities

The integration of diarization capabilities into ASR systems has been significantly influenced by the advent of transformer models, particularly in the context of OpenAI's Whisper and NVIDIA's NeMo frameworks. These advancements have improved the accuracy of ASR systems and introduced new methodologies to handle speaker diarization tasks. Let's delve into the similarities and differences between Whisper diarization using Pyannote and diarization using NVIDIA's NeMo, focusing on speech activity detection, speaker change detection, and overlapped speech detection. Understanding the differences between these two approaches to speaker diarization is crucial for making informed decisions when choosing a solution for your specific use case. By examining how each system handles critical aspects of the diarization process, such as speech activity detection, speaker change detection, and overlapped speech detection, you can better assess which approach aligns with your accuracy, efficiency, and ease of integration requirements:

| Diarization feature | Whisper with Pyannote | NVIDIA NeMo |
|---|---|---|
| Detecting speech activity | Whisper does not inherently perform VAD as part of its diarization process. However, when combined with Pyannote, an external VAD model from the Pyannote toolkit can segment the audio into speech and non-speech intervals before applying diarization. This approach requires integrating Whisper's ASR capabilities with Pyannote's VAD models, based on deep learning techniques and fine-tuning, for accurate speech/non-speech segmentation. | NeMo's speaker diarization pipeline includes a dedicated VAD module that is trainable and optimized as part of the diarization system. This VAD model is designed to detect the presence or absence of speech and generate timestamps for speech activity within an audio recording. Integrating VAD within NeMo's diarization pipeline allows for a more streamlined process, directly feeding the VAD results into subsequent diarization steps. |

| Diarization feature | Whisper with Pyannote | NVIDIA NeMo |
|---|---|---|
| Detecting speaker change | The integration of Whisper with Pyannote for diarization purposes relies on Pyannote's speaker change detection capabilities. Pyannote employs neural network models to identify points in audio where a speaker change occurs. This process is crucial for segmenting the audio into homogeneous segments attributed to individual speakers. Speaker change detection in Pyannote is a separate module that works with its diarization pipeline. | NeMo's approach to speaker change detection is implicitly handled within its diarization pipeline, including modules for extracting and clustering speaker embeddings. While NeMo does not explicitly mention a standalone speaker change detection module, identifying speaker changes is integrated into the overall diarization workflow, mainly through analyzing speaker embeddings and their temporal distribution across audio. |
| Detecting overlapped speech | Overlapped speech detection is another area where Pyannote complements Whisper's capabilities. Pyannote's toolkit includes models designed to detect and handle overlapping speech, a challenging aspect of speaker diarization. This functionality is crucial for accurately diarizing conversations where multiple speakers simultaneously talk. | Like speaker change detection, NeMo's treatment of overlapped speech is integrated into its diarization pipeline rather than being addressed by a separate module. The system's ability to handle overlapped speech results from its sophisticated speaker embedding and clustering techniques, which can identify and separate speakers even in challenging overlapping scenarios. |

| Diarization feature | Whisper with Pyannote | NVIDIA NeMo |
|---|---|---|
| Integrating speaker embeddings in the diarization pipeline | Whisper's combination with Pyannote relies on external modules for these tasks, offering flexibility and modularity. In contrast, NeMo's diarization pipeline directly integrates these functionalities, providing a streamlined and cohesive workflow. These advancements underscore the transformative impact of transformer models on speech processing, paving the way for more accurate and efficient diarization systems. | NVIDIA's NeMo toolkit includes a more integrated approach to speaker diarization. It provides a complete diarization pipeline that includes VAD, speaker embedding extraction, and clustering. NeMo's speaker embeddings are extracted using models explicitly trained for this purpose, and these embeddings are then used within the same framework to perform the clustering necessary for diarization. |
| Clustering and assigning speaker embeddings | After extracting speaker embeddings, Pyannote uses various clustering algorithms, such as hierarchical clustering, to group and assign the embeddings to the respective speakers. This clustering process is crucial for determining which audio segments belong to which speaker. | NeMo also uses clustering algorithms to group speaker embeddings. However, NeMo employs a multiscale, auto-tuning, spectral clustering approach, reportedly more resilient than the Pyannote version. This approach involves segmenting the audio file with different window lengths and calculating embeddings for multiple scales, which are then clustered to label each segment with a speaker. |

Table 8.1 – How different diarization approaches handle critical diarization features

While both Whisper augmented with Pyannote and NVIDIA's NeMo use speaker embeddings as a core part of their diarization pipelines, their approaches have notable differences. Whisper requires an external toolkit (`pyannote.audio`) to perform diarization, whereas NeMo offers an all-in-one solution with its speaker embedding extraction and clustering modules. NeMo's multiscale clustering approach is a distinctive feature, differentiating it from the Pyannote implementation used with Whisper. These differences reflect the diverse methodologies and innovations present in the field of speaker diarization research.

---

**Blending Whisper and PyAnnote – WhisperX**

WhisperX (`https://replicate.com/dinozoiddev/whisperx`) provides fast ASR (70x faster than OpenAI's `Whisper large-v2`) with word-level timestamps and speaker diarization, a feature not natively supported by Whisper. WhisperX builds upon the foundational strengths of Whisper by addressing some of its limitations, particularly in timestamp accuracy and speaker diarization. While Whisper provides utterance-level timestamps, WhisperX advances this by offering word-level timestamps, crucial for applications requiring precise synchronization between text and audio, such as subtitling and detailed audio analysis. This is achieved through combining techniques, including VAD, pre-segmentation of audio into manageable chunks, and forced alignment with an external phoneme model to provide accurate word-level timestamps.

The implementation of WhisperX supports transcription in all languages supported by Whisper, with alignment currently available for English audio. It has been upgraded to incorporate the latest Whisper models and diarization technologies powered by Pyannote to enhance its performance further. At the time of writing, WhisperX incorporates `whisper-large-v3` along with diarization upgrades to speaker-diarization-3.1 and segmentation-3.0, powered by Pyannote. WhisperX demonstrates significant improvements over Whisper in word segmentation precision and recall, as well as reductions in WER and increases in transcription speed, especially when employing batched transcription with VAD preprocessing.

In summary, WhisperX is a significant evolution of OpenAI's Whisper, offering enhanced functionality through word-level timestamps and speaker diarization. These advancements make WhisperX a powerful tool for applications requiring detailed and accurate speech transcription and analysis.

---

With this solid theoretical foundation, it's time to put our knowledge into practice. The next hands-on section will explore a practical implementation that combines WhisperX, NeMo, and other supporting Python libraries to perform speech diarization on real-world audio data.

# Performing hands-on speech diarization

Transitioning from the theoretical context of speech diarization, let's immerse ourselves in the practical implementation that combines WhisperX, NeMo, and other supporting Python libraries, all from the comfort of our trusty Google Colaboratory. I encourage you to visit the book's GitHub repository, find the `LOAIW_ch08_diarizing_speech_with_WhisperX_and_NVIDIA_NeMo.ipynb` notebook (`https://github.com/PacktPublishing/Learn-OpenAI-Whisper/blob/main/Chapter08/LOAIW_ch08_diarizing_speech_with_WhisperX_and_NVIDIA_NeMo.ipynb`), and run the Python code yourself; feel free to experiment by modifying parameters and observe the results. The notebook provides a detailed walk-through to integrate Whisper's transcription capabilities with NeMo's diarization framework, offering a robust solution to analyze speech in audio recordings.

The notebook is structured into several key sections, each focusing on a specific aspect of the diarization process.

## Setting up the environment

The first section of the notebook outlines the installation of several Python libraries and tools essential for the diarization process:

```
!pip install git+https://github.com/m-bain/whisperX.
git@78dcfaab51005aa703ee21375f81ed31bc248560
!pip install --no-build-isolation nemo_toolkit[asr]==1.22.0
!pip install --no-deps git+https://github.com/facebookresearch/
demucs#egg=demucs
!pip install dora-search "lameenc>=1.2" openunmix
!pip install deepmultilingualpunctuation
!pip install wget pydub
```

Let's review each to understand their role in diarization:

- whisperX: An extension of OpenAI's Whisper model, tailored for enhanced functionality. Notably, WhisperX installs faster-whisper (https://github.com/SYSTRAN/faster-whisper), a reimplementation of OpenAI's Whisper model using CTranslate2 (https://github.com/OpenNMT/CTranslate2/). This implementation is up to four times faster than OpenAI's Whisper with the same accuracy, while using less memory. The efficiency can be improved with 8-bit quantization on both the CPU and GPU.

- nemo_toolkit[asr]: NVIDIA's NeMo toolkit for ASR, providing the foundation for speaker diarization.

- demucs: A library for music source separation, functional for preprocessing audio files by isolating speech from background music.

- dora-search, lameenc, and openunmix: Tools and libraries for audio processing, enhancing the quality and compatibility of audio data for diarization tasks.

- deepmultilingualpunctuation: A library for adding punctuation to transcriptions, improving the readability and structure of the generated text.

- wget and pydub: Utilities for downloading and manipulating audio files, facilitating audio data handling within the Python environment.

These libraries collectively form the foundation for processing audio files, transcribing speech, and performing speaker diarization. Each tool plays a specific role, from preparing the audio data to generating accurate transcriptions and identifying distinct speakers within the audio.

## Streamlining the diarization workflow with helper functions

The notebook defines several supporting functions to simplify the process of diarizing speech with Whisper and NeMo. These functions are instrumental in managing audio data, aligning transcriptions with speaker identities, and enhancing the workflow. The following is a concise description of each function:

- `create_config()`: Initializes and returns a configuration object, setting up essential parameters for the diarization process:

```python
def create_config(output_dir):
    DOMAIN_TYPE = "telephonic"  # Can be meeting, telephonic, or
general based on domain type of the audio file
    CONFIG_FILE_NAME = f"diar_infer_{DOMAIN_TYPE}.yaml"
    CONFIG_URL = f"https://raw.githubusercontent.com/
NVIDIA/NeMo/main/examples/speaker_tasks/diarization/conf/
inference/{CONFIG_FILE_NAME}"
    MODEL_CONFIG = os.path.join(output_dir, CONFIG_FILE_NAME)
    if not os.path.exists(MODEL_CONFIG):
        MODEL_CONFIG = wget.download(CONFIG_URL, output_dir)

    config = OmegaConf.load(MODEL_CONFIG)

    data_dir = os.path.join(output_dir, "data")
    os.makedirs(data_dir, exist_ok=True)

    meta = {
        "audio_filepath": os.path.join(output_dir, "mono_file.
wav"),
        "offset": 0,
        "duration": None,
        "label": "infer",
        "text": "-",
        "rttm_filepath": None,
        "uem_filepath": None,
    }
    with open(os.path.join(data_dir, "input_manifest.json"),
"w") as fp:
        json.dump(meta, fp)
        fp.write("\n")

    pretrained_vad = "vad_multilingual_marblenet"
    pretrained_speaker_model = "titanet_large"
    config.num_workers = 0  # Workaround for multiprocessing
hanging with ipython issue
    config.diarizer.manifest_filepath = os.path.join(data_dir,
```

```
"input_manifest.json")
    config.diarizer.out_dir = (
        output_dir  # Directory to store intermediate files and
prediction outputs
    )

    config.diarizer.speaker_embeddings.model_path = pretrained_
speaker_model
    config.diarizer.oracle_vad = (
        False  # compute VAD provided with model_path to vad
config
    )
    config.diarizer.clustering.parameters.oracle_num_speakers =
False

    # Here, we use our in-house pretrained NeMo VAD model
    config.diarizer.vad.model_path = pretrained_vad
    config.diarizer.vad.parameters.onset = 0.8
    config.diarizer.vad.parameters.offset = 0.6
    config.diarizer.vad.parameters.pad_offset = -0.05
    config.diarizer.msdd_model.model_path = (
        "diar_msdd_telephonic"  # Telephonic speaker diarization
model
    )

    return config
```

- `get_word_ts_anchor()`: Determines the anchor timestamp for words, facilitating accurate alignment between spoken words and their timestamps in the audio:

```
def get_word_ts_anchor(s, e, option="start"):
    if option == "end":
        return e
    elif option == "mid":
        return (s + e) / 2
    return s
```

- `get_words_speaker_mapping()`: Maps each word in the transcription to the corresponding speaker based on the diarization results, ensuring that every word is attributed to the correct speaker:

```
def get_words_speaker_mapping(wrd_ts, spk_ts, word_anchor_
option="start"):
    s, e, sp = spk_ts[0]
    wrd_pos, turn_idx = 0, 0
    wrd_spk_mapping = []
```

```
    for wrd_dict in wrd_ts:
        ws, we, wrd = (
            int(wrd_dict["start"] * 1000),
            int(wrd_dict["end"] * 1000),
            wrd_dict["word"],
        )
        wrd_pos = get_word_ts_anchor(ws, we, word_anchor_option)
        while wrd_pos > float(e):
            turn_idx += 1
            turn_idx = min(turn_idx, len(spk_ts) - 1)
            s, e, sp = spk_ts[turn_idx]
            if turn_idx == len(spk_ts) - 1:
                e = get_word_ts_anchor(ws, we, option="end")
        wrd_spk_mapping.append(
            {"word": wrd, "start_time": ws, "end_time": we,
"speaker": sp}
        )
    return wrd_spk_mapping
```

- `get_first_word_idx_of_sentence()`: Identifies the index of the first word in a sentence, crucial for processing sentences in the context of speaker attribution and alignment:

```
def get_first_word_idx_of_sentence(word_idx, word_list, speaker_
list, max_words):
    is_word_sentence_end = (
        lambda x: x >= 0 and word_list[x][-1] in sentence_
ending_punctuations
    )
    left_idx = word_idx
    while (
        left_idx > 0
        and word_idx - left_idx < max_words
        and speaker_list[left_idx - 1] == speaker_list[left_idx]
        and not is_word_sentence_end(left_idx - 1)
    ):
        left_idx -= 1

    return left_idx if left_idx == 0 or is_word_sentence_
end(left_idx - 1) else -1
```

- `get_last_word_idx_of_sentence()`: Finds the index of the last word in a sentence, aiding in delineating sentence boundaries within the transcribed text:

```
def get_last_word_idx_of_sentence(word_idx, word_list, max_
words):
    is_word_sentence_end = (
```

```
        lambda x: x >= 0 and word_list[x][-1] in sentence_
ending_punctuations
    )
    right_idx = word_idx
    while (
        right_idx < len(word_list)
        and right_idx - word_idx < max_words
        and not is_word_sentence_end(right_idx)
    ):
        right_idx += 1

    return (
        right_idx
        if right_idx == len(word_list) - 1 or is_word_sentence_
end(right_idx)
        else -1
    )
```

- `get_realigned_ws_mapping_with_punctuation()`: Adjusts the word-to-speaker mapping by considering punctuation, enhancing the accuracy of speaker attribution, especially in complex conversational scenarios:

```
def get_realigned_ws_mapping_with_punctuation(
    word_speaker_mapping, max_words_in_sentence=50
):
    is_word_sentence_end = (
        lambda x: x >= 0
        and word_speaker_mapping[x]["word"][-1] in sentence_
ending_punctuations
    )
    wsp_len = len(word_speaker_mapping)

    words_list, speaker_list = [], []
    for k, line_dict in enumerate(word_speaker_mapping):
        word, speaker = line_dict["word"], line_dict["speaker"]
        words_list.append(word)
        speaker_list.append(speaker)

    k = 0
    while k < len(word_speaker_mapping):
        line_dict = word_speaker_mapping[k]
        if (
            k < wsp_len - 1
            and speaker_list[k] != speaker_list[k + 1]
            and not is_word_sentence_end(k)
```

```
        ):
            left_idx = get_first_word_idx_of_sentence(
                k, words_list, speaker_list, max_words_in_
sentence
            )
            right_idx = (
                get_last_word_idx_of_sentence(
                    k, words_list, max_words_in_sentence - k +
left_idx - 1
                )
                if left_idx > -1
                else -1
            )
            if min(left_idx, right_idx) == -1:
                k += 1
                continue

            spk_labels = speaker_list[left_idx : right_idx + 1]
            mod_speaker = max(set(spk_labels), key=spk_labels.
count)
            if spk_labels.count(mod_speaker) < len(spk_labels)
// 2:
                k += 1
                continue

            speaker_list[left_idx : right_idx + 1] = [mod_
speaker] * (
                right_idx - left_idx + 1
            )
            k = right_idx

        k += 1

    k, realigned_list = 0, []
    while k < len(word_speaker_mapping):
        line_dict = word_speaker_mapping[k].copy()
        line_dict["speaker"] = speaker_list[k]
        realigned_list.append(line_dict)
        k += 1

    return realigned_list
```

- get_sentences_speaker_mapping(): Generates a mapping of entire sentences to speakers, providing a higher-level view of speaker contributions throughout the audio:

```python
def get_sentences_speaker_mapping(word_speaker_mapping, spk_ts):
    sentence_checker = nltk.tokenize.PunktSentenceTokenizer().
text_contains_sentbreak
    s, e, spk = spk_ts[0]
    prev_spk = spk

    snts = []
    snt = {"speaker": f"Speaker {spk}", "start_time": s, "end_
time": e, "text": ""}

    for wrd_dict in word_speaker_mapping:
        wrd, spk = wrd_dict["word"], wrd_dict["speaker"]
        s, e = wrd_dict["start_time"], wrd_dict["end_time"]
        if spk != prev_spk or sentence_checker(snt["text"] + " "
+ wrd):
            snts.append(snt)
            snt = {
                "speaker": f"Speaker {spk}",
                "start_time": s,
                "end_time": e,
                "text": "",
            }
        else:
            snt["end_time"] = e
        snt["text"] += wrd + " "
        prev_spk = spk

    snts.append(snt)
    return snts
```

- `get_speaker_aware_transcript()`: Produces a transcript aware of speaker identities, integrating both the textual content and the speaker information into a cohesive format:

```python
def get_speaker_aware_transcript(sentences_speaker_mapping, f):
    previous_speaker = sentences_speaker_mapping[0]["speaker"]
    f.write(f"{previous_speaker}: ")

    for sentence_dict in sentences_speaker_mapping:
        speaker = sentence_dict["speaker"]
        sentence = sentence_dict["text"]

        # If this speaker doesn't match the previous one, start
a new paragraph
        if speaker != previous_speaker:
            f.write(f"\n\n{speaker}: ")
```

```
        previous_speaker = speaker

    # No matter what, write the current sentence
    f.write(sentence + " ")
```

- `format_timestamp()`: Converts timestamps into a human-readable format, essential for annotating the transcript with precise timing information:

```
def format_timestamp(
    milliseconds: float, always_include_hours: bool = False,
decimal_marker: str = "."
):
    assert milliseconds >= 0, "non-negative timestamp expected"

    hours = milliseconds // 3_600_000
    milliseconds -= hours * 3_600_000

    minutes = milliseconds // 60_000
    milliseconds -= minutes * 60_000

    seconds = milliseconds // 1_000
    milliseconds -= seconds * 1_000

    hours_marker = f"{hours:02d}:" if always_include_hours or
hours > 0 else ""
    return (
        f"{hours_marker}{minutes:02d}:{seconds:02d}{decimal_
marker}{milliseconds:03d}"
    )
```

- `write_srt()`: Outputs the diarization results in the **SubRip Text (SRT)** format, suitable for subtitles or detailed analysis, including speaker labels and timestamps:

```
def write_srt(transcript, file):
    """
    Write a transcript to a file in SRT format.

    """
    for i, segment in enumerate(transcript, start=1):
        # write srt lines
        print(
            f"{i}\n"
            f"{format_timestamp(segment['start_time'], always_
include_hours=True, decimal_marker=',')} --> "
            f"{format_timestamp(segment['end_time'], always_
include_hours=True, decimal_marker=',')}\n"
```

```
            f"{segment['speaker']}: {segment['text'].strip().
    replace('-->', '->')}\n",
            file=file,
            flush=True,
        )
```

- `find_numeral_symbol_tokens()`: Identifies tokens within the transcription that represent numeral symbols, aiding in processing numerical data within text:

```
def find_numeral_symbol_tokens(tokenizer):
    numeral_symbol_tokens = [
        -1,
    ]
    for token, token_id in tokenizer.get_vocab().items():
        has_numeral_symbol = any(c in "0123456789%$£" for c in
    token)
        if has_numeral_symbol:
            numeral_symbol_tokens.append(token_id)
    return numeral_symbol_tokens
```

- `_get_next_start_timestamp()`: Calculates the start timestamp for the next word, ensuring continuity in the sequence of timestamps across a transcription:

```
def _get_next_start_timestamp(word_timestamps, current_word_
index, final_timestamp):
    # if current word is the last word
    if current_word_index == len(word_timestamps) - 1:
        return word_timestamps[current_word_index]["start"]

    next_word_index = current_word_index + 1
    while current_word_index < len(word_timestamps) - 1:
        if word_timestamps[next_word_index].get("start") is
    None:
            # if next word doesn't have a start timestamp
            # merge it with the current word and delete it
            word_timestamps[current_word_index]["word"] += (
                " " + word_timestamps[next_word_index]["word"]
            )

            word_timestamps[next_word_index]["word"] = None
            next_word_index += 1
            if next_word_index == len(word_timestamps):
                return final_timestamp

        else:
```

```
                    return word_timestamps[next_word_index]["start"]
```

- `filter_missing_timestamps()`: Filters and corrects any missing or incomplete timestamps in transcription data, maintaining the integrity of temporal information:

```python
def filter_missing_timestamps(
    word_timestamps, initial_timestamp=0, final_timestamp=None
):
    # handle the first and last word
    if word_timestamps[0].get("start") is None:
        word_timestamps[0]["start"] = (
            initial_timestamp if initial_timestamp is not None
else 0
        )
        word_timestamps[0]["end"] = _get_next_start_timestamp(
            word_timestamps, 0, final_timestamp
        )

    result = [
        word_timestamps[0],
    ]

    for i, ws in enumerate(word_timestamps[1:], start=1):
        # if ws doesn't have a start and end
        # use the previous end as start and next start as end
        if ws.get("start") is None and ws.get("word") is not
None:
            ws["start"] = word_timestamps[i - 1]["end"]
            ws["end"] = _get_next_start_timestamp(word_
timestamps, i, final_timestamp)

        if ws["word"] is not None:
            result.append(ws)
    return result
```

- `cleanup()`: Cleans up temporary files or directories created during diarization, ensuring a tidy working environment:

```python
def cleanup(path: str):
    """path could either be relative or absolute."""
    # check if file or directory exists
    if os.path.isfile(path) or os.path.islink(path):
        # remove file
        os.remove(path)
    elif os.path.isdir(path):
```

```
        # remove directory and all its content
        shutil.rmtree(path)
    else:
        raise ValueError("Path {} is not a file or
dir.".format(path))
```

- process_language_arg(): Processes the language argument to ensure compatibility with models, facilitating accurate transcription across different languages:

```
def process_language_arg(language: str, model_name: str):
    """
    Process the language argument to make sure it's valid and
convert language names to language codes.
    """
    if language is not None:
        language = language.lower()
    if language not in LANGUAGES:
        if language in TO_LANGUAGE_CODE:
            language = TO_LANGUAGE_CODE[language]
        else:
            raise ValueError(f"Unsupported language:
{language}")

    if model_name.endswith(".en") and language != "en":
        if language is not None:
            logging.warning(
                f"{model_name} is an English-only model but
received '{language}'; using English instead."
            )
        language = "en"
    return language
```

- transcribe(): Utilizes Whisper to transcribe audio into text, providing foundational textual data for the diarization process:

```
def transcribe(
    audio_file: str,
    language: str,
    model_name: str,
    compute_dtype: str,
    suppress_numerals: bool,
    device: str,
):
    from faster_whisper import WhisperModel
    from helpers import find_numeral_symbol_tokens, wav2vec2_
langs
```

```
    # Faster Whisper non-batched
    # Run on GPU with FP16
    whisper_model = WhisperModel(model_name, device=device,
compute_type=compute_dtype)

    # or run on GPU with INT8
    # model = WhisperModel(model_size, device="cuda", compute_
type="int8_float16")
    # or run on CPU with INT8
    # model = WhisperModel(model_size, device="cpu", compute_
type="int8")

    if suppress_numerals:
        numeral_symbol_tokens = find_numeral_symbol_
tokens(whisper_model.hf_tokenizer)
    else:
        numeral_symbol_tokens = None

    if language is not None and language in wav2vec2_langs:
        word_timestamps = False
    else:
        word_timestamps = True

    segments, info = whisper_model.transcribe(
        audio_file,
        language=language,
        beam_size=5,
        word_timestamps=word_timestamps,  # TODO: disable this
if the language is supported by wav2vec2
        suppress_tokens=numeral_symbol_tokens,
        vad_filter=True,
    )
    whisper_results = []
    for segment in segments:
        whisper_results.append(segment._asdict())
    # clear gpu vram
    del whisper_model
    torch.cuda.empty_cache()
    return whisper_results, language
```

- `transcribe_batched()`: Offers a batch processing capability to transcribe audio files, optimizing the transcription process for efficiency and scalability:

```
def transcribe_batched(
```

```
        audio_file: str,
        language: str,
        batch_size: int,
        model_name: str,
        compute_dtype: str,
        suppress_numerals: bool,
        device: str,
    ):
        import whisperx

        # Faster Whisper batched
        whisper_model = whisperx.load_model(
            model_name,
            device,
            compute_type=compute_dtype,
            asr_options={"suppress_numerals": suppress_numerals},
        )
        audio = whisperx.load_audio(audio_file)
        result = whisper_model.transcribe(audio, language=language,
    batch_size=batch_size)
        del whisper_model
        torch.cuda.empty_cache()
        return result["segments"], result["language"]
```

These functions collectively form the notebook foundation of the diarization workflow, enabling seamless integration of Whisper's transcription capabilities with NeMo's advanced diarization features.

## Separating music from speech using Demucs

As we explore the notebook, let's focus on the preprocessing step, which is crucial for enhancing speech clarity before diarization. This section introduces **Demucs**, a deep-learning model for separating music source vocals from complex audio tracks.

Separating music from speech is essential, mainly when dealing with recordings containing background music or other non-speech elements. By extracting the vocal component, the diarization system can more effectively analyze and attribute speech to the correct speakers, as the spectral and temporal characteristics of their speech signals become more pronounced and less obscured by music:

```
if enable_stemming:
    # Isolate vocals from the rest of the audio

    return_code = os.system(
        f'python3 -m demucs.separate -n htdemucs --two-stems=vocals
    "{audio_path}" -o "temp_outputs"'
```

```
        )

    if return_code != 0:
        logging.warning("Source splitting failed, using original audio
file.")
        vocal_target = audio_path
    else:
        vocal_target = os.path.join(
            "temp_outputs",
            "htdemucs",
            os.path.splitext(os.path.basename(audio_path))[0],
            "vocals.wav",
        )
else:
    vocal_target = audio_path
```

Demucs operates by leveraging a neural network trained to distinguish between different audio sources within a mixture. When applied to an audio file, it can separate the vocal track from the instrumental, allowing subsequent tools such as Whisper and NeMo to process the speech without the interference of background music.

This separation step is beneficial for the accuracy of speaker diarization and any downstream tasks that require clean speech input, such as transcription and speech recognition. By using Demucs as part of the preprocessing pipeline, the notebook ensures that the input to the diarization system is optimized for the best possible performance.

## Transcribing audio using WhisperX

The next step is to leverage WhisperX to transcribe the audio content. The transcription process involves processing the audio file through Whisper to generate a set of text segments, each accompanied by timestamps indicating when the segment was spoken:

```
compute_type = "float16"
# or run on GPU with INT8
# compute_type = "int8_float16"
# or run on CPU with INT8
# compute_type = "int8"

if batch_size != 0:
    whisper_results, language = transcribe_batched(
        vocal_target,
        language,
        batch_size,
        whisper_model_name,
```

```
        compute_type,
        suppress_numerals,
        device,
    )
else:
    whisper_results, language = transcribe(
        vocal_target,
        language,
        whisper_model_name,
        compute_type,
        suppress_numerals,
        device,
    )
```

This foundational step provides the textual content necessary for speaker diarization and further analysis. I hope you've noticed that both functions, `transcribe()` and `transcribe_batch()`, were previously defined in the notebook.

## Aligning the transcription with the original audio using Wav2Vec2

Following transcription, the notebook introduces the use of **Wav2Vec2** for forced alignment, a process that refines the alignment between the transcribed text and the original audio. Wav2Vec2, a large-scale neural network model, excels at learning representations of speech that are beneficial for speech recognition and alignment tasks. By employing Wav2Vec2, we demonstrate how to fine-tune the alignment of transcription segments with the audio signal, ensuring that the text is accurately synchronized with the spoken words:

```
if language in wav2vec2_langs:
    device = "cuda"
    alignment_model, metadata = whisperx.load_align_model(
        language_code=language, device=device
    )
    result_aligned = whisperx.align(
        whisper_results, alignment_model, metadata, vocal_target,
device
    )
    word_timestamps = filter_missing_timestamps(
        result_aligned["word_segments"],
        initial_timestamp=whisper_results[0].get("start"),
        final_timestamp=whisper_results[-1].get("end"),
    )

    # clear gpu vram
```

```
        del alignment_model
        torch.cuda.empty_cache()
else:
        assert batch_size == 0, (  # TODO: add a better check for word
timestamps existence
            f"Unsupported language: {language}, use --batch_size to 0"
            " to generate word timestamps using whisper directly and fix
this error."
        )
        word_timestamps = []
        for segment in whisper_results:
            for word in segment["words"]:
                word_timestamps.append({"word": word[2], "start": word[0],
"end": word[1]})
```

This alignment is essential for diarization, as it allows for a more precise segmentation of audio based on speaker changes. The combined output of Whisper and Wav2Vec2 offers a fully aligned transcription, which is instrumental for tasks such as speaker diarization, sentiment analysis, and language identification. This section in the notebook emphasizes that if a Wav2Vec2 model is not available for a specific language, the word timestamps generated by Whisper will be utilized, showcasing the flexibility of the approach.

By integrating Whisper's transcription capabilities with Wav2Vec2's alignment precision, we set the stage for accurate speaker diarization, enhancing the overall quality and reliability of the diarization process.

## Using NeMo's MSDD model for speaker diarization

At the core of the notebook, the focus shifts toward the intricate process of speaker diarization, leveraging the advanced capabilities of NVIDIA's NeMo MSDD. This section in the notebook is pivotal, as it addresses distinguishing between different speakers within an audio signal, a task essential for accurately attributing speech segments to individual speakers:

```
# Initialize NeMo MSDD diarization model
msdd_model = NeuralDiarizer(cfg=create_config(temp_path)).to("cuda")
msdd_model.diarize()

del msdd_model
torch.cuda.empty_cache()
```

The NeMo MSDD model stands at the forefront of this process, employing a sophisticated approach to diarization that considers multiple temporal resolutions of speaker embeddings. This multiscale strategy enhances the model's ability to discern between speakers, even in challenging audio environments with overlapping speech or background noise.

## Mapping speakers to sentences according to timestamps

After successfully separating speech from music, transcribing the audio using Whisper, and performing speaker diarization with the NeMo MSDD model, the next challenge is to accurately map each sentence in the transcription to its corresponding speaker. This involves analyzing the timestamps associated with each word or segment in the transcription and the speaker labels assigned during the diarization process:

```
speaker_ts = []
with open(os.path.join(temp_path, "pred_rttms", "mono_file.rttm"),
"r") as f:
    lines = f.readlines()
    for line in lines:
        line_list = line.split(" ")
        s = int(float(line_list[5]) * 1000)
        e = s + int(float(line_list[8]) * 1000)
        speaker_ts.append([s, e, int(line_list[11].split("_")[-1])])

wsm = get_words_speaker_mapping(word_timestamps, speaker_ts, "start")
```

The preceding code ensures that each sentence in the transcription is correctly attributed to a speaker, considering the start and end times of spoken segments. This meticulous mapping is crucial for applications where understanding conversation dynamics, such as who said what and when, is essential. It enables a more granular analysis of dialogues, meetings, interviews, and audio content involving multiple speakers.

## Enhancing speaker attribution with punctuation-based realignment

The following code snippet demonstrates how punctuation determines the predominant speaker for each sentence in a transcription. It employs a pre-trained punctuation model, `kredor/punctuate-all`, to predict punctuation marks for the transcribed words. The code then processes the words and their predicted punctuation, handling exceptional cases such as acronyms (e.g., USA) to avoid incorrect punctuation. This approach ensures that the speaker attribution remains consistent within each sentence, even in the presence of background comments or brief interjections from other speakers. This is particularly useful in scenarios where the transcription may not indicate speaker changes, such as when a speaker's utterance is interrupted or overlapped by another's. By analyzing the distribution of speaker labels for each word in a sentence, the code can assign a consistent speaker label to the entire sentence, enhancing the coherence of the diarization output:

```
if language in punct_model_langs:
    # restoring punctuation in the transcript to help realign the
sentences
    punct_model = PunctuationModel(model="kredor/punctuate-all")
```

```
words_list = list(map(lambda x: x["word"], wsm))

labled_words = punct_model.predict(words_list)

ending_puncts = ".?!"
model_puncts = ".,;:!?"

# We don't want to punctuate U.S.A. with a period. Right?
is_acronym = lambda x: re.fullmatch(r"\b(?:[a-zA-Z]\.){2,}", x)

for word_dict, labeled_tuple in zip(wsm, labled_words):
    word = word_dict["word"]
    if (
        word
        and labeled_tuple[1] in ending_puncts
        and (word[-1] not in model_puncts or is_acronym(word))
    ):
        word += labeled_tuple[1]
        if word.endswith(".."):
            word = word.rstrip(".")
        word_dict["word"] = word

else:
    logging.warning(
        f"Punctuation restoration is not available for {language}
language. Using the original punctuation."
    )

wsm = get_realigned_ws_mapping_with_punctuation(wsm)
ssm = get_sentences_speaker_mapping(wsm, speaker_ts)
```

This approach also addresses instances where background comments or brief interjections occur while a primary speaker delivers a monologue. The code effectively attributes the main body of speech to the dominant speaker, disregarding sporadic remarks from others. This results in a more accurate and reliable mapping of speech segments to the appropriate speakers, ensuring that the diarization process reflects the actual structure of the conversation.

# Finalizing the diarization process

In this final section, the code performs essential cleanup tasks, exports the diarization results for further use, and replaces speaker IDs with their corresponding names. The main steps include the following:

1. **Saving the speaker-aware transcript**: The get_speaker_aware_transcript function generates a transcript incorporating textual content and speaker information. This transcript is then saved as a file with the same name as the input audio file but with a .txt extension:

```
with open(f"{os.path.splitext(audio_path)[0]}.txt", "w",
encoding="utf-8-sig") as f:
    get_speaker_aware_transcript(ssm, f)
```

2. **Exporting the diarization results in SRT format**: The write_srt function is employed to export the diarization results in the SRT format. This format is commonly used for subtitles and includes speaker labels and precise timestamps for each utterance. The SRT file is saved with the same name as the input audio file but with a .srt extension:

```
with open(f"{os.path.splitext(audio_path)[0]}.srt", "w",
encoding="utf-8-sig") as srt:
    write_srt(ssm, srt)
```

3. **Cleaning up temporary files**: The cleanup function removes any temporary files or directories created during the diarization process. This step ensures a clean and organized working environment, freeing storage space and maintaining system efficiency:

```
cleanup(temp_path)
```

4. **Mapping speaker identifiers to speaker names**: The code reads the content of the previously saved speaker-aware transcript file and replaces the generic speaker IDs (e.g., Speaker 0, Speaker 1, and Speaker 2) with the actual names of the speakers:

```
# Open the file
with open(f"{os.path.splitext(audio_path)[0]}.txt", 'r') as f:
    text = f.read()

# Replace the speaker IDs with names
text = text.replace('Speaker 0','Ewa Jasiewicz')
text = text.replace('Speaker 1','Chris Faulkner')
text = text.replace('Speaker 2','Matt Frei')

# Write the file to disk
with open(audio_path[:-4] + '-with-speakers-names.txt', 'w') as
f:
    f.write(text)
```

By completing these final steps, the diarization process is concluded, and the results are made available for further analysis, post-processing, or integration with other tools and workflows. The exported speaker-aware transcript, SRT file, and transcript with mapped speaker names provide valuable insights into the content and structure of the audio recording, enabling a wide range of applications, such as content analysis, speaker identification, and subtitle generation.

After diving into the notebook, we uncovered a treasure trove of insights into the nuanced world of speech diarization using cutting-edge AI tools. The notebook was a hands-on guide, meticulously walking us through separating and transcribing speech from complex audio files.

One of the first lessons was setting up the right environment. The notebook emphasized the need to install specific dependencies, such as Whisper and NeMo, which were pivotal for the tasks. This step was crucial, laying the groundwork for all subsequent operations.

As we delved deeper, we learned about the utility of helper functions. These functions were the unsung heroes that streamlined the workflow, from processing audio files to handling timestamps and cleaning up resources. They exemplified the principle of writing clean, reusable code that significantly reduced the project's complexity.

The notebook also introduced us to separating music from speech using Demucs. This step was a testament to the power of preprocessing in enhancing the accuracy of diarization. By isolating vocals, we focused on speech's spectral and temporal characteristics, which are essential for identifying different speakers.

Another key takeaway was the integration of multiple models to achieve better results. The notebook showcased how Whisper was used for transcription and Wav2Vec2 for aligning the transcription with the original audio. This synergy between models was a brilliant example of how combining different AI tools leads to a more robust solution.

Mapping speakers into sentences and realigning speech segments using punctuation was particularly enlightening. It demonstrated the intricacies of diarization and the need for attention to detail to ensure that each speaker was accurately represented in the transcript.

In essence, the notebook was a masterclass in the practical application of AI for speech diarization. It not only taught us the technical steps involved but also imparted broader lessons on the importance of preprocessing, the power of combining different AI models, and the need for meticulous post-processing to ensure the integrity of the final output.

# Summary

In this chapter, we embarked on an exciting exploration of the advanced voice capabilities of OpenAI's Whisper. We delved into powerful techniques that enhance Whisper's performance, such as quantization, and uncovered its potential for speaker diarization and real-time speech recognition.

We augmented Whisper with speaker diarization capabilities, allowing it to identify and attribute speech segments to different speakers within an audio recording. By integrating Whisper with the NVIDIA NeMo framework, we discovered how to perform accurate speaker diarization, opening new possibilities for analyzing multispeaker conversations. Our hands-on experience with WhisperX and NVIDIA NeMo showcased the power of combining Whisper's transcription capabilities with advanced diarization techniques.

Throughout the chapter, we acquired a solid understanding of advanced techniques to optimize Whisper's performance and expand its capabilities with speaker diarization. The hands-on coding examples and practical insights equipped us with the knowledge and skills to apply these techniques in our projects, pushing the boundaries of what is possible with Whisper.

As we conclude this chapter, we will look ahead to *Chapter 9, Harnessing Whisper for Personalized Voice Synthesis*. In that chapter, we will gain the knowledge and skills to preprocess audio data, fine-tune voice models, and generate realistic speech using a personal voice synthesis model. The hands-on coding examples and practical insights will empower you to apply these techniques in your projects, pushing the boundaries of what is possible with personalized voice synthesis.

Join me as we continue our journey with Whisper, ready to embrace the exciting possibilities in the rapidly evolving world of voice-synthesis technology.

# 9

# Harnessing Whisper for Personalized Voice Synthesis

Welcome to *Chapter 9*, where we'll delve into **personalized voice synthesis** (**PVS**). This field encompasses various applications and technologies that create synthetic voices tailored to individual preferences or needs. PVS is a versatile process that can be customized for various purposes, including assistive technologies, virtual assistant development, and digital content creation. In this context, OpenAI's Whisper tool enables voice synthesis by providing accurate speech data transcriptions during preprocessing and registration.

As we begin, it's crucial to distinguish between voice cloning and PVS. Voice cloning involves creating a digital replica of a natural person's voice. While this technology has valid applications, it also raises significant ethical concerns. PVS, however, focuses on creating unique voices inspired by specific characteristics without directly copying an individual's voice. This distinction is vital in discussions about the ethical use of voice synthesis technologies. In this chapter, we will guide you on harnessing Whisper's power to create PVS models, ensuring you have the knowledge to use this technology responsibly.

We'll begin by exploring **speech synthesis** and **text-to-speech** (**TTS**) fundamentals. You will gain insights into the role of neural networks, audio processing, and voice synthesis in this domain. Building on this foundation, we will guide you through converting audio files to the **LJSpeech** format, a standardized dataset structure commonly used in TTS tasks.

Next, we will introduce you to the **Deep Learning Art School** (**DLAS**) toolkit, a robust framework for fine-tuning PVS models. This is where your learning journey will truly begin. You will discover how to set up the training environment, prepare the dataset, and configure the model architecture. By leveraging the power of Whisper's accurate transcriptions, you can align audio segments with their corresponding text, creating a dataset suitable for training PVS models. This tutorial is not just a guide but your gateway to mastering the art of PVS with Whisper. Get ready to be inspired and motivated!

Hands-on examples and code snippets will give you practical experience fine-tuning a pre-trained PVS model using your LJSpeech dataset. You will discover how to customize the training process, select appropriate hyperparameters, and evaluate the model's performance.

Finally, we will test your fine-tuned PVS model by synthesizing realistic and expressive speech. You will learn how to generate natural-sounding speech by providing text input to the model, bringing the PVS voice to life.

In this chapter, we will cover the following main topics:

- Understanding TTS in PVS

- Converting audio files into LJSpeech format

- Fine-tuning a PVS model using the DLAS toolkit

- Synthesizing speech using a fine-tuned PVS model

By the end of this chapter, you will have a comprehensive understanding of how to utilize Whisper for PVS. You will possess the knowledge and skills to preprocess audio data, fine-tune voice models, and generate realistic speech using PVS frameworks. Whether you are a researcher, developer, or enthusiast in speech technology, this chapter will equip you with valuable insights and practical techniques to unlock the potential of PVS using OpenAI's Whisper.

## Technical requirements

To harness the capabilities of OpenAI's Whisper for advanced applications, this chapter leverages Python and Google Colab for ease of use and accessibility. The Python environment setup includes the Whisper library for transcription tasks.

**Key requirements**:

- **Google Colab notebooks**: The notebooks are set to run our Python code with the minimum required memory and capacity. If the **T4 GPU** runtime type is available, select it for better performance.

- **Python environment**: Each notebook contains directives to load the required Python libraries.

- **Hugging Face account**: Some notebooks require a Hugging Face account and login API key. The Colab notebooks include information about this topic.

- **Audacity**: Audacity is a free and open source digital audio editor and recording application available for Windows, macOS, Linux, and other Unix-like operating systems. It is an excellent choice if you want to synthesize your voice.

- **Microphone and speakers**: Some notebooks implement audio with voice recording and audio playback. A microphone and speakers connected to your computer might help you experience the interactive voice features.

- **GitHub repository access**: All Python code, including examples, is available in this chapter's GitHub repository (`https://github.com/PacktPublishing/Learn-OpenAI-Whisper/tree/main/Chapter09`). These Colab notebooks are ready to run, providing a practical and hands-on approach to learning.

By meeting these technical requirements, you will be prepared to explore Whisper in different contexts while enjoying the streamlined experience of Google Colab and the comprehensive resources available on GitHub.

# Understanding text-to-speech in voice synthesis

TTS is a crucial component in the voice synthesis process, enabling speech to be generated from written text using the synthesized voice. Understanding the fundamentals of TTS is essential to grasp how voice synthesizing works and how it can be applied in various scenarios. *Figure 9.1* illustrates a high-level overview of how TTS works in the context of voice synthesis without delving too deeply into technical specifics:

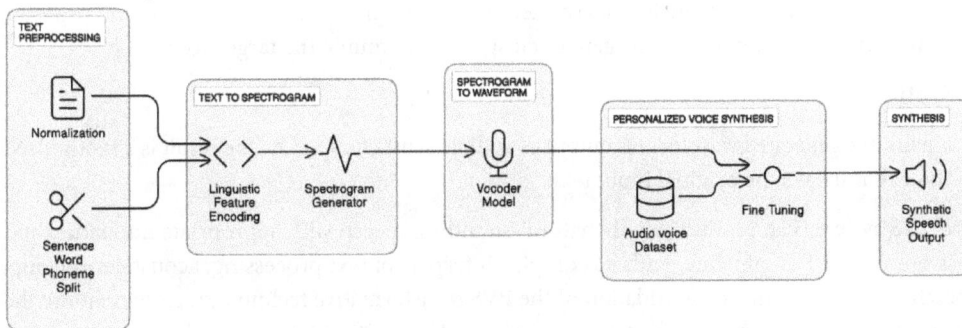

Figure 9.1 – The TTS voice synthesis pipeline

There are five components in the TTS voice synthesis pipeline:

1. **Text preprocessing**:

    I.    The input text is first normalized and preprocessed.

    II.   Numbers, abbreviations, and special characters are expanded into full words.

    III.  The text is divided into individual sentences, words, and phonemes (distinct sound units).

2. **Text-to-spectrogram**:

   I.   The normalized text is converted into a sequence of linguistic features and encoded into a vector representation.

   II.  A spectrogram generator model, usually a deep learning model, takes this encoded text and generates a spectrogram.

   III. The spectrogram visually represents the frequencies and intensities of the speech sounds over time.

3. **Spectrogram-to-waveform**:

   The spectrogram is then fed into a vocoder model. The vocoder is a generative model trained to convert spectrograms into audible waveforms. It reconstructs the speech signal from the frequency information in the spectrogram.

4. **Voice synthesis**:

   To synthesize a specific person's voice, the TTS models are fine-tuned on a dataset of that person's speech. This allows the models to learn their voices' unique characteristics, tone, and prosody. With sufficient training data, the generated speech will mimic the target voice.

5. **Synthesis**:

   Finally, the generated waveform is output as audible synthetic speech. The result is a synthesized voice that speaks the original input text.

Modern TTS systems can produce highly natural-sounding speech with appropriate intonation and expressiveness. The TTS pipeline, with its complex interplay of text processing, acoustic modeling, and speech synthesis, forms the foundation of the PVS transformative technology. As we explore the intricacies of voice synthesis, it is essential to understand how TTS systems can be leveraged to create personalized voices.

One such robust TTS implementation is **TorToiSe-TTS-Fast**, a high-performance TTS system that harnesses the power of neural networks to generate realistic and expressive speech. The following sections will delve into TorToiSe-TTS-Fast's capabilities and demonstrate how it can synthesize voices with remarkable accuracy and naturalness.

## Introducing TorToiSe-TTS-Fast

In *Chapter 5*, we used the gTTS Python library, an interface to Google Translate's TTS API. gTTS lets you generate spoken audio from text using Google's TTS engine. This time, we will explore the TorToiSe-TTS-Fast project, a high-performance TTS system that leverages neural networks to synthesize realistic speech without fine-tuning. Next, we will learn how to initialize the `TextToSpeech` model, which is the core component of the TTS system. We will explore the `TextToSpeech` class and understand its role in converting text into speech.

One of the exciting features of the TorToiSe-TTS-Fast project is its ability to generate speech using different audio clip samples of a given voice. The project provides a collection of pre-packaged voices as audio clips organized in separate folders. These audio clips are used to determine many properties of the voice synthesized output, such as the pitch and tone of the voice, speaking speed, and even speaking defects, such as a lisp or stuttering. We will delve into selecting a voice from that collection of pre-existing voice samples. *Figure 9.2* shows the TorToiSe-TTS-Fast voice processing:

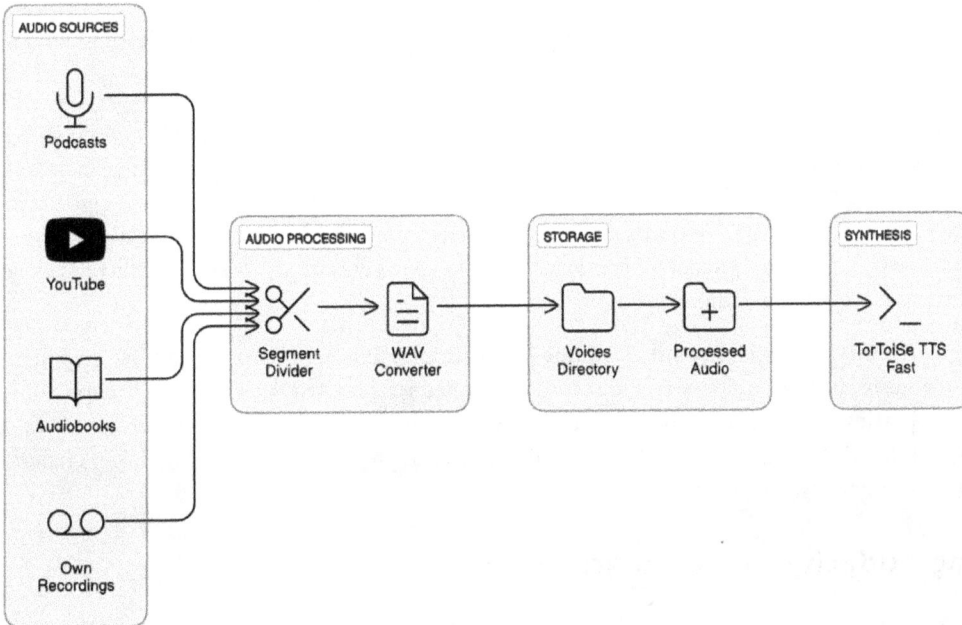

Figure 9.2 – TorToiSe-TTS-Fast voice processing pipeline

By following the steps in *Figure 9.2*, you can incorporate additional voices into TorToiSe and enhance its versatility:

1.  Collect audio samples featuring the desired voice(s). Interviews on YouTube (which can be downloaded using `youtube-dl` or the `pytube` Python library, as we did in *Chapter 6*), audiobooks, and podcasts are excellent sources. I recommend the **Audacity** tool as a viable option for recording your voice and processing audio files.

2.  Divide the collected audio into segments of approximately 10 seconds each. A minimum of 3 clips is required, but more clips are recommended for better results. During testing, I experimented with up to 5 clips.

3.  Convert the audio segments into WAV format with floating-point encoding and a sample rate of 22,050 Hz.

4. Once you run the `LOAIW_ch09_1_Synthesizing_voices_with_tortoise_tts_fast.ipynb` notebook later in this chapter, you will see a directory structure called `/tortoise/voices/` with audio clip samples in it. This is the default folder TorToiSe uses to store and retrieve audio samples. If you create your samples, create a folder in that `/tortoise/voices/` directory and save your files there. For example, I made the `/tortoise/voices/josue` folder to store my audio files.

5. Transfer the processed audio segments into the newly created subdirectory.

6. To utilize the new voice, execute the `tortoise` utilities with the `--voice` flag, followed by the name of your subdirectory.

After exploring the TorToiSe-TTS-Fast pipeline, it should be clear that high-quality audio data is foundational to creating convincing, natural-sounding synthesized voices. Preparing this audio data involves creating new recordings or manipulating existing audio files to ensure they are suitable for voice synthesis. This is where Audacity comes into play as a powerful tool for audio creation, editing, and refinement. Of course, I encourage you to use other tools you are already using for audio processing; Audacity and creating an audio file is optional.

Audacity is a versatile tool for creating, editing, and manipulating audio files, an essential step in the voice synthesis pipeline. It allows you to record your voice samples, trim and split audio clips, adjust audio properties such as pitch and speed, and export files in various formats compatible with voice synthesis tools. By leveraging Audacity's capabilities, you can prepare high-quality audio data tailored to your voice synthesis requirements.

## Using Audacity for audio processing

At its core, Audacity is a multitrack audio editor and recorder that supports many operating systems, including Windows, macOS, GNU/Linux, and other Unix-like systems. Its open source nature ensures it remains free for all users and fosters a vibrant community of developers and audio enthusiasts who continuously contribute to its development and enhancement. This collaborative effort has equipped Audacity with various capabilities, from basic recording and editing to more advanced features such as noise reduction, spectral analysis, and support for different audio formats. If you prefer another audio editor, go for it. The use of Audacity is optional. If you want to install it, here is a step-by-step guide.

The installation process of Audacity is straightforward, regardless of your operating system. Detailed instructions are available on the Audacity website (`https://support.audacityteam.org/basics/downloading-and-installing-audacity`). Here, we'll cover the basic steps to get Audacity up and running on your machine.

## Installing Audacity for Windows

Follow these steps:

1.  **Download the installer**: Navigate to the official Audacity website (`https://www.audacityteam.org/`) and click on the download link for the Windows version. The site will automatically detect your operating system, but you can manually select the version if needed.

2.  **Run the installer**: Once the download is complete, locate the installer file (usually in your `Downloads` folder) and double-click to initiate installation. You might encounter a security prompt asking for permission to allow the installer to change your system; click **Yes** to proceed.

3.  **Follow the installation wizard**: The installer will guide you through several steps. You'll select your preferred language, agree to the license terms, choose the installation directory, and decide on additional tasks, such as creating a desktop shortcut.

4.  **Complete the installation**: After configuring your preferences, click **Install** to begin the installation. Once completed, you can launch Audacity directly from the installer or find it in your Start menu.

## Installing Audacity for macOS

Follow these steps:

1.  **Download the DMG file**: Visit the Audacity website and download the macOS version. The site should automatically provide the correct version of your system.

2.  **Install Audacity**: Open the downloaded DMG file and drag the Audacity icon to your `Applications` folder to install the software. You might need to authenticate using your administrator password.

3.  **Launch Audacity**: Open it in your `Applications` folder. macOS might prompt you to confirm that you trust the application, especially if you're running it for the first time.

## Installing Audacity for Linux

Linux users can download AppImage from the Audacity website or install Audacity using their distribution's package manager. For AppImage, follow these steps:

1.  **Make the AppImage executable**: After downloading, right-click the file, navigate to **Properties | Permissions**, and check the option to make the file executable.

2.  **Run Audacity**: Double-click AppImage to launch Audacity.

Alternatively, use commands such as `sudo apt install audacity` for Debian-based distributions or `sudo yum install audacity` for Fedora/RHEL to install Audacity through the Terminal.

## Running the notebook with TorToiSe-TTS-Fast

With a more detailed understanding of Audacity as an audio creation, manipulation, and management tool, let's do some hands-on work with TorToiSe-TTS-Fast. Please find and open the Colab notebook called LOAIW_ch09_1_Synthesizing_voices_with_tortoise_tts_fast.ipynb (https://github.com/PacktPublishing/Learn-OpenAI-Whisper/blob/main/Chapter09/LOAIW_ch09_1_Synthesizing_voices_with_tortoise_tts_fast.ipynb). This notebook is based on the TorToiSe-TTS-Fast (https://github.com/152334H/tortoise-tts-fast) TTS project, which drastically boosts the performance of TorToiSe (https://github.com/neonbjb/tortoise-tts), without modifying the base models.

Using the notebook, we will develop speech from a given text with the TextToSpeech model initialized and a chosen voice. Furthermore, we will investigate the flexibility of the TorToiSe-TTS-Fast project by generating speech using random and even custom voices. We can create personalized voices for speech synthesis by uploading and preprocessing our WAV files.

Lastly, we will explore the fascinating capability of combining multiple voices to generate speech with blended characteristics. We can create unique and intriguing voice combinations by loading voice samples and conditioning latents from different voices.

By the end of this section, you will have a solid understanding of TTS in the context of voice synthesis. You will have the knowledge and practical skills to set up the environment, initialize the TextToSpeech model, select voices, generate speech, and create custom and combined voices using the TorToiSe-TTS-Fast project. This understanding will serve as a foundation for further exploring the potential of voice synthesis and its applications in various domains.

Let's open the notebook and run the cells to better understand the voice synthesis pipeline in the TorToiSe-TTS-Fast project:

1. **Setting up the environment**: Here, we will install and instantiate several libraries, each serving a distinct purpose in the project setup:

```
!git clone https://github.com/152334H/tortoise-tts-fast
%cd tortoise-tts-fast
!pip3 install -r requirements.txt --no-deps
!pip3 install -e .
!pip3 install git+https://github.com/152334H/BigVGAN.git
!pip install transformers==4.29.2
!pip install voicefixer==0.1.2
%cd tortoise-tts-fast

from huggingface_hub import notebook_login
notebook_login()

from huggingface_hub import whoami
whoami()
```

Let's briefly review each library:

- `torch`: This is the PyTorch library, a popular open source machine learning library for computer vision and natural language processing applications. In the context of this project, PyTorch provides the foundational framework for building and training the neural networks that underpin the voice synthesis capabilities of TorToiSe-TTS-Fast.

- `torchaudio`: As an extension to PyTorch, `torchaudio` offers easy access to audio processing tools within the PyTorch framework. It is used for loading and saving audio files and performing transformations and augmentations on audio data, which are essential tasks in voice synthesis.

- `huggingface_hub`: This library from Hugging Face allows users to easily download and upload models and other files to the Hugging Face Hub, which may include the pre-trained models or components required by the `TextToSpeech` class for voice synthesis. The `huggingface_hub` library also provides a function for authenticating with the Hugging Face Hub via `notebook_login()` and managing user information via `whoami()`, facilitating access to the models and resources stored on the Hub required for voice synthesis.

- `transformers (version 4.29.2)`: The `transformers` library, also from Hugging Face, provides thousands of pre-trained models for various natural language processing tasks, including TTS. This library supports the underlying NLP and TTS functionalities of the TorToiSe-TTS-Fast project by providing access to state-of-the-art models and utilities.

- `voicefixer (version 0.1.2)`: This tool is designed for repairing and enhancing human voice recordings. `voicefixer` improves the quality of voice samples before they are processed by the voice synthesis system, ensuring higher fidelity in the synthesized voices.

- `BigVGAN`: The TTS model uses a **voice generative adversarial network** (**VGAN**) to generate or modify voice data. The `BigVGAN` library plays a role in the voice synthesis process, enhancing the realism or quality of the generated voices.

Each of these libraries contributes to the overall functionality of the TorToiSe-TTS-Fast project by providing essential tools and frameworks for machine learning, audio processing, model management, and voice enhancement. These enable the efficient and practical synthesis of human voices.

2. **Initializing the TextToSpeech model**: Here, we'll set up the `TextToSpeech` model that will be used for synthesizing voices. The steps involved in this section are designed to initialize the TTS model so that it is ready to process text input and generate corresponding speech output using the selected voice:

```
from tortoise.api import TextToSpeech
from tortoise.utils.audio import load_audio, load_voice, load_
voices
# This will download all the models Tortoise uses from the
HuggingFace hub.
tts = TextToSpeech()
```

Here are the steps that are outlined in the preceding code:

- **Importing the** `TextToSpeech` **class**: The first step is to import the `TextToSpeech` class from the `tortoise.api` module. This class is the primary interface for the TTS functionality provided by the TorToiSe-TTS-Fast project.

- **Creating an instance of** `TextToSpeech`: After importing the class, an instance of `TextToSpeech` is created by simply calling the class constructor without any arguments. This instance is assigned to the `tts` variable.

- **Downloading the required models**: Upon creating an instance of `TextToSpeech`, the necessary models are automatically downloaded from the Hugging Face Hub. This step ensures that all the required components for voice synthesis are available locally.

- The `TextToSpeech` class encapsulates the functionality needed to convert text into speech. Initializing it is critical in preparing the system for voice synthesizing tasks. Once the model has been initialized, it can be used in subsequent steps to generate speech from text using various voices.

3.  **Selecting a voice**: This section is crucial for personalizing the voice synthesis process by allowing you to choose from various pre-existing voice samples or uploaded voice clips:

```
import os
from ipywidgets import Dropdown
voices_dir = "tortoise/voices"
# Get a list of all directories in the voices directory
voice_names = os.listdir(voices_dir)

voice_folder = Dropdown(
    options=sorted(voice_names),
    description='Select a voice:',
    value='tom',
    disabled=False,
    style={'description_width': 'initial'},
)
voice_folder

import os
from ipywidgets import Dropdown
voices_dir = f"tortoise/voices/{voice_folder.value}"

# Get a list of all directories in the voices directory
voice_files = os.listdir(voices_dir)
voice = Dropdown(
    options=sorted(voice_files),
    description='Select a voice:',
```

```
    # value='tom',
    disabled=False,
    style={'description_width': 'initial'},
)
Voice

#Pick one of the voices from the output above
IPython.display.Audio(filename=f'tortoise/voices/{voice_folder.
value}/{voice.value}')
```

The steps involved in guiding you through the process of selecting a specific voice to synthesize are as follows:

- **Listing the available voice folders**: The code utilizes the `os` module to interact with the operating system and lists all the available voice folders in the `tortoise/voices` directory. This step is essential for identifying which voices are available for synthesis.

- **Creatinge a dropdown widget for voice folder selection**: A dropdown widget is created using the `Dropdown` class from the `ipywidgets` library. This widget allows you to select a voice folder from the list of available folders. The `Dropdown` widget is configured with options populated from the list of voice folders, a description prompt (`"Select a voice:"`), and other settings to ensure usability.

- **Selecting a specific voice file**: After selecting a voice folder, another dropdown widget is created to select a particular voice file within the chosen folder. This step is similar to the previous one but focuses on the individual voice files within the selected folder. The list of voice files is obtained using the `os.listdir` function, and the `Dropdown` widget is again used to present these options to you.

- **Playing the selected voice**: Once a specific voice file is selected, the `IPython.display.Audio` class plays the voice file chosen directly in the Colab notebook. This feature provides immediate auditory feedback, enabling you to confirm that the selected voice is the desired one for synthesis.

These steps collectively enable a user-friendly and interactive approach to selecting a voice for synthesis. They ensure that you can easily navigate the available options and make an informed choice based on your preferences or project requirements.

4. **Generating speech with a selected voice**: This section of the notebook generates speech audio from text using the voice previously specified by you:

```
text = " Words, once silent, now dance on digital breath,
speaking volumes through the magic of text-to-speech."
preset = "ultra_fast"
voice = voice_folder.value
voice_samples, conditioning_latents = load_voice(voice)
gen = tts.tts_with_preset(text, voice_samples=voice_samples,
```

```
conditioning_latents=conditioning_latents, preset=preset)
torchaudio.save(generated_filename, gen.squeeze(0).cpu(), 24000)
IPython.display.Audio(generated_filename)
```

The steps in this section are designed to convert your input text into spoken words in the chosen voice's style. Here are the steps outlined in the code:

- **Defining the text**: Here, you define the text that they want to be spoken by assigning it to a variable named `text`. This text will be synthesized into speech.

- **Setting the quality preset**: The quality of the generated speech is set by assigning a value to the preset variable. The options for the preset include `"ultra_fast"`, `"fast"`, `"standard"`, and `"high_quality"`. This determines the trade-off between generation speed and audio quality.

- **Loading the selected voice**: The `load_voice` function from `tortoise.utils.audio` is used to load the selected voice. This function returns two items: `voice_samples` and `conditioning_latents`. These condition the TTS model to generate speech in the selected voice's style.

- **Generating the speech**: The `tts_with_preset` method of the `tts` object (an instance of the `TextToSpeech` class) is called with the text, voice samples, conditioning latents, and preset. This method synthesizes the speech based on the given parameters.

- **Saving and playing the speech**: The generated speech is saved as a WAV file using the `torchaudio.save` function. The file is then played using `IPython.display.Audio` to allow you to hear the synthesized speech.

These steps enable you to create a spoken version of their text using the specific characteristics of the chosen voice, effectively using the PVS model for speech synthesis.

5. **Generating speech with a random voice**: This is similar to the previous step, *Generating speech with a selected voice*, but with `voice_samples` and `conditioning_latents` set to None, which generates speech using a random voice:

```
gen = tts.tts_with_preset(text, voice_samples=None,
conditioning_latents=None, preset=preset)
torchaudio.save(' synthetized_voice_sample.wav', gen.squeeze(0).
cpu(), 24000)
IPython.display.Audio('synthetized_voice_sample.wav')
```

6. **Using a custom voice**: The following code allows users to upload their WAV files (6-10 seconds long) to create a custom voice:

```
CUSTOM_VOICE_NAME = "custom"
import os
from google.colab import files
```

```
custom_voice_folder = f"tortoise/voices/{CUSTOM_VOICE_NAME}"
os.makedirs(custom_voice_folder)
for i, file_data in enumerate(files.upload().values()):
  with open(os.path.join(custom_voice_folder, f'{i}.wav'), 'wb')
as f:
    f.write(file_data)
# Generate speech with the custom voice.
voice_samples, conditioning_latents = load_voice(CUSTOM_VOICE_
NAME)
gen = tts.tts_with_preset(text, voice_samples=voice_samples,
conditioning_latents=conditioning_latents,
                         preset=preset)
torchaudio.save(f'generated-{CUSTOM_VOICE_NAME}.wav', gen.
squeeze(0).cpu(), 24000)
IPython.display.Audio(f'generated-{CUSTOM_VOICE_NAME}.wav')
```

It creates a custom voice folder using `os.makedirs` and saves the uploaded files in that folder. The custom voice is then loaded and used to generate speech, similar to *steps 4* and *5*.

7. **Combining voices**: The `load_voices` function loads multiple voices (in this case, `'pat'` and `'william'`). The `tts_with_preset` method combines voice samples and conditioning latents to generate speech with traits from both voices:

```
voice_samples, conditioning_latents = load_voices(['freeman',
'deniro'])
gen = tts.tts_with_preset("Words, once silent, now dance on
digital breath, speaking volumes through the magic of text-to-
speech.",
                         voice_samples=voice_samples,
conditioning_latents=conditioning_latents,
                         preset=preset)
torchaudio.save('freeman_deniro.wav', gen.squeeze(0).cpu(),
24000)
IPython.display.Audio('freeman_deniro.wav')
```

Having gained a foundational understanding of TTS in voice synthesis and explored the powerful capabilities of the TorToiSe-TTS-Fast project, we'll turn our attention to a crucial step in preparing our data for the voice synthesizing process: converting audio files into the LJSpeech format.

# PVS step 1 – Converting audio files into LJSpeech format

This section and the accompanying notebook, `LOAIW_ch09_2_Processing_audio_to_LJ_format_with_Whisper_OZEN.ipynb`, represent the initial step in the three-step PVS process outlined in this chapter. This step takes an audio sample of the target voice as input and processes it into the LJSpeech dataset format. The notebook demonstrates using the OZEN Toolkit and OpenAI's Whisper to extract speech, transcribe it, and organize the data according to the LJSpeech structure. The resulting LJSpeech-formatted dataset, consisting of segmented audio files and corresponding transcriptions, serves as the input for the second step, *PVS step 2 – Fine-tuning a discrete variational autoencoder using the DLAS toolkit*, where a PVS model will be fine-tuned using this dataset.

An LJSpeech-formatted dataset is crucial in TTS models as it provides a standardized structure for organizing audio files and their corresponding transcriptions. By following the LJSpeech format, researchers and developers can ensure compatibility with various TTS tools and facilitate training.

An LJSpeech-formatted dataset refers to a specific structure and organization of audio files and their corresponding transcriptions modeled after the **LJSpeech dataset** (`https://keithito.com/LJ-Speech-Dataset/`). The LJSpeech dataset is a public domain speech dataset that includes 13,100 short audio clips of a single speaker reading passages from seven non-fiction books, with a transcription of each clip. The audio clips vary in length and have a total duration of approximately 24 hours. When formatting a dataset for training a TTS model in the style of LJSpeech, the following structure is recommended:

- Audio clips should be divided into separate files with a corresponding transcription.
- The WAV file format should be used for the audio to avoid compression artifacts.
- The audio clips and their transcriptions are collected in a folder named `wavs`.
- A metadata text file maps each audio clip to its transcription. This file should have columns delimited by a special character, typically a pipe ( | ), to separate the audio filename, the transcription, and the normalized transcription.
- The delimiter used in the metadata file should not appear in the transcription text itself.
- If normalized transcriptions are unavailable, the same transcription can be used for both columns, with normalization applied later in the pipeline.

The folder structure for an LJSpeech-formatted dataset would look like this:

```
/MyDataset
---├── train.txt
---├── valid.txt
---├── /wavs
---------├── 0.wav
---------├── 1.wav
        ...
```

In the text files, entries would be formatted as follows:

```
wavs/0.wav|This is Josue Batista.
wavs/1.wav|I am the author of the book Learn OpenAI Whisper, Transform
Your Understanding of Generative AI
wavs/2.wav|through robust and accurate speech processing solutions.
...
```

The LJSpeech format is widely used because it is supported by various TTS tools, such as TorToiSe, which provides tooling for the LJSpeech dataset. Formatting a dataset this way allows for immediate model training without additional formatting steps.

Now that we understand the LJSpeech format and why it's used, let's convert our audio files into this format. By doing so, we'll ensure that our dataset is compatible with various TTS tools and ready for training our PVS models.

Once you have a recording of the voice you would like to synthesize, the next step is to preprocess the audio files using the OZEN Toolkit. This toolkit simplifies extracting speech, transcribing it using Whisper, and saving the results in the LJSpeech format. It can handle both single audio files and entire folders of audio files.

Leveraging the OZEN Toolkit and Whisper allows us to efficiently convert our audio data into the LJSpeech format. The toolkit automates segmenting audio files, generating corresponding WAV files, and creating the necessary metadata files (`train.txt` and `valid.txt`) that map the audio files to their transcriptions.

Converting audio files into the LJSpeech format is a crucial skill in the voice synthesis pipeline as it ensures data compatibility and facilitates the training process. Mastering this technique will prepare you to tackle the subsequent steps, such as fine-tuning PVS models and synthesizing speech.

Please find and open the Colab notebook called `LOAIW_ch09_2_Processing_audio_to_LJ_format_with_Whisper_OZEN.ipynb` (`https://github.com/PacktPublishing/Learn-OpenAI-Whisper/blob/main/Chapter09/LOAIW_ch09_2_Processing_audio_to_LJ_format_with_Whisper_OZEN.ipynb`). This notebook is based on the OZEN Toolkit project (`https://github.com/devilismyfriend/ozen-toolkit`). Given a folder of files or a single audio file, it will extract the speech, transcribe using Whisper, and save it in LJ format (segmented audio files in WAV format go in the `wavs` folder, while transcriptions go in the `train` and `valid` folders). Let's walk through the code while explaining the steps and providing code samples:

1. **Cloning the OZEN Toolkit repository**: The following command clones the OZEN Toolkit repository from GitHub, which contains the necessary scripts and utilities for processing audio files:

   ```
   !git clone https://github.com/devilismyfriend/ozen-toolkit
   ```

2. **Installing the required libraries**: The following commands install the necessary libraries for audio processing, speech recognition, and text formatting. After installing the dependencies, restarting the session is recommended to ensure the installed packages are initialized adequately:

   ```
   !pip install transformers
   !pip install huggingface
   !pip install pydub
   !pip install yt-dlp
   ```

```
!pip install pyannote.audio
!pip install colorama
!pip install termcolor
!pip install pyfiglet
```

3. **Changinge the working directory**: The following command changes the working directory to the cloned `ozen-toolkit` directory:

```
%cd ozen-toolkit
```

4. **Downloading a sample audio file**: If you do not have an audio file for synthesis, this command downloads a sample audio file from the specified URL for demonstration purposes:

```
!wget -nv https://github.com/PacktPublishing/Learn-OpenAI-
Whisper/raw/main/Chapter01/Learn_OAI_Whisper_Sample_
Audio01.mp3
```

5. **Uploading custom audio files**: If you have your audio file, this code block allows users to upload their audio files to the Colab environment. It creates a directory in `/content/ozen-toolkit` to store the uploaded files and saves them in that directory:

```
import os
from google.colab import files

custom_voice_folder = "./myaudiofile"

os.makedirs(custom_voice_folder, exist_ok=True)   # Create the
directory if it doesn't exist

for filename, file_data in files.upload().items():
    with open(os.path.join(custom_voice_folder, filename), 'wb')
as f:
        f.write(file_data)

%ls -l "$PWD"/{*,.*}
```

6.  **Creating a configuration file**: The following code section creates a configuration file named `"config.ini"` using the `configparser` library. It defines various settings, such as the Hugging Face API key, Whisper model, device, diarization and segmentation models, validation ratio, and segmentation parameters:

```
import configparser

config = configparser.ConfigParser()

config['DEFAULT'] = {
    'hf_token': '<Your HF API key>',
    'whisper_model': 'openai/whisper-medium',
    'device': 'cuda',
    'diaization_model': 'pyannote/speaker-diarization',
    'segmentation_model': 'pyannote/segmentation',
    'valid_ratio': '0.2',
    'seg_onset': '0.7',
    'seg_offset': '0.55',
    'seg_min_duration': '2.0',
    'seg_min_duration_off': '0.0'
}

with open('config.ini', 'w') as configfile:
    config.write(configfile)
```

7.  **Running the OZEN script**: This command runs the `ozen.py` script with the sample audio file as an argument (or the file you uploaded):

```
!python ozen.py Learn_OAI_Whisper_Sample_Audio01.mp3
```

**IMPORTANT**: `ozen.py` requires Hugging Face's `pyannote/segmentation` model. This is a gated model; you MUST request access before attempting to run the next cell. Thankfully, getting access is relatively straightforward and fast. Here are the steps:

I.   You must already have a Hugging Face account; if you do not have one, see the instructions in the notebook for *Chapter 3*: `LOAIW_ch03_working_with_audio_data_via_Hugging_Face.ipynb` (`https://github.com/PacktPublishing/Learn-OpenAI-Whisper/blob/main/Chapter03/LOAIW_ch03_working_with_audio_data_via_Hugging_Face.ipynb`)

II.   Visit `https://hf.co/pyannote/segmentation` to accept the user conditions:

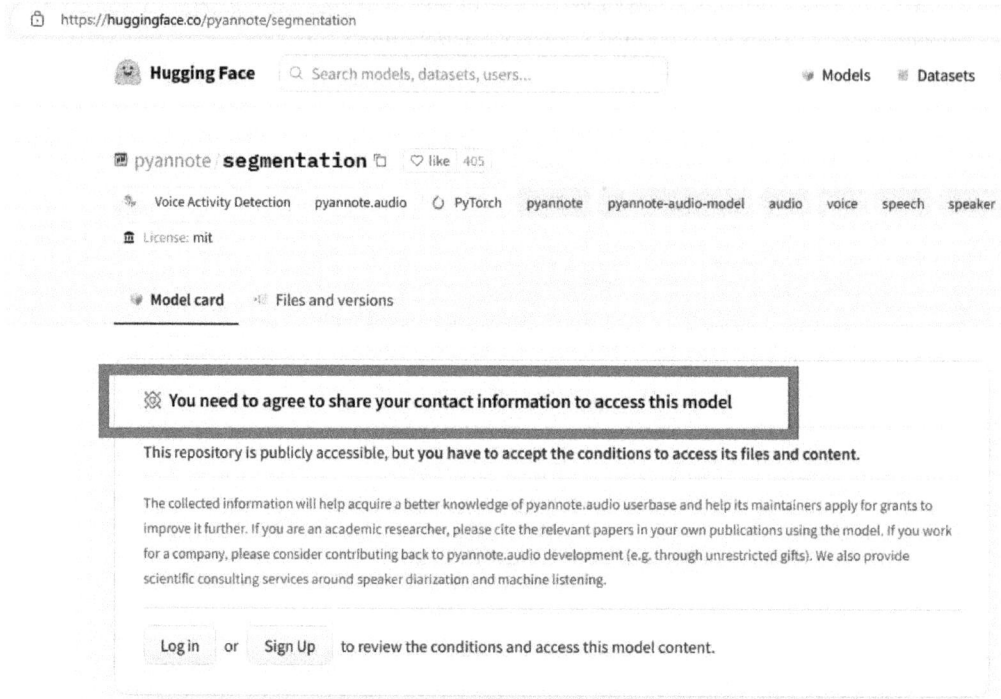

Figure 9.3 – The pyannote/segmentation gated model on Hugging Face

III.  Run the cell after ensuring you have access to the pyannote/segmentation model. The `ozen.py` script processes the audio file, extracts speech, transcribes it using Whisper, and saves the output in the LJSpeech format. The script saves the DJ format files in a folder called `ozen-toolkit/output/<audio file name + timestamp>/`. Here is an example of the expected file structure:

```
ozen-toolkit/output/
---├── Learn_OAI_Whisper_Sample_Audio01.mp3_2024_03_16-16_36/
-----------------├── valid.txt
-----------------├── train.txt
-----------------├── wavs/
---------------------------├── 0.wav
---------------------------├── 1.wav
---------------------------├── 2.wav
```

8. **Mounting Google Drive**: The following lines mount your Google Drive to the Colab environment, allowing access to the drive for saving checkpoints and loading datasets:

```
from google.colab import drive
drive.mount('/content/gdrive')
```

9. **Copying the output to Google Drive**: The following command copies the processed output files from the `ozen-toolkit/output` directory to your Google Drive. After running the cell, go to your Google Drive using a web browser, as shown in *Figure 9.4*; you will see a directory called `output` with the DJ format dataset files in it:

```
%cp -r /content/ozen-toolkit/output/ /content/gdrive/MyDrive/
ozen-toolkit/output/
```

Here's the output:

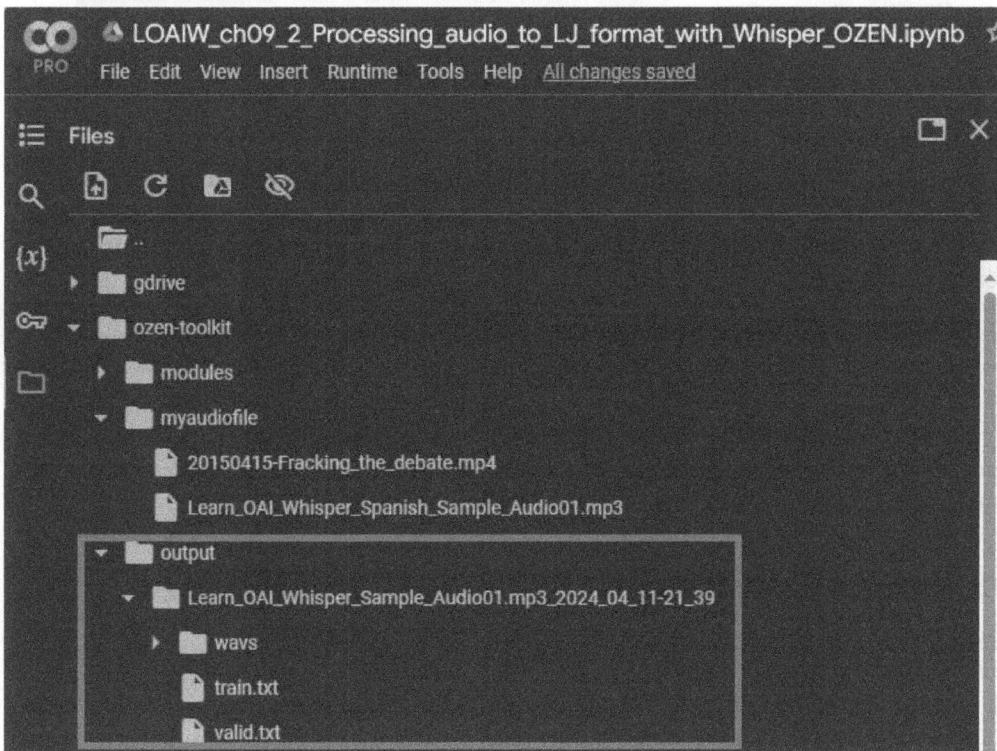

Figure 9.4 – Identifying the location of the DJ format files from the ozen.py script

After running the cell, go to your Google Drive using a web browser; you will see a directory called `output` with the DJ format dataset files in it:

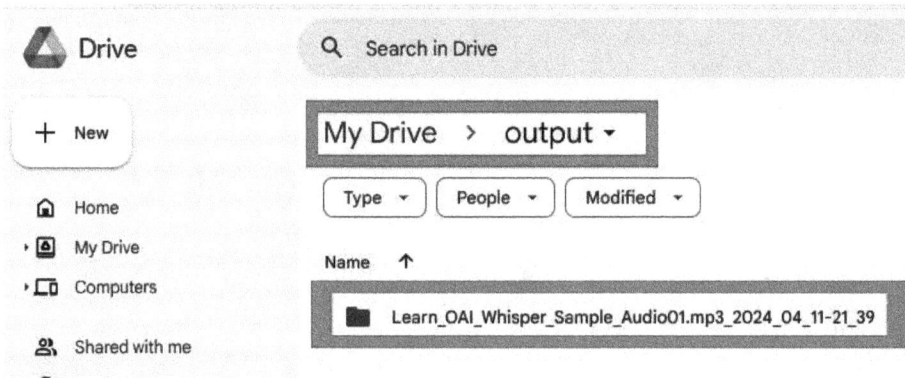

Figure 9.5 – Example of the DJ format output folder in Google Colab

The Python code in the `LOAIW_ch09_2_Processing_audio_to_LJ_format_with_Whisper_OZEN.ipynb` notebook demonstrates setting up the environment, installing dependencies, configuring the OZEN Toolkit, processing audio files using Whisper, and saving the output in the LJSpeech format. It provides a streamlined workflow for preparing audio data for further analysis or use in downstream tasks.

With our audio data now converted into the LJSpeech format, we are well-prepared to embark on the following critical stage of the voice synthesis journey: fine-tuning a PVS model using the powerful DLAS toolkit. The `LOAIW_ch09_3_Fine-tuning_PVS_models_with_DLAS.ipynb` notebook will cover this process in detail in the next section. By leveraging the DLAS toolkit's comprehensive features and the structured LJSpeech dataset, we can create a personalized voice model that captures the unique characteristics of our target voice with remarkable accuracy and naturalness.

## PVS step 2 – Fine-tuning a PVS model with the DLAS toolkit

Fine-tuning a PVS model is a critical step in creating personalized voices that capture the unique characteristics of a voice. To achieve high-quality results, utilizing a robust framework that leverages state-of-the-art techniques and provides flexibility in customizing the training process is essential. The DLAS toolkit emerges as a comprehensive solution for fine-tuning PVS models, offering a range of features and capabilities.

Before starting the fine-tuning process, ensuring that the necessary components and resources are in place is crucial. This includes setting up a suitable training environment, such as Google Colab, which provides access to powerful GPUs and sufficient RAM to handle the computational demands of PVS models. Checking the availability and compatibility of NVIDIA GPUs is vital to ensuring optimal performance during training.

The dataset preparation phase is another essential aspect of fine-tuning a PVS model. The DLAS toolkit requires a specific repository structure and dependencies, which must be cloned and installed before proceeding. Additionally, pre-trained model checkpoints, such as the **discrete variational autoencoder (dVAE)**, play a crucial role in learning a discrete latent representation of the speech data. Verifying the integrity of these checkpoints is necessary to accelerate the fine-tuning process and achieve better results.

Selecting appropriate hyperparameters based on the dataset's size is critical in fine-tuning a PVS model. The DLAS toolkit offers intelligent suggestions for hyperparameters, such as batch sizes, learning rate decay steps, and validation frequencies, all of which consider the specific characteristics of the dataset. Understanding how these hyperparameters are calculated and their impact on the training process is essential for achieving optimal results.

Customization is another critical aspect of fine-tuning a PVS model using the DLAS toolkit. Researchers and developers often have specific requirements and preferences for training settings, such as experiment names, dataset names, and turning certain features on or off. The DLAS toolkit provides flexibility in modifying these settings, allowing for tailored fine-tuning processes that align with specific needs and goals.

The DLAS toolkit utilizes a YAML configuration file to ensure the fine-tuning process is configured according to the desired specifications. This file serves as a blueprint for the training process, specifying various parameters and settings. The toolkit applies the customized training settings to the YAML file using sophisticated `sed` commands, ensuring that the fine-tuning process is tailored to the specific requirements and enables the reproducibility of the experiments (`sed` stands for Stream Editor, a powerful command-line utility that's used for parsing and transforming text using a simple, compact programming language).

With the configuration file ready, the training process can be initiated by running the `train.py` script provided by the DLAS toolkit. This script leverages the power of GPUs to efficiently fine-tune the PVS model, utilizing optimization algorithms and loss functions to guide the learning process. Monitoring the training progress and evaluating the model's performance using appropriate metrics is crucial for ensuring the quality of the fine-tuned PVS model.

Finally, saving and exporting the fine-tuned PVS model is essential for future use and deployment. The DLAS toolkit provides convenient methods to store the trained model checkpoints and experiment files, ensuring data persistence and facilitating research collaboration. Proper management and organization of the fine-tuned models are critical for seamless integration into various applications, such as virtual assistants, audiobook narration, and personalized voice interfaces.

Researchers and developers can create personalized voices that capture the nuances and characteristics by understanding the components, processes, and considerations involved in fine-tuning a PVS model using the DLAS toolkit. The ability to customize the training process, select appropriate hyperparameters, and leverage pre-trained checkpoints empowers users to achieve high-quality results and explore exciting possibilities in voice synthesis.

Please find and open the Colab notebook called `LOAIW_ch09_3_Fine-tuning_PVS_models_ with_DLAS.ipynb` (`https://github.com/PacktPublishing/Learn-OpenAI- Whisper/blob/main/Chapter09/LOAIW_ch09_3_Fine-tuning_PVS_models_ with_DLAS.ipynb`).

This notebook fine-tunes a PVS model using the DLAS toolkit. It is based on the *TorToiSe fine-tuning with DLAS* project by James Betker (`https://github.com/152334H/DL-Art-School`). I cloned and modified the code to run on Google Colab and leveraged an NVIDIA GPU for training.

Let's walk through the steps in the `LOAIW_ch09_3_Fine-tuning_PVS_models_with_DLAS. ipynb` notebook:

1.  **Checking the GPU**: The code first checks whether an NVIDIA GPU is available using the `nvidia-smi` command. It prints out the GPU's information if it's connected:

    ```
    gpu_info = !nvidia-smi
    gpu_info = '\n'.join(gpu_info)
    if gpu_info.find('failed') >= 0:
      print('Not connected to a GPU')
    else:
      print(gpu_info)
    ```

2.  **Checking virtual memory**: It then checks the available RAM on the runtime using the `psutil` library. It prints a message if it's using a high-RAM runtime:

    ```
    from psutil import virtual_memory
    ram_gb = virtual_memory().total / 1e9
    print('Your runtime has {:.1f} gigabytes of available RAM\n'.
    format(ram_gb))

    if ram_gb < 20:
      print('Not using a high-RAM runtime')
    else:
      print('You are using a high-RAM runtime!')
    ```

3.  **Mounting Google Drive**: The following code mounts your Google Drive to save trained checkpoints and load the dataset:

    ```
    from google.colab import drive
    drive.mount('/content/gdrive')
    ```

4. **Installing the requirements**: It clones the DLAS repository, downloads pre-trained model checkpoints, and installs the required dependencies:

```
!git clone https://github.com/josuebatista/DL-Art-School.git
%cd DL-Art-School
!wget https://huggingface.co/Gatozu35/tortoise-tts/resolve/main/dvae.pth -O experiments/dvae.pth
!wget https://huggingface.co/jbetker/tortoise-tts-v2/resolve/main/.models/autoregressive.pth -O experiments/autoregressive.pth
!pip install -r codes/requirements.laxed.txt
```

After installing the dependencies, restarting the session is recommended to ensure the installed packages are properly initialized. We must also verify the integrity of the dVAE checkpoint. A dVAE is a component of the PVS model that helps learn a discrete latent representation of the data. The `sha256sum` command calculates the checksum. If the `grep` command does not find a match to the expected checksum value, the integrity check fails and reports an error message. Thus, the entire command line ensures that the checkpoint file has not been corrupted or altered:

```
!sha256sum /content/DL-Art-School/experiments/dvae.pth | grep
a990825371506c16bcf0e8167bf24ccf82f65bb6a1dbcbfcf058d76f9b197e35
|| echo "SOMETHING IS WRONG WITH THE CHECKPOINT; REPORT THIS AS
A GITHUB ISSUE AND DO NOT PROCEED"
```

The preceding command calculates the SHA-256 checksum of the checkpoint file and compares it with the expected checksum. If the checksums match, the integrity of the file is confirmed. If they do not match, the user is advised to report the issue on GitHub and not proceed further.

5. **Calculating hyperparameters**: This section automatically calculates suggested hyperparameters for training based on the provided dataset sizes. It adjusts the batch sizes to minimize leftover samples in each epoch, calculates the number of steps per epoch, and determines the frequencies for learning rate decay, validation, and checkpoint saving. Hyperparameters are crucial as they directly control the training algorithm's behavior and significantly impact the model's performance.

To find the path to `Dataset_Training_Path` and `ValidationDataset_Training_Path`, click on Google Colab's **Files** option and search Google Drive for the directory where the DJ-format datasets were stored in the previous notebook. *Figure 9.6* shows an example of where the DJ-format dataset is found. Keep in mind that *Figure 9.6* is just an example. Do not search for that literal name. Instead, you must search for the directory name you set while creating the DJ-formatted files:

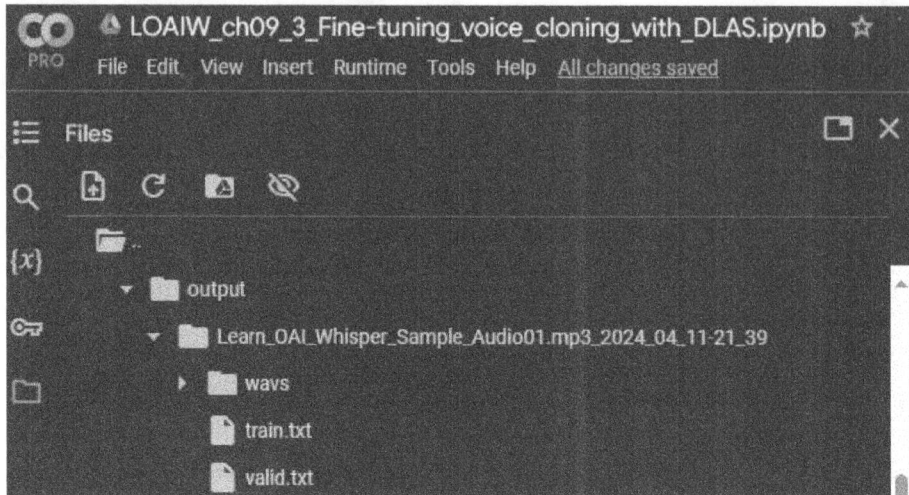

Figure 9.6 – Example of searching for the DJ format dataset in the
output directory created in the previous notebook

The following is the entire script that performs the hyperparameter calculation. An explanation of how it works is provided after the code listing:

```
from pathlib import Path
from math import ceil
DEFAULT_TRAIN_BS = 64
DEFAULT_VAL_BS = 32
Dataset_Training_Path = "/content/gdrive/MyDrive/Generative_
AI/Deep_Fakes_Voice/output/Learn_OAI_Whisper_Sample_Audio01.
mp3_2024_03_16-16_36/train.txt" #@param {type:"string"}
ValidationDataset_Training_Path = "/content/gdrive/MyDrive/
Generative_AI/Deep_Fakes_Voice/output/Learn_OAI_Whisper_Sample_
Audio01.mp3_2024_03_16-
if Dataset_Training_Path == ValidationDataset_Training_Path:
  print("WARNING: training dataset path == validation dataset
path!!!")
  print("\tThis is technically okay but will make all of the
validation metrics useless. ")
  print("it will also SUBSTANTIALLY slow down the rate of
training, because validation datasets are supposed to be much
smaller than training ones.")

def txt_file_lines(p: str) -> int:
  return len(Path(p).read_text().strip().split('\n'))
training_samples = txt_file_lines(Dataset_Training_Path)
val_samples = txt_file_lines(ValidationDataset_Training_Path)
```

```
if training_samples < 128: print("WARNING: very small dataset!
the smallest dataset tested thus far had ~200 samples.")
if val_samples < 20: print("WARNING: very small validation
dataset! val batch size will be scaled down to account")


def div_spillover(n: int, bs: int) -> int: # returns new batch
size
  epoch_steps,remain = divmod(n,bs)
  if epoch_steps*2 > bs: return bs # don't bother optimising
this stuff if epoch_steps are high
  if not remain: return bs # unlikely but still

  if remain*2 < bs: # "easier" to get rid of remainder -- should
increase bs
    target_bs = n//epoch_steps
  else: # easier to increase epoch_steps by 1 -- decrease bs
    target_bs = n//(epoch_steps+1)
  assert n%target_bs < epoch_steps+2 # should be very few extra
  return target_bs


if training_samples < DEFAULT_TRAIN_BS:
  print("WARNING: dataset is smaller than a single batch. This
will almost certainly perform poorly. Trying anyway")
  train_bs = training_samples
else:
  train_bs = div_spillover(training_samples, DEFAULT_TRAIN_BS)
if val_samples < DEFAULT_VAL_BS:
  val_bs = val_samples
else:
  val_bs = div_spillover(val_samples, DEFAULT_VAL_BS)


steps_per_epoch = training_samples//train_bs
lr_decay_epochs = [20, 40, 56, 72]
lr_decay_steps = [steps_per_epoch * e for e in lr_decay_epochs]
print_freq = min(100, max(20, steps_per_epoch))
val_freq = save_checkpoint_freq = print_freq * 3


print("===CALCULATED SETTINGS===")
print(f'{train_bs=} {val_bs=}')
print(f'{val_freq=} {lr_decay_steps=}')
print(f'{print_freq=} {save_checkpoint_freq=}')
```

Let's break down the purpose and steps in this section:

- The code imports the necessary libraries: `Path` from `pathlib` to work with files and directories, and `ceil`, a function in the built-in `math` module for rounding numbers up to the nearest integer.

- It defines default training and validation batch size values: `DEFAULT_TRAIN_BS = 64` and `DEFAULT_VAL_BS = 32`.

- You are prompted to provide the paths for the training and validation datasets: `Dataset_Training_Path` and `ValidationDataset_Training_Path`.

- The code checks whether the training and validation dataset paths are identical. If they are, it prints a warning message indicating that validation metrics will be useless and the training rate will be substantially slowed.

- The code defines a helper function, `txt_file_lines`, that takes a file path as input and returns the number of lines in the file.

- It calculates the training and validation samples by calling `txt_file_lines` with the respective dataset paths.

- The code prints warning messages if the dataset sizes are small: less than 128 for training and less than 20 for validation.

- It defines a helper function called `div_spillover` that takes the number of samples (n) and batch size (`bs`) as input and returns an adjusted batch size to minimize the number of leftover samples in each epoch.

- The code calculates the training batch size (`train_bs`) based on the number of training samples. If the number of training samples is smaller than `DEFAULT_TRAIN_BS`, it sets `train_bs` to the number of training samples and prints a warning message. Otherwise, it calls `div_spillover` with the number of training samples and `DEFAULT_TRAIN_BS` to calculate an adjusted batch size.

- Similarly, it calculates the validation batch size (`val_bs`) based on the number of validation samples. If the number of validation samples is smaller than `DEFAULT_VAL_BS`, it sets `val_bs` to the number of validation samples. Otherwise, it calls `div_spillover` with the number of validation samples and `DEFAULT_VAL_BS` to calculate an adjusted batch size.

- The code calculates the number of steps per epoch (`steps_per_epoch`) by dividing the number of training samples by the training batch size.

- It defines the epochs at which learning rate decay should occur via `lr_decay_epochs = [20, 40, 56, 72]`.

- The code calculates the corresponding steps for learning rate decay (`lr_decay_steps`) by multiplying `steps_per_epoch` with each value in `lr_decay_epochs`.

- It calculates the frequency for printing training progress (`print_freq`) based on the number of steps per epoch, with a minimum of 20 and a maximum of 100.

- The code sets the frequency for validation and saving checkpoints (`val_freq` and `save_checkpoint_freq`) to three times the `print_freq` value.

- Finally, it prints the calculated settings: `train_bs`, `val_bs`, `val_freq`, `lr_decay_steps`, `print_freq`, and `save_checkpoint_freq`.

- From the creator of the DLAS trainer, the values of `print_freq`, `val_freq`, and `save_checkpoint_freq` should all be adjusted to the dataset's size. The Python code states the recommended values: `val_freq == save_checkpoint_freq == print_freq*3`; `print_freq == min(epoch_steps,100)`. Again, these are recommendations; I encourage you to experiment with different ones and compare results for optimal hyperparameter settings.

By calculating these hyperparameters, the code aims to provide reasonable default values that can be used to train the PVS model. However, we can override these calculated values in the subsequent sections if needed.

6. **Training settings**: This section allows us to customize the training settings according to their requirements and available resources. It provides flexibility in naming the experiment, specifying dataset names, turning certain features on or off, and overriding calculated settings. The code also includes notes and warnings to guide you in making appropriate choices based on the system's storage and computational capabilities:

```
Experiment_Name = "Learn_OAI_Whisper_20240316"
Dataset_Training_Name= "TestDataset"
ValidationDataset_Name = "TestValidation"
SaveTrainingStates = False
Keep_Last_N_Checkpoints = 0
Fp16 = False
Use8bit = True
TrainingRate = "1e-5"
TortoiseCompat = False
TrainBS = ""
ValBS = ""
ValFreq = ""
LRDecaySteps = ""
PrintFreq = ""
SaveCheckpointFreq = ""

def take(orig, override):
  if override == "": return orig
  return type(orig)(override)
```

```
train_bs = take(train_bs, TrainBS)
val_bs = take(val_bs, ValBS)
val_freq = take(val_freq, ValFreq)
lr_decay_steps = eval(LRDecaySteps) if LRDecaySteps else lr_
decay_steps
print_freq = take(print_freq, PrintFreq)
save_checkpoint_freq = take(save_checkpoint_freq,
SaveCheckpointFreq)
assert len(lr_decay_steps) == 4
gen_lr_steps = ', '.join(str(v) for v in lr_decay_steps)
```

Let's break down the purpose and steps in this section:

- You can specify the following training settings:

  - Experiment_Name: A string to name the experiment.

  - Dataset_Training_Name: A string to name the training dataset.

  - ValidationDataset_Name: A string to name the validation dataset.

  - SaveTrainingStates: A Boolean to indicate whether to save training states.

  - Keep_Last_N_Checkpoints: An integer slider to specify the number of checkpoints to keep. Setting it to 0 means keeping all saved models.

- The code provides notes and warnings:

  - It mentions that keeping all saved models (setting Keep_Last_N_Checkpoints to 0) could potentially cause out-of-storage issues.

  - Without training states, **each model takes up approximately 1.6 GB** in Google Drive, giving you around 50 GB of free space.

  - With training states, each model (including the state) takes up approximately 4.9 GB, and Colab may crash if there are around 10 undeleted checkpoints.

- Other training parameters:

  - Fp16: A Boolean to turn 16-bit floating-point precision on or off.

  - Use8bit: A Boolean to turn 8-bit precision on or off.

  - TrainingRate: A string to specify the learning rate.

  - TortoiseCompat: A Boolean to turn compatibility with the TorToiSe model on or off. Enabling it to introduce breaking changes to the training process and then disabling it is recommended to reproduce older models.

- Calculated settings override:

  - You can manually override the calculated settings from the previous cell by specifying values for `TrainBS`, `ValBS`, `ValFreq`, `LRDecaySteps`, `PrintFreq`, and `SaveCheckpointFreq`

  - If left blank, the calculated defaults from the previous cell will be used

- The code defines a `take` function to override the calculated settings. If the override is an empty string, it returns the original value; otherwise, it returns the overridden value.

- The code assigns the overridden or default values to the corresponding variables: `train_bs`, `val_bs`, `val_freq`, `lr_decay_steps`, `print_freq`, and `save_checkpoint_freq`.

- Finally, the code prompts you to run the cell after editing the settings.

7. **Applying the the settings**: This section applies the user-defined settings from *step 6* to a fresh YAML configuration file using `sed` commands:

```
%cd /content/DL-Art-School
# !wget https://raw.githubusercontent.com/152334H/DL-Art-
School/master/experiments/EXAMPLE_gpt.yml -O experiments/
EXAMPLE_gpt.yml
!wget https://raw.githubusercontent.com/josuebatista/
DL-Art-School/master/experiments/EXAMPLE_gpt.yml -O
experiments/EXAMPLE_gpt.yml
import os
%cd /content/DL-Art-School
!sed -i 's/batch_size: 128/batch_size: '"$train_bs"'/g' ./
experiments/EXAMPLE_gpt.yml
!sed -i 's/batch_size: 64/batch_size: '"$val_bs"'/g' ./
experiments/EXAMPLE_gpt.yml
!sed -i 's/val_freq: 500/val_freq: '"$val_freq"'/g' ./
experiments/EXAMPLE_gpt.yml
!sed -i 's/500, 1000, 1400, 1800/'"$gen_lr_steps"'/g' ./
experiments/EXAMPLE_gpt.yml
!sed -i 's/print_freq: 100/print_freq: '"$print_freq"'/g' ./
experiments/EXAMPLE_gpt.yml
!sed -i 's/save_checkpoint_freq: 500/save_checkpoint_freq:
'"$save_checkpoint_freq"'/g' ./experiments/EXAMPLE_gpt.yml

!sed -i 's+CHANGEME_validation_dataset_
name+'"$ValidationDataset_Name"'+g' ./experiments/EXAMPLE_gpt.
yml
!sed -i 's+CHANGEME_path_to_validation_
dataset+'"$ValidationDataset_Training_Path"'+g' ./experiments/
EXAMPLE_gpt.yml
if(Fp16==True):
  os.system("sed -i 's+fp16: false+fp16: true+g' ./experiments/
```

```
EXAMPLE_gpt.yml")
!sed -i 's/use_8bit: true/use_8bit: '"$Use8bit"'/g' ./
experiments/EXAMPLE_gpt.yml

!sed -i 's/disable_state_saving: true/disable_state_saving:
'"$SaveTrainingStates"'/g' ./experiments/EXAMPLE_gpt.yml
!sed -i 's/tortoise_compat: True/tortoise_compat:
'"$TortoiseCompat"'/g' ./experiments/EXAMPLE_gpt.yml
!sed -i 's/number_of_checkpoints_to_save: 0/number_of_
checkpoints_to_save: '"$Keep_Last_N_Checkpoints"'/g' ./
experiments/EXAMPLE_gpt.yml

!sed -i 's/CHANGEME_training_dataset_name/'"$Dataset_Training_
Name"'/g' ./experiments/EXAMPLE_gpt.yml
!sed -i 's/CHANGEME_your_experiment_name/'"$Experiment_Name"'/g'
./experiments/EXAMPLE_gpt.yml
!sed -i 's+CHANGEME_path_to_training_dataset+'"$Dataset_
Training_Path"'+g' ./experiments/EXAMPLE_gpt.yml

if (not TrainingRate=="1e-5"):
  os.system("sed -i 's+!!float 1e-5 # CHANGEME:+!!float '" +
TrainingRate + "' #+g' ./experiments/EXAMPLE_gpt.yml")
```

Let's break down the purpose and steps in this section:

- The code changes the current directory to /content/DL-Art-School using the %cd magic command

- It downloads a fresh YAML configuration file named EXAMPLE_gpt.yml from the GitHub repository, 152334H/DL-Art-School, using the wget command and saves it in the experiments directory

- The code then uses a series of sed commands to modify the values in the EXAMPLE_gpt. yml file based on the user-defined settings:

  - It replaces the batch_size values for training and validation with the values stored in the $train_bs and $val_bs variables, respectively

  - It updates the val_freq value with the value stored in $val_freq

  - It replaces the learning rate decay steps with the values stored in $gen_lr_steps

  - It updates the print_freq and save_checkpoint_freq values with the corresponding values stored in the $print_freq and $save_checkpoint_freq variables

- The code replaces placeholders in the YAML file with the user-defined values

- Finally, if TrainingRate is not equal to the default value of 1e-5, the code uses sed to replace the placeholder of CHANGEME: with the user-defined TrainingRate value in the YAML file

The training process can be customized according to your requirements by modifying the YAML file with the user-specified values. This ensures the training process is configured based on your preferences and dataset specifications.

8.   **Training**: The code starts training by running the `train.py` script with the configured YAML file. Press the stop button for this cell when you are satisfied with the results and have seen the following output:

```
INFO:base:Saving models and training states
```

*Figure 9.7* shows an example of the output and the `60_gtp.pht` checkpoint as it looks in Google Colab's **Files** interface:

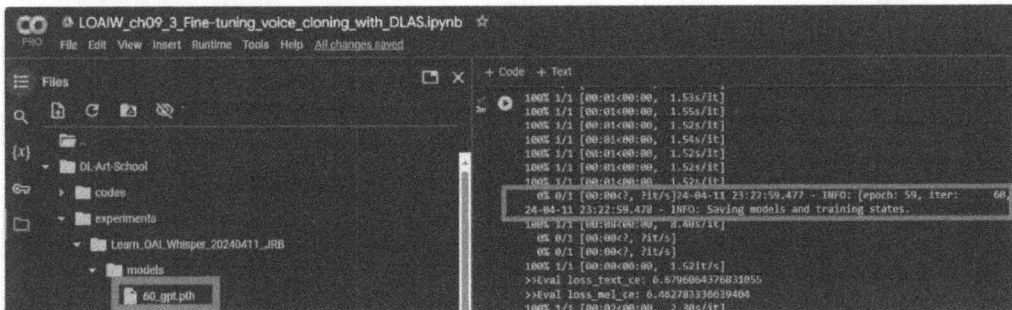

Figure 9.7 – Example of the checkpoint file at the 60-epoch mark

If your training run saves many models, you might exceed the storage limits on the Google Colab runtime. To prevent this, try to delete old checkpoints in `/content/DL-Art-School/ experiments/$Experiment_Name/(models|training_state)/` via the file explorer panel as the training runs. Resuming training after a crash requires config editing, so try not to let that happen:

```
%cd /content/DL-Art-School/codes
!python3 train.py -opt ../experiments/EXAMPLE_gpt.yml
```

9.   **Exporting to Google Drive**: After training, the code allows you to copy the `experiments` folder to Google Drive for persistence:

```
!cp -r /content/DL-Art-School/experiments/$Experiment_Name /
content/gdrive/MyDrive/
```

After running the cell, go to your Google Drive using a web browser; you will see a directory with the same name as the value of the `Experiment_Name` variable:

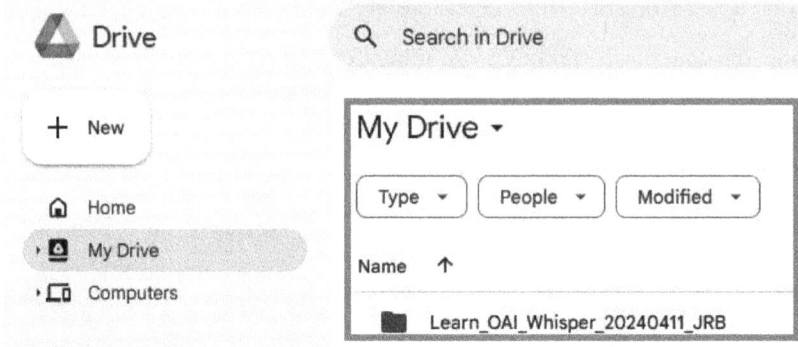

Figure 9.8 – Example of the folder in Google Drive containing DLAS checkpoint files

You will find the model checkpoints in the `<Experiment_Name>/models` folder – that is, the files with the `.pth` extension:

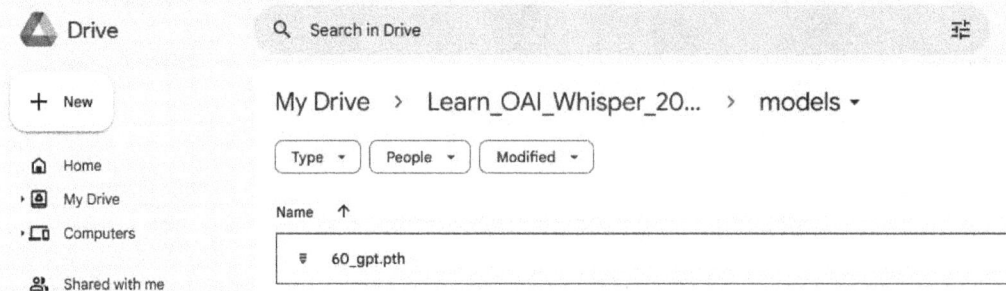

Figure 9.9 – Example of the DLAS checkpoint file. In the next step, you will
need a checkpoint file to synthesize a fine-tuned PVS model

This concludes our overview of the PVS fine-tuning process using the DLAS toolkit. That `.pth` file is the fine-tuned PVS model we just created with DLAS. In the next step, we will use that file to synthesize the voice using TorToiSe-TTS-Fast.

**Fine-tuning PVS models – Hyperparameters versus dataset size**

When fine-tuning PVS models, it's essential to consider the relationship between hyperparameters and dataset size. The Google Colab training notebook automatically suggests appropriate hyperparameters based on the provided dataset size to ensure optimal performance and results.

One key aspect to remember is that the number of steps per epoch (`epoch_steps`) is calculated as `dataset_size // batch_size`. The trainer discards partial batches, so selecting a batch size that evenly divides the dataset is crucial.

If your dataset is relatively small (50-500 samples), consider making the following adjustments to the hyperparameters:

- **Reduce the batch size**: To minimize discarded samples, choose a batch size that is a clean divisor of your dataset size.

- **Set the learning rate decay** (`gen_lr_steps`): Decay the learning rate faster for smaller datasets. Aim for no more than 10 epochs before the first decay – that is, the first value in `gen_lr_steps` should be less than `epoch_steps * 10`. Experiments show that the loss may increase if there is no decay by epoch 20.

- **Adjust the validation and checkpoint frequencies**: Adjust `print_freq`, `val_freq`, and `save_checkpoint_freq` based on the dataset's size. A recommended setting is `val_freq == save_checkpoint_freq == print_freq*3; print_freq == min(epoch_steps,100)`.

By carefully tuning these hyperparameters according to your dataset's size, you can optimize the fine-tuning process and achieve better results with your PVS models.

Having successfully fine-tuned our PVS model using the DLAS toolkit, we are now ready to test our personalized voice by synthesizing speech that captures the essence of our target voice. In the next section, we will explore generating realistic and expressive speech using our fine-tuned model, bringing the synthesized voice to life and unlocking the potential for a wide range of exciting applications.

# PVS step 3 – Synthesizing speech using a fine-tuned PVS model

Synthesizing speech using a fine-tuned PVS model is the culmination of the voice synthesizing process, where the personalized voice is brought to life. It is the stage where the fine-tuned model is tested, generating realistic and natural-sounding speech. The ability to synthesize speech using a fine-tuned PVS model opens up various applications, from creating virtual assistants and audiobook narration to personalized voice interfaces.

Several key components and considerations come into play when embarking on the journey of speech synthesis. Firstly, it is essential to have a suitable computing environment that can handle the computational demands of speech synthesis. This often involves leveraging the power of GPUs, particularly NVIDIA GPUs, which can significantly accelerate the synthesis process. Checking the availability and compatibility of the GPU is crucial to ensure smooth and efficient speech generation.

In addition to the hardware requirements, the speech synthesis process relies on a robust software stack. The TorToiSe-TTS-Fast project, a high-performance TTS system, emerges as a powerful tool. To utilize TorToiSe-TTS-Fast, it is necessary to clone the project repository and install the required dependencies, ensuring that all the necessary libraries and packages are available.

Loading the fine-tuned PVS model is a critical step in speech synthesis. During the fine-tuning phase, we stored the model in **Google Drive** as a checkpoint file or a serialized model object. The location and format of the fine-tuned model may vary depending on the specific locations you used during the fine-tuning process.

With the fine-tuned PVS model loaded, the next step is to prepare the text input that will be synthesized into speech. Depending on the desired output, this text input can be a single sentence, a paragraph, or a script. It is essential to ensure that the text input is formatted correctly and free of any errors or inconsistencies that could impact the quality of the synthesized speech.

The speech synthesis process involves feeding the text input into the fine-tuned PVS model and generating the corresponding audio output. This is where the magic happens as the model applies its learned knowledge to convert the text into speech that mimics the voice we used as the foundation. The synthesis process may involve various techniques, such as neural vocoding, to generate high-quality audio waveforms.

Depending on the specific requirements and preferences, various parameters and settings can be adjusted during speech synthesis. These can include factors such as the speaking rate, pitch, volume, and emotional tone of the generated speech. Fine-tuning these parameters allows greater control over the final output and can help achieve the desired expressiveness and naturalness in the synthesized speech.

Once the speech synthesis process is complete, the generated audio can be saved to a file format suitable for playback, such as WAV or MP3. This allows us to integrate synthesized speech into various applications and platforms easily. Consider the desired audio quality and compatibility when choosing the output file format – that is, `"ultra_fast"`, `"fast"` (default), `"standard"`, or `"high_quality"`.

Finally, evaluating the quality and naturalness of the synthesized speech is a crucial step in the process. This can involve subjective assessments, such as listening tests conducted by human evaluators, and objective metrics that measure various aspects of the generated speech, such as intelligibility, naturalness, and similarity to the target voice. Iterative refinement and fine-tuning based on the evaluation results can help improve the overall quality of the synthesized speech.

Researchers and developers can unlock the potential of personalized speech generation by understanding the components, considerations, and steps involved in synthesizing speech using a fine-tuned PVS model. The ability to create realistic and expressive speech that captures the essence of a voice opens exciting possibilities in various domains, from entertainment and education to accessibility and beyond.

With the right tools, techniques, and attention to detail, the speech synthesis process using fine-tuned PVS models can be a powerful and transformative technology in speech and audio processing.

Please find and open the Colab notebook called `LOAIW_ch09_4_Synthesizing_speech_using_fine-tuned_PVS_models.ipynb` (https://github.com/PacktPublishing/Learn-OpenAI-Whisper/blob/main/Chapter09/LOAIW_ch09_4_Synthesizing_speech_using_fine-tuned_PVS_models.ipynb).

This notebook demonstrates how to check the GPU and RAM, install the necessary libraries, load a fine-tuned PVS model, synthesize speech using the model, and play the generated audio. The code relies on the tortoise-TTS-Fast project to achieve high-performance voice synthesis. Let's walk through the code in the `LOAIW_ch09_4_Synthetizing_speech_using_fine-tuned_PVS_models.ipynb` file, with explanations and code samples:

1. **Checking the NVIDIA GPU**: The notebook starts by checking if an NVIDIA GPU is available using the `nvidia-smi` command. It prints the GPU information if connected. Otherwise, it indicates that no GPU is connected:

```
gpu_info = !nvidia-smi
gpu_info = '\n'.join(gpu_info)
if gpu_info.find('failed') >= 0:
  print('Not connected to a GPU')
else:
  print(gpu_info)
```

2. **Checking the virtual memory**: Next, we check the available RAM using the `psutil` library. It prints the amount of available RAM in GB and indicates if a high-RAM runtime is being used:

```
from psutil import virtual_memory
ram_gb = virtual_memory().total / 1e9
print('Your runtime has {:.1f} gigabytes of available RAM\n'.
format(ram_gb))

if ram_gb < 20:
  print('Not using a high-RAM runtime')
else:
  print('You are using a high-RAM runtime!')
```

3. **Cloning and installing tortoise-tts-fast**: This section of the notebook clones the `tortoise-tts-fast` repository from GitHub and installs the required dependencies using `pip3`:

```
!git clone https://github.com/152334H/tortoise-tts-fast
%cd tortoise-tts-fast
!pip3 install -r requirements.txt --no-deps
!pip3 install -e .
```

4. **Installing additional supporting libraries**: Next, we install additional libraries, such as `transformers`, `voicefixer`, and `BigVGAN`, using `pip3`:

```
!pip3 install transformers==4.29.2
!pip3 uninstall voicefixer
!pip3 install voicefixer==0.1.2
!pip3 install git+https://github.com/152334H/BigVGAN.git
```

5. **Mounting Google Drive**: We must mount Google Drive to load the fine-tuned PVS model we created earlier:

```
from google.colab import drive
drive.mount('/content/gdrive')
```

6. **Loading a fine-tuned PVS voice model**: Here, we set the path to the fine-tuned PVS model (`gpt_path`) and the text to be synthesized (`text`):

```
gpt_path = '/content/gdrive/MyDrive/<filepath/ filename_gpt.pth'
text = "Benny, bring me everyone. EVERYONE!"
```

In Google Colab, use the **Files** interface to navigate and find the checkpoint you created in the previous notebook using DLAS. *Figure 9.10* shows an example of a `.pth` checkpoint file in Google Colab's **Files** interface. Right-click on that checkpoint file; you will be presented with a menu option to copy the location of such a file using `Copy Path`:

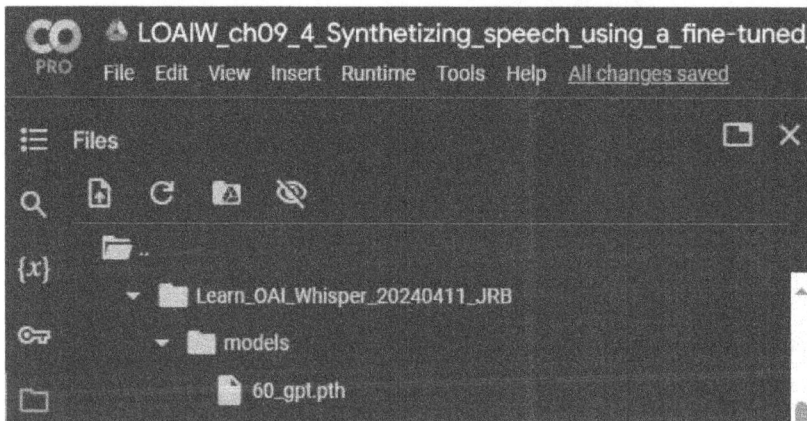

Figure 9.10 – Example of a checkpoint file created by DLAS in Google Colab

7. **Running tortoise_tts.py:** We are now ready to run the `tortoise_tts.py` script with the specified arguments, including the `--preset` option for inference speed, the `--ar_checkpoint` option for the fine-tuned model path, the `-o` option for an output filename, and the text to be synthesized:

```
!python tortoise_tts.py --preset fast --ar_checkpoint $gpt_path
-o "152.wav" $text
```

8. **Playing the synthesized audio:** Finally, we search in Google Colab's **Files** interface for the `tortoise-tts-fast/scripts/results/` directory. You will find the generated audio from the voice synthesis model in that directory. We use `IPython` to display and play the synthesized audio file.

```
import IPython
IPython.display.Audio('/content/tortoise-tts-fast/scripts/
results/random_00_00.wav')
```

*Figure 9.11* shows an example of the directory structure and files created by TorToiSe-TTS-Fast:

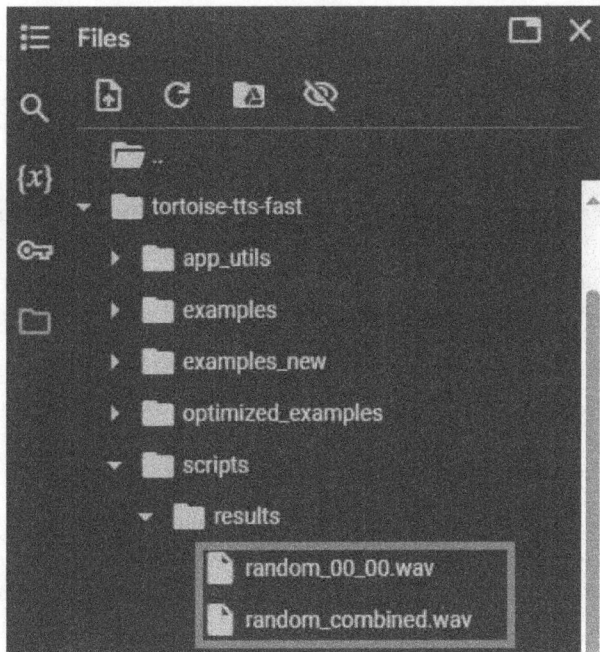

Figure 9.11 – Example of audio files created with TorToiSe-TTS-
Fast using a fine-tuned PVS model's checkpoint .pth file

I encourage you to examine and run all the notebooks in this chapter. They provide an end-to-end practical understanding of leveraging fine-tuned PVS models for high-quality, efficient TTS synthesis. This knowledge will enable you to create PVS applications and explore the potential of personalized speech generation using the TorToiSe-TTS-Fast project in conjunction with the techniques you've learned throughout this book.

## Summary

In this chapter, we explored PVS using OpenAI's Whisper. We discovered how to harness its power to create customized voice models that capture the unique characteristics of a voice or entirely new voices, opening a wide range of exciting applications.

We began by exploring the fundamentals of TTS in voice synthesis, gaining insights into the role of neural networks, audio processing, and voice synthesis. We learned how to convert audio files into the LJSpeech format, a standardized dataset structure commonly used in TTS tasks, using the OZEN Toolkit and Whisper. This hands-on experience provided a solid foundation for the subsequent steps in the voice synthesizing process.

Next, we delved into the DLAS toolkit, a robust framework for fine-tuning PVS models. We learned how to set up the training environment, prepare the dataset, and configure the model architecture. By leveraging Whisper's accurate transcriptions, we aligned audio segments with their corresponding text, creating a dataset suitable for training personalized PVS models.

Through practical examples and code snippets, we gained hands-on experience fine-tuning a pre-trained PVS model using our LJSpeech dataset. We discovered how to customize the training process, select appropriate hyperparameters, and evaluate the model's performance. This experience gave us the knowledge and skills to create high-quality personalized PVS models.

Finally, we tested our fine-tuned PVS model by synthesizing realistic and expressive speech. We learned how to generate natural-sounding speech by providing text input to the model, bringing our synthesized voice to life. The ability to create personalized speech opened a wide range of applications, from virtual assistants and audiobook narration to personalized voice interfaces.

As we conclude this chapter, we look ahead to *Chapter 10, Shaping the Future with Whisper*. In this final chapter, we will explore the evolving landscape of ASR and Whisper's role in shaping its future. We will delve into upcoming trends, anticipated features, ethical considerations, and the general direction of voice technologies, including advanced voice TTS fine-tuning techniques. This forward-looking perspective will provide us with the knowledge and foresight to prepare for and adapt to the future of ASR and voice technology.

# 10

# Shaping the Future with Whisper

Welcome to this book's final chapter, where we will delve into the future of ASR and its thrilling possibilities. This chapter will explore the advancements and trends shaping the ASR landscape and illuminate the potential impact of OpenAI's groundbreaking Whisper model, sparking excitement for what lies ahead.

We'll begin by exploring the ongoing efforts to enhance Whisper's accuracy and robustness. This includes techniques such as increasing training data, domain-specific fine-tuning, and model architecture optimization. These theoretical efforts have practical implications for Whisper's performance across diverse languages, accents, and acoustic conditions, which we will delve into in this chapter.

Furthermore, this chapter will underscore the crucial role of ethical considerations and responsible AI practices in developing and deploying ASR technologies. We will explore strategies for ensuring fairness, mitigating bias, protecting user privacy, and establishing guidelines for using Whisper and other ASR systems responsibly.

Finally, we will look ahead to the future of ASR and the evolving voice technology landscape. You will learn about emerging architectures, training techniques, and the potential of multimodal interfaces and textless NLP in revolutionizing how we interact with spoken language.

In this chapter, we will cover the following topics:

- Anticipating future trends, features, and enhancements in Whisper
- Considering the ethical implications of ASR technologies
- Preparing for the evolving ASR and voice technologies landscape

By the end of this chapter, as an ASR professional, researcher, and enthusiast, you will have a comprehensive understanding of the cutting-edge advancements and future directions in ASR. This will empower you to leverage Whisper and other state-of-the-art models to build innovative, inclusive, and responsible voice-enabled applications. Get ready to shape the future of ASR and unlock new possibilities in the exciting world of voice technology.

# Anticipating future trends, features, and enhancements

This section will explore the ongoing efforts to improve OpenAI Whisper's accuracy, robustness, and performance. We will discuss techniques such as increasing training data, leveraging domain-specific fine-tuning, optimizing model architecture, and implementing strategies to address bias and fairness challenges. These advancements enhance Whisper's capabilities, making it an even more powerful tool for various ASR applications.

## Improving accuracy and robustness

OpenAI Whisper has already demonstrated impressive capabilities in transcribing and translating speech across multiple languages. However, there is always room for improvement in accuracy, robustness, and efficiency. This section will explore the key areas where Whisper's performance can be enhanced, including optimizing model architecture and inference processes to deliver even more precise and reliable results.

### Optimizing model architecture and inference

Building upon the success of Whisper's encoder-decoder transformer model, researchers are exploring novel architectures and techniques that can potentially enhance the model's performance while reducing computational costs.

One notable example of architectural optimization is the work done by Hugging Face, which has achieved up to 40% speed improvements in Whisper's inference code through two key technical enhancements:

First, they integrated native **scaled dot product attention (SDPA)** into Whisper's architecture. SDPA is a mechanism in transformer-based neural networks such as Whisper to process speech inputs. By optimizing how Whisper handles the computational load of attention mechanisms through native SDPA integration, Hugging Face enabled the model to process speech data more efficiently without compromising accuracy.

Second, Hugging Face switched to using the highly GPU-optimized Torch backend for computing the **short-term Fourier transform (STFT)**, a crucial component in preprocessing audio signals for speech recognition. Leveraging the Torch backend allows faster computation of STFT on GPUs, further contributing to the overall speed boost.

The impact of these optimizations is reflected in a significant reduction of Whisper's **real-time factor (RTF)**, a measure of speech processing speed relative to real time. The `large-v3` model's RTF decreased from 10.3 to 7.45, while the `distil-v2` model's RTF improved from 4.93 to 2.08. By streamlining the computational processes and leveraging more efficient algorithms, these optimizations boost Whisper's performance and make it more accessible and cost-effective to deploy at scale.

Applying quantization techniques, which reduce the precision of model weights and activations, can further enhance Whisper's efficiency without compromising accuracy. By intelligently compressing the model's parameters, quantization allows for faster inference times and reduced memory footprint, enabling Whisper to be deployed on a broader range of devices and platforms.

### Enhancing punctuation, diarization, and non-speech detection

More advanced models based on Whisper's foundation are being developed to deliver more precise and reliable punctuation predictions. These models aim to generate transcripts that closely mirror human-like sentence structures and readability by incorporating contextual understanding, linguistic rules, and techniques such as part-of-speech tagging, dependency parsing, and language modeling. As a result, these advanced models can better understand the syntactic and semantic structure of the transcribed text, enabling them to predict more accurate and contextually relevant punctuation, which is essential for creating easily understandable transcripts and helpful for downstream applications.

Additionally, ongoing research focuses on improving Whisper's speaker diarization capabilities, enabling it to accurately identify and differentiate between multiple speakers within a single audio recording. This advancement is particularly valuable for transcribing meetings, interviews, and multiparticipant conversations, where attributing speech to the correct speaker is crucial for clarity and context.

Enhancing Whisper's detection and handling of non-speech sounds, such as laughter, music, or background noise, contributes to the overall robustness of the generated transcripts. By accurately identifying and labeling these non-speech elements, Whisper can provide a more comprehensive and nuanced representation of the audio content, making the transcripts more useful for analysis and interpretation.

### Addressing bias and fairness challenges

As we strive to improve Whisper's accuracy and robustness, it is crucial to address the potential bias and fairness challenges that may arise. To ensure that Whisper's performance improvements benefit all users equally, it is essential to prioritize the collection of diverse and representative training datasets.

This involves actively seeking and including speech data from various demographic groups, accents, and linguistic backgrounds. By incorporating diverse voices and speech patterns into the training data, Whisper can learn to recognize and transcribe speech more accurately across different populations.

Data augmentation, stratified sampling, and ensemble methods can mitigate bias and promote fairness in Whisper's predictions:

- Data augmentation involves creating additional training examples by applying various transformations to the existing data, such as pitch shifting, speed alteration, or adding background noise. This helps to increase the diversity and robustness of the training data.

- Stratified sampling ensures that the training data is representative of different demographic groups and linguistic variations. By carefully balancing the composition of the training set, Whisper can learn to perform equitably across different subpopulations.

- Ensemble methods involve combining multiple models trained on different subsets of the data or using other architectures. By aggregating the predictions of numerous models, ensemble methods can help mitigate individual model biases and improve overall accuracy and fairness.

We can develop a highly accurate, robust, inclusive, and equitable ASR system by proactively addressing these bias and fairness considerations. This is essential for ensuring that the benefits of Whisper's advancements are accessible to all users, regardless of their background or linguistic characteristics.

While Whisper has made significant strides in multilingual ASR, there is still substantial potential for expanding its language support. Increasing the model's ability to accurately transcribe and translate a broader range of languages is crucial for making ASR technology more inclusive and accessible globally.

## Expanding language support in OpenAI Whisper

OpenAI Whisper has already demonstrated impressive multilingual capabilities, supporting transcription and translation in various languages. However, there is still significant room for improvement and expansion, particularly in low-resource and underrepresented languages.

### Increasing training data for low-resource languages

One of the primary factors influencing Whisper's accuracy and performance in a given language is the availability of training data. As noted in the research paper *Robust Speech Recognition via Large-Scale Weak Supervision*, known as the *Whisper paper* (`https://cdn.openai.com/papers/whisper.pdf`), the model's performance is strongly correlated with the volume of training data. About a third of Whisper's training data is non-English, with most languages having less than 1,000 hours of data compared to English. This disparity limits the model's potential in low-resource languages.

Over the past year, a targeted effort to increase training data for low-resource languages is crucial to address the disparity in performance and improve Whisper's accuracy across a broader range of languages. By actively collecting and curating diverse speech data from these languages, Whisper can learn and adapt to their unique linguistic patterns, phonemes, grammatical structures, accents, and dialects. This expansion of training data will boost the model's performance, making it more robust, equitable, and inclusive, ensuring that the benefits of ASR technology are accessible to a broader range of language communities.

### Leveraging transfer learning and self-supervised pre-training

In addition to increasing training data, advanced techniques such as transfer learning and self-supervised pre-training can significantly boost Whisper's performance in low-resource languages. Transfer learning involves leveraging knowledge gained from high-resource languages to improve the model's understanding of low-resource languages. By identifying shared linguistic features and patterns, Whisper can learn and adapt to new languages more efficiently with limited supervised data.

On the other hand, self-supervised pre-training enables Whisper to learn from vast amounts of unlabeled speech data. By exposing the model to diverse speech samples without explicit transcriptions, it can discover inherent structures and representations that are transferable across languages. This unsupervised learning approach is precious for low-resource languages, where labeled data is scarce, as it allows Whisper to capture language-agnostic features and improve its overall language understanding.

## Developing more inclusive and diverse datasets

To ensure that Whisper's language expansion is broad but also unbiased and robust, it is essential to prioritize the development of inclusive and diverse datasets. This involves actively seeking and incorporating speech data from various sources, accents, and domains. By training in diverse voices and linguistic variations, Whisper can learn to recognize and transcribe speech more accurately across different demographics and regional dialects.

Initiatives such as Mozilla's Common Voice project are vital in democratizing access to speech technology by crowdsourcing multilingual speech datasets. By engaging language communities and volunteers worldwide, these efforts aim to create large-scale, open source datasets representing human speech's rich diversity. Incorporating such datasets into Whisper's training pipeline can significantly improve its language coverage and fairness.

## Optimizing multilingual model architectures

Optimizing its model architecture becomes increasingly crucial as Whisper scales to support hundreds or thousands of languages. Researchers are exploring various techniques to improve the efficiency and performance of multilingual models, such as parameter-efficient architectures, language-specific components, and enhanced cross-lingual representations.

One promising approach is using adapter modules, lightweight neural networks that can be plugged into pre-trained models to specialize them for specific languages or tasks. By fine-tuning these adapters while keeping the core model parameters fixed, Whisper can more efficiently adapt to new languages without extensive retraining. This modular approach enables faster and more scalable language expansion.

Another area of research focuses on developing language-specific components, such as language embeddings or language-dependent attention mechanisms. These components allow Whisper to capture each language's unique characteristics and nuances more effectively, improving recognition accuracy and linguistic understanding.

## Enhancing language identification and code-switching

In real-world multilingual environments, identifying spoken language and handling code-switching (mixing multiple languages in speech) are critical challenges. To make Whisper more practical and valuable in these scenarios, ongoing research aims to enhance its language identification capabilities and improve its robustness to code-switching.

Advanced language identification techniques that leverage phonetic and prosodic features can enable Whisper to accurately determine an utterance's language, even for short speech segments. By leveraging a combination of acoustic and linguistic cues, Whisper can more reliably detect and switch between languages on the fly, ensuring seamless transcription and translation in multilingual conversations.

Moreover, developing specialized models and training strategies for code-switched speech can significantly improve Whisper's performance in handling language mixing. By exposing the model to a diverse range of code-switched examples and incorporating language-specific constraints and transition probabilities, Whisper can learn to navigate the complexities of multilingual speech more effectively.

### The importance of expanding language support

Expanding Whisper's language support is a technical challenge and a crucial step toward making ASR technology more inclusive and accessible worldwide. With over 7,000 languages spoken globally, most of which are currently underserved by existing speech technologies, Whisper has the potential to bridge the language divide. By supporting a more comprehensive range of languages, Whisper can enable greater access to information, services, and communication for multilingual populations, empowering essential applications in education, healthcare, government services, and entertainment. Moreover, Whisper's language expansion efforts can facilitate cross-lingual communication and break language barriers, fostering more natural and seamless interactions between individuals from different linguistic backgrounds and promoting cultural exchange, collaboration, and understanding on a global scale. Furthermore, a multilingual Whisper model is a foundation for other advanced speech AI tasks, such as language understanding, dialogue systems, and speech-to-speech translation, driving innovation and progress in natural language processing.

Speech interfaces powered by an expanded Whisper model can improve accessibility for individuals with limited literacy or those who prefer to interact with technology using their native language. This inclusivity ensures that the benefits of ASR are not limited to speakers of high-resource languages but are available to communities across the globe. By providing high-quality ASR capabilities in a wide range of languages, Whisper enables the development of more sophisticated and inclusive speech-based applications, unlocking the full potential of speech technology for users worldwide.

Generating accurate and readable transcripts involves more than just converting speech to text. As Whisper expands its language support and improves its ability to handle diverse linguistic structures, the importance of accurate punctuation, formatting, and speaker diarization becomes even more critical. These elements play crucial roles in creating human-readable outputs that capture the nuances of spoken language across various languages and cultural contexts. By enhancing Whisper's capabilities in these areas, we can ensure that the benefits of expanded language support are fully realized, enabling the creation of high-quality, accessible transcripts for a broader range of users worldwide.

## Achieving better punctuation, formatting, and speaker diarization in OpenAI Whisper

OpenAI Whisper has demonstrated remarkable capabilities in transcribing speech across multiple languages. While the current version of Whisper already shows competence in punctuation, speaker diarization, and non-speech detection, further enhancements are needed to bridge the gap between raw speech recognition output and human-readable transcripts. Enhancing these capabilities is crucial for several reasons. First, it dramatically improves the readability and usability of transcripts as punctuation conveys the structure, intonation, and intended meaning of spoken words, making the transcript easier for human readers to follow and comprehend. Second, accurate speaker diarization clearly explains who said what in a conversation, providing essential context for the transcribed content. Finally, detecting and properly handling non-speech elements, such as laughter, silence, or background noise, contributes to a more comprehensive and nuanced audio representation. This section will explore the fundamental techniques and approaches developed to refine Whisper's ability to handle these crucial aspects and improve its accuracy and robustness.

## Enhancing punctuation and capitalization

Accurate punctuation and capitalization make transcripts more readable, understandable, and valuable for downstream tasks. One approach to achieve this is by integrating separate punctuation and capitalization models as a post-processing step after Whisper has generated the initial transcript. These models can restore punctuation marks such as periods, commas, question marks, and exclamation points, capitalize proper nouns, and begin sentences.

However, it is essential to note that models based solely on text may sometimes produce less optimal outputs than punctuation derived directly from the audio. Future research aims to develop more sophisticated methods that leverage textual and acoustic cues to achieve more accurate and contextually relevant punctuation.

## Improving timestamp accuracy for better alignment

Whisper outputs timestamps for each transcribed segment, indicating the corresponding audio snippet. However, these timestamps may have a lead or lag of a few seconds, negatively impacting speaker diarization. Tools such as WhisperX and stable-ts are being developed to force-align the transcription with the audio using phoneme-based models such as wav2vec2.0 to address this issue. By achieving more precise word-level timestamps, these techniques enable more accurate mapping of speaker labels to the transcribed segments.

## Enhancing speaker embedding and clustering

Speaker diarization involves identifying and attributing speech segments to individual speakers within a conversation. After obtaining precise timestamps, the next step is to map each segment to a speaker label. This process typically involves extracting speaker embeddings using a model such as TitaNet and then clustering the embeddings to group segments from the same speaker.

Ongoing research focuses on improving the speaker embedding models and clustering algorithms to achieve lower diarization error rates. By leveraging more robust and discriminative speaker representations, Whisper can more accurately distinguish between speakers and assign the correct labels to each segment.

## Handling speaker overlap and interruptions

Overlapping speech from multiple speakers remains a significant challenge for diarization systems. When various speakers talk simultaneously or interrupt each other, it becomes difficult to accurately attribute the speech to the correct individuals. Future research aims to develop advanced methods to detect interruption points and handle speaker overlap more effectively.

One promising approach is to employ source separation techniques to isolate individual speaker audio streams. By separating the overlapping speech into distinct channels, Whisper can more easily identify and transcribe each speaker's contributions. This capability will be precious in real-world scenarios where conversations often involve dynamic turn-taking and interruptions.

### Leveraging punctuation for speaker change detection

Punctuation marks predicted by Whisper can provide valuable cues for refining speaker diarization. Since speaker changes often occur at natural pauses or sentence boundaries, the presence of punctuation can indicate potential transition points between speakers. Minor errors can be corrected by realigning the diarization output based on punctuation, resulting in a more coherent and accurate speaker segmentation.

Firstly, accurate speaker diarization enables a more precise mapping of speech to individual speakers, which is essential for downstream tasks such as sentiment analysis, conversation summarization, and speaker-attributed machine translation. By correctly identifying who said what, Whisper can provide a richer and more contextualized representation of the conversation.

Moreover, improved punctuation and formatting facilitate the segmentation of transcripts into coherent sentences and paragraphs. This semantic structuring enhances the performance of various NLP models, such as those used for information extraction, named entity recognition, and machine translation.

Lastly, transcripts with accurate punctuation, capitalization, and speaker labels are more amenable to automated processing and analysis. By reducing the need for manual post-editing, Whisper can save significant time and effort, enabling faster and more efficient utilization of speech data across various applications.

As OpenAI Whisper continues to evolve, the integration of advanced techniques for punctuation restoration, timestamp alignment, speaker embedding, and overlap handling will significantly elevate the quality and usability of its transcripts. Combining these enhancements, Whisper aims to generate transcripts that capture conversational speech's full nuance and dynamics, bridging the gap between raw speech recognition output and human-readable documents.

Accurate punctuation, formatting, and speaker diarization cannot be overstated. These elements are essential for unlocking the true potential of speech data, enabling more sophisticated analysis, summarization, and translation tasks. As Whisper pushes the boundaries of what's possible in these areas, it will empower various applications, from virtual assistants and customer service to media analysis and educational content creation.

Ongoing research and development efforts in punctuation modeling, speaker diarization, and related fields hold immense promise for Whisper's future and the broader ASR landscape. By staying at the forefront of these advancements, Whisper is poised to revolutionize the way we interact with and utilize speech data, making it more accessible, actionable, and valuable than ever before.

As the demand for instant speech recognition grows across various applications, improving the speed and real-time performance of OpenAI Whisper has become a critical focus area. Accelerating Whisper's inference and enabling smooth, low-latency speech-to-text conversion will open up new possibilities for interactive and time-sensitive use cases.

# Accelerating performance and enabling real-time capabilities in OpenAI Whisper

As the demand for instant speech recognition grows across various applications, improving the speed and real-time performance of OpenAI Whisper has become a critical focus area. This section will explore the essential techniques and optimizations being developed to accelerate Whisper's inference and enable smooth, low-latency speech-to-text conversion.

## Enabling streaming inference for real-time transcription

Whisper must support streaming inference to achieve accurate real-time speech recognition, a crucial emerging capability. In streaming inference, the model processes audio chunks as they arrive instead of waiting for the entire audio file. This generates transcripts with minimal latency, keeping pace with the speaker. Optimizing chunk size, sampling rate, and buffer management ensures smooth, uninterrupted real-time performance. By carefully balancing these parameters, Whisper can strike an optimal balance between responsiveness and accuracy, significantly improving its efficiency and real-time capabilities.

## Reducing model size through compression techniques

Another effective strategy for accelerating Whisper's inference speed is compressing the model size using quantization, distillation, and parameter sharing. Quantization reduces the precision of model weights and activations, while distillation involves training a smaller, more efficient *student* model to mimic the behavior of the more significant *teacher* model. Parameter sharing, however, aims to identify and exploit redundancies within the model architecture. By reducing the model's memory footprint and computational complexity, these compression techniques can significantly speed up inference, making deployment on resource-constrained edge devices more feasible for real-time use cases.

## Leveraging hardware acceleration

Harnessing the power of specialized hardware such as GPUs, TPUs, and dedicated AI accelerators can significantly speed up the computation of Whisper's neural networks. Substant performance gains can be achieved by optimizing the model for specific hardware architectures and using frameworks such as NVIDIA TensorRT or Intel OpenVINO. These hardware acceleration technologies leverage techniques such as parallel processing, mixed-precision arithmetic, and custom instruction sets to maximize throughput and minimize latency. Deploying Whisper on high-performance computing infrastructure can enable faster-than-real-time transcription and quick processing of large volumes of audio data.

## Improving accuracy and robustness in challenging conditions

While speed is crucial for real-time speech recognition, it should not come at the cost of accuracy and robustness. Enhancing Whisper's performance in challenging real-world conditions, such as noisy environments, diverse accents, and spontaneous speech, is essential for reliable real-time operation.

Techniques such as spectral augmentation, which introduces random perturbations to the audio spectrogram during training, can improve the model's resilience to noise. Speaker adaptation methods, such as fine-tuning the model on a small amount of speaker-specific data, can help Whisper better handle individual variations in speech patterns. Furthermore, incorporating contextual language models that leverage the surrounding text to refine predictions can boost accuracy in complex linguistic scenarios.

Speeding up Whisper's performance and enabling real-time capabilities cannot be overstated. It opens up a wide range of transformative applications across domains. In the realm of accessibility, real-time transcription empowers individuals who are deaf or hard of hearing to participate fully in lectures, meetings, and live events. For conversational AI interfaces such as voice assistants and chatbots, instant speech recognition is essential for natural, human-like interaction. Real-time ASR enables time-sensitive use cases, including live broadcast captioning, emergency response transcription, and simultaneous interpretation, where even a few seconds of delay can be disruptive.

Moreover, faster-than-real-time performance brings significant computational efficiency and scalability benefits. By minimizing the time audio data spends in buffers, large-scale transcription workloads' overall latency and processing costs can be significantly reduced. This is particularly valuable for organizations dealing with massive archives of audio content as it allows for rapid indexing, searching, and analysis.

Looking ahead, the advancements in Whisper's speed and real-time capabilities will pave the way for a new era of ubiquitous, seamless voice interaction. As edge computing becomes more prevalent, the ability to perform real-time speech recognition on devices such as smartphones, smart speakers, and IoT sensors will enable privacy-preserving, offline voice interfaces. This decentralized approach to ASR will unlock novel ambient computing applications where voice commands and conversations can be processed locally without reliance on cloud connectivity.

In conclusion, the ongoing efforts to optimize Whisper's architecture, compress its model size, leverage hardware acceleration, and improve its accuracy in challenging conditions are crucial for achieving real-time, low-latency speech recognition. As these advancements come to fruition, Whisper will become an even more powerful and versatile tool, enabling a wide range of transformative applications across accessibility, conversational AI, live interpretation, and edge computing. The future of voice interaction is where instant, accurate, and ubiquitous speech recognition is the norm, and OpenAI Whisper is at the forefront of making this vision a reality.

The advancements in Whisper's speed and real-time capabilities, combined with its expanded language support, open exciting possibilities for seamless multilingual communication. As Whisper becomes capable of instantly transcribing and translating speech across a wide range of languages, it can break down language barriers and facilitate more natural and efficient cross-lingual interactions. Integrating Whisper with other AI technologies, such as machine translation and natural language understanding, can further enhance its ability to bridge linguistic gaps and enable more sophisticated multilingual applications.

# Enhancing Whisper's integration with other AI systems

As AI continues to advance rapidly, integrating OpenAI Whisper with other cutting-edge AI technologies is emerging as a critical area of focus for future development. By combining Whisper's state-of-the-art speech recognition capabilities with complementary AI models, researchers and developers aim to create more powerful, versatile, and user-friendly applications that can understand and respond to human speech in increasingly natural and contextual ways.

## Combining Whisper with large language models

One of the most promising avenues for integrating Whisper with other AI systems is its combination with large language models, such as GPT-4, Mistral, Claude, and Llama. These advanced language models have demonstrated remarkable natural language understanding, generation, and reasoning abilities, making them ideal partners for Whisper's speech recognition functionality.

Integrating Whisper with these language models makes it possible to create end-to-end speech-to-text-to-action pipelines that can process spoken input, understand its meaning, and generate appropriate responses or actions. For example, Whisper could transcribe a user's speech, which could then be fed into GPT-3 to summarize the content, extract relevant action items, generate coherent replies, or even translate the message into another language. This tight integration between speech recognition and language understanding enables the development of more natural and efficient voice-based interfaces that can truly grasp the user's intent and provide helpful, contextually relevant assistance.

## Enabling multimodal AI applications with Whisper

Another exciting frontier for Whisper's integration with other AI systems lies in multimodal AI applications. By combining Whisper's speech recognition capabilities with computer vision models for image and video understanding, sensor fusion models for robotics, and other specialized AI components, developers can create intelligent systems that simultaneously process and make sense of multiple input modalities.

For instance, a virtual assistant powered by Whisper and computer vision AI could understand and respond to voice commands and interpret visual cues and gestures from the user, enabling a more intuitive and immersive interaction experience. Similarly, an autonomous vehicle equipped with Whisper and sensor fusion AI could process spoken instructions from passengers while simultaneously analyzing real-time visual and spatial data from cameras and LiDAR to navigate safely and efficiently.

Integrating Whisper with multimodal AI technologies opens up a wide range of possibilities for creating intelligent systems that can perceive, understand, and interact with the world in a more human-like manner. By leveraging the strengths of different AI models across various modalities, these systems can provide more comprehensive and contextually aware assistance, leading to enhanced user experiences and improved decision-making in complex real-world scenarios.

### Powering conversational AI with Whisper and dialogue models

Conversational AI is another critical area where Whisper's integration with other AI models holds immense promise. Researchers and developers can create highly sophisticated conversational agents engaging in fluid, contextually relevant, and human-like interactions by combining Whisper's accurate speech-to-text capabilities with advanced dialogue management models, intent recognition systems, and natural-sounding text-to-speech engines.

The integration of Whisper with these conversational AI components enables the development of voice assistants, chatbots, and other conversational interfaces that can not only understand and transcribe user speech with high accuracy but also interpret the underlying intent, maintain coherent dialogue context, and generate natural, dynamically adapted responses. This level of integration allows for more engaging and productive human-machine interactions where users can communicate their needs, queries, and instructions through natural speech and receive intelligent, personalized assistance in return.

### The importance of integrating Whisper with other AI systems

Integrating Whisper with other AI technologies is crucial for realizing the full potential of speech-based interfaces and unlocking a wide range of transformative applications across various domains. By combining the strengths of multiple complementary AI models, developers can create more powerful, natural, and user-friendly systems that can understand, process, and respond to human speech in increasingly sophisticated and contextually relevant ways.

One key benefit of tighter integration between Whisper and other AI models is enabling multimodal AI applications that can process and combine speech, vision, language, and other input modalities to understand and interact with the world more human-likely. This level of multimodal intelligence is essential for developing AI systems that can operate effectively in complex, real-world environments and provide comprehensive, context-aware assistance to users.

Moreover, the integration of Whisper with language models and machine translation systems paves the way for real-time speech-to-speech translation applications that can break down language barriers and facilitate more natural cross-lingual communication. These integrated systems can foster better understanding, collaboration, and cultural exchange in an increasingly globalized world by seamlessly converting spoken words from one language to another.

Another significant advantage of integrating Whisper with other AI models is the potential for efficiency gains in reduced latency and computational costs. By optimizing the interaction and data flow between different AI components, developers can create more streamlined and resource-efficient pipelines to process speech input and generate responses with minimal delay, making powerful AI applications more practically feasible for real-world deployment.

Integrating OpenAI Whisper with other cutting-edge AI technologies is crucial in developing more advanced, natural, and accessible speech-based interfaces. By combining Whisper's state-of-the-art speech recognition capabilities with language models, computer vision, dialogue systems, and other AI components, researchers and developers can create intelligent systems that understand, process, and respond to human speech in increasingly sophisticated and contextually relevant ways. As Whisper continues to evolve and integrate more closely with other AI models, it will play a pivotal role in shaping the future of human-machine interaction, enabling a wide range of transformative applications that can enhance productivity, accessibility, and quality of life for people worldwide.

# Considering ethical implications

As ASR technologies such as OpenAI Whisper become more advanced and widely adopted, addressing the ethical considerations and implications associated with their development and deployment is crucial. This section will delve into critical issues such as ensuring fairness, mitigating bias, protecting user privacy and data security, and establishing guidelines and safeguards for their responsible use. By proactively establishing a framework for responsible ASR deployment, we can harness the benefits of these technologies while mitigating potential risks and negative consequences.

## Ensuring fairness and mitigating bias in ASR

ASR systems such as Whisper may exhibit performance bias against certain types of speech, such as non-native accents, dialects, age groups, and genders. Studies have shown that these biases can result in higher error rates and unequal user experiences, leading to discrimination against underserved populations. Addressing performance disparities across different demographics is one of the primary ethical considerations in ensuring fairness and mitigating bias in ASR, which is critical for providing equitable access and preventing discrimination. Developing inclusive and diverse training datasets is crucial to minimize these performance disparities. By incorporating a wide range of accents, dialects, and demographics in the training data, ASR models can learn to recognize and transcribe speech more accurately across different user groups. Additionally, implementing techniques such as data augmentation and transfer learning can help improve the robustness of ASR models in handling diverse speech patterns.

### Responsible data collection and usage practices

Responsible data collection and usage practices ensure fairness in ASR systems. The models can amplify biases in the training data, leading to skewed performance and perpetuating societal inequities. To address this, it is essential to prioritize diversity, inclusivity, and privacy in data collection processes.

A crucial ethical consideration is obtaining representative speech data across various demographics while respecting user consent and data protection. This involves implementing transparent data collection policies, obtaining informed consent from participants, and ensuring the secure storage and handling of sensitive speech data. By adhering to responsible data practices, ASR developers can build more inclusive and unbiased models that serve the needs of diverse user populations.

## *Promoting transparency and accountability*

Transparency and accountability are crucial ethical considerations in developing and deploying ASR systems such as Whisper. Clear documentation of the development process, training data, and potential limitations of ASR models enables informed decision-making and helps build trust among users.

To promote transparency, it is essential to provide detailed information about the appropriate use cases, performance characteristics across different demographics, and known biases of the ASR system. This allows users and stakeholders to understand the capabilities and limitations of the technology and make informed decisions about its deployment.

Accountability measures, such as regular audits and fairness evaluations, are essential for identifying and mitigating biases in ASR models. These assessments should involve diverse stakeholders, including AI researchers, ethicists, policymakers, and affected communities, to comprehensively understand the system's impact on different user groups.

## *Fostering interdisciplinary collaboration*

Addressing fairness and mitigating bias in ASR requires a collaborative and interdisciplinary approach. Engaging diverse stakeholders, including AI researchers, ethicists, policymakers, and affected communities, helps identify blind spots, understand societal implications, and develop inclusive solutions.

Interdisciplinary research on bias mitigation strategies, such as data augmentation, model selection techniques, and fairness constraints, is crucial for progress in this area. By bringing together experts from various fields, including computer science, linguistics, sociology, and ethics, we can develop a more comprehensive understanding of the challenges and devise practical solutions to ensure fairness in ASR systems.

Ensuring fairness and mitigating bias in ASR systems such as OpenAI Whisper is a multifaceted challenge that requires addressing performance disparities, responsible data practices, transparency and accountability, and interdisciplinary collaboration. By prioritizing these ethical considerations, we can build more inclusive and equitable ASR technologies that benefit all users while upholding fairness and social justice principles.

Another critical ethical consideration in developing and deploying ASR systems is protecting user privacy and data security. As Whisper and other ASR technologies become more prevalent, robust safeguards and best practices for handling sensitive voice data are essential.

# Protecting privacy and data

Training ASR systems such as Whisper requires vast amounts of speech data, which can raise privacy concerns regarding the collection, storage, and use of potentially sensitive audio recordings. Preserving data privacy, obtaining informed consent, and implementing robust data protection measures are critical ethical priorities. Federated learning has been proposed as a privacy-preserving approach to training ASR models.

## *Obtaining informed consent and ensuring transparency*

One of the key ethical considerations in protecting privacy and data when using ASR systems such as OpenAI Whisper is obtaining informed consent from individuals whose voice data is collected. Informed consent involves fully disclosing the extent of data collection, the purposes for which the speech data will be used, and the parties with access to it. This transparency allows users to make informed decisions about whether they are comfortable with their voice data being collected and used by the ASR system.

To ensure transparency, ASR system providers should communicate their data practices in their privacy policies. These policies should be easily accessible and written in plain language, enabling users to understand how their voice data will be handled. By being transparent about data practices and obtaining informed consent, ASR system providers can build trust with their users and demonstrate their commitment to upholding user privacy rights.

## *Implementing robust data security measures*

Protecting the privacy of voice data collected by ASR systems requires implementing robust data security measures. Voice data can be susceptible, potentially revealing personal information, biometric details, and intimate aspects of an individual's life. Therefore, it is crucial to safeguard this data against unauthorized access, misuse, or breaches.

ASR system providers should employ techniques such as data encryption, secure storage, and access controls to ensure the confidentiality and integrity of user voice data. Encryption helps protect data in transit and at rest, making it unreadable to unauthorized parties. Secure storage involves using protected databases and servers with strict access controls, ensuring that only authorized personnel can access the data.

Regular security assessments and updates to data protection practices are essential to address evolving threats and maintain compliance with privacy regulations. This includes conducting vulnerability scans, penetration testing, and security audits to identify and address potential weaknesses in the system's security posture. By implementing robust data security measures, ASR system providers can minimize the risk of data breaches and protect user privacy.

## *Adhering to data minimization and purpose limitation principles*

Responsible data management in ASR systems involves adhering to data minimization and purpose limitations.

Data minimization means collecting and retaining only the minimum voice data necessary for specific, legitimate purposes. ASR systems should be designed to limit data collection to what is essential for functionality and avoid gathering excessive or irrelevant information.

Purpose limitation involves clearly defining the purposes for which voice data is collected and used and ensuring that the data is not used for other purposes without user consent. This helps prevent misuse of voice data and aligns with user expectations. ASR system providers should be transparent about the specific purposes for collecting and using voice data. They should obtain user consent to use the data for additional purposes.

By adhering to data minimization and purpose limitation principles, ASR system providers can demonstrate their commitment to responsible data management and protect user privacy. This approach helps build trust with users and ensures compliance with data protection regulations.

### Empowering users with control over their data

Giving users control over their voice data is critical to protecting privacy in ASR systems. Users should be able to access, review, correct, or delete their data. This empowers users to manage their personal information and ensures they control how their voice data is used.

ASR system providers should implement mechanisms for users to exercise their data rights, such as providing user-friendly interfaces for accessing and managing their data. Users should also be able to opt out of data collection or withdraw their consent at any time. This allows users to make informed decisions about their privacy preferences and gives them the power to control their voice data.

Respecting user rights, such as the right to be forgotten under GDPR, is another important aspect of empowering users. ASR system providers should have processes to honor user requests for data deletion and ensure that voice data is securely erased when no longer needed or requested by the user.

By giving users control over their voice data and respecting their data rights, ASR system providers can demonstrate their commitment to user privacy and build trust with their user base. This approach aligns with ethical principles and legal requirements for protecting personal data.

Protecting privacy and data in the context of ASR systems such as OpenAI Whisper is a critical ethical consideration. Obtaining informed consent, implementing robust data security measures, adhering to data minimization and purpose limitation principles, and empowering users with control over their data are vital strategies for safeguarding user privacy.

By prioritizing these ethical principles, ASR system providers can mitigate privacy risks, comply with data protection regulations, and foster trust in these technologies' responsible development and deployment. Ensuring the privacy and security of user voice data is an ethical obligation and essential for the widespread adoption and beneficial use of ASR systems in various domains.

As ASR technologies continue to advance and become more prevalent, developers, researchers, and organizations must remain vigilant in addressing privacy concerns and implementing best practices for data protection. By doing so, we can harness the power of ASR systems such as Whisper while respecting individual privacy rights and promoting the responsible use of these technologies for the benefit of society.

In addition to addressing fairness and privacy concerns, the responsible development and deployment of ASR systems such as Whisper requires clear guidelines and safeguards. Establishing a framework for the ethical use of these technologies is crucial to prevent misuse, protect vulnerable populations, and ensure transparency and accountability.

# Establishing guidelines and safeguards for responsible use

By proactively establishing a framework for responsible ASR deployment, we can harness the benefits of these technologies while mitigating potential risks and negative consequences.

## Defining appropriate use cases and preventing misuse

One of the primary goals of establishing guidelines for responsible ASR use is to define appropriate use cases and prevent misuse or malicious applications. Clear guidelines should outline the intended purposes for which ASR systems such as Whisper can be employed, such as transcription, language learning, or accessibility. These guidelines should also explicitly prohibit using ASR for unauthorized recording, surveillance, or generating fake audio content.

To prevent misuse, ASR system providers should implement access controls and authentication measures to ensure only authorized users can utilize the technology. This may require user registration, API key authentication, or other forms of access management. Additionally, providers should educate users about the ethical boundaries and responsible use of ASR, emphasizing the importance of obtaining consent and respecting privacy rights.

## Implementing transparency and accountability measures

Transparency and accountability are essential components of responsible ASR use. ASR system providers should be transparent about their technologies' capabilities, limitations, and potential biases. Users should be informed about how their voice data is collected, processed, and protected. This transparency helps users make informed decisions about whether to engage with ASR systems and enables them to understand the implications of their participation.

To ensure accountability, ASR system providers should establish oversight mechanisms and regular audits to verify compliance with ethical principles and legal requirements. This may involve third-party assessments, public reporting, or establishing an ethics board to monitor and review ASR practices. Accountability measures help build trust with users and stakeholders, demonstrating a commitment to responsible and ethical ASR deployment.

## Safeguarding vulnerable populations

Guidelines for responsible ASR use must prioritize the protection of vulnerable populations, such as children, elderly individuals, and those with disabilities. Special considerations should be given to obtaining informed consent from these groups as they may have unique needs or challenges in understanding the implications of ASR technology.

ASR systems should be designed with accessibility and inclusivity in mind, ensuring that individuals with diverse abilities and backgrounds can use them. This may involve incorporating features such as voice commands, text-to-speech output, or language support for non-native speakers. By prioritizing the needs of vulnerable populations, ASR guidelines can help prevent harm, discrimination, and exclusion.

### Promoting a culture of ethical innovation

Finally, establishing guidelines for responsible ASR use should be accompanied by efforts to promote a culture of ethical innovation within the ASR community. This involves encouraging researchers and developers to prioritize ethical considerations throughout the design, development, and deployment processes.

ASR system providers should provide training and resources to help their teams navigate ethical challenges and make responsible decisions. Sharing best practices, case studies, and lessons learned can help build a collective understanding of responsible ASR use in practice. By fostering a culture of ethical innovation, the ASR community can proactively address potential risks and ensure that the technology is developed and used to benefit society.

Establishing guidelines and safeguards for the responsible use of ASR systems such as OpenAI Whisper is critical in ensuring these technologies are deployed ethically and beneficially. By defining appropriate use cases, implementing transparency and accountability measures, safeguarding vulnerable populations, fostering collaboration, and promoting a culture of ethical innovation, the ASR community can mitigate risks and realize the full potential of these powerful tools. As ASR continues to advance and integrate into various aspects of our lives, prioritizing responsible use will be essential for building trust, protecting rights, and driving positive social impact.

# Preparing for the evolving ASR and voice technologies landscape

The field of ASR is rapidly evolving, with new architectures, training techniques, and applications emerging at an unprecedented pace. To stay at the forefront of this dynamic landscape, organizations must adopt strategies that enable them to capitalize on the latest advancements and prepare for future breakthroughs. This section will discuss the importance of focusing on data quality and diversity, embracing emerging architectures and training techniques, and preparing for multimodal interfaces and textless NLP. By investing in these areas, organizations can build robust, flexible, and future-proof ASR systems that can adapt to the changing needs of users and industries.

## Focusing on data quality and diversity

The foundation of any successful ASR system lies in the quality and diversity of its training data. Whisper's remarkable robustness to accents, background noise, and technical language can be attributed to its training on an extensive dataset comprising 680,000 hours of multilingual and multitask supervised data. To achieve similar performance, organizations must prioritize collecting high-quality speech data covering a wide range of speakers (gender, age, race, socioeconomic status), accents and dialects (native and non-native, regional variations), domains (conversational, read, spontaneous, command and control), acoustic environments (clean studio, noisy conditions, far-field), and languages (high and low resource, language families). This diverse data collection is essential for developing ASR models that accurately transcribe speech in various real-world scenarios. For example, Mozilla's Common Voice project crowdsources a diverse dataset by having volunteers record voice samples in their native language. At the same time, Switchboard and Fisher are conversational telephone speech corpora with speakers from various US regions.

## Ensuring responsible and ethical data practices

While data quality and diversity cannot be overstated, ensuring that data collection and usage adhere to responsible and ethical practices is equally crucial. The development of Whisper serves as a cautionary tale, as training on web-scraped data without proper consent or attribution can raise significant moral concerns, mainly when dealing with underrepresented languages and communities. Organizations must establish clear guidelines for data collection, respect intellectual property rights, obtain necessary permissions, and maintain transparency throughout the process. Engaging with relevant stakeholders and following the principles of **Indigenous Data Sovereignty (IDSov)** is vital when working with sensitive linguistic data. This involves doing the following:

- Defining precise data requirements, quality metrics, and documentation standards
- Implementing validation and cleaning pipelines to catch errors
- Conducting manual spot checks and listening tests
- Monitoring quality KPIs and addressing issues promptly

For instance, the MALACH corpus of Holocaust survivor interviews illustrates the importance of implementing rigorous quality assurance processes to ensure the accuracy and integrity of speech datasets. By employing a combination of automatic checks, manual spot checks, and listening tests, the creators of the MALACH corpus demonstrated their commitment to data quality and responsible practices in handling sensitive audio content. This multistep QA process helps identify and correct errors in the alignment between audio and transcripts. It exemplifies how organizations can prioritize data quality and ethical considerations in their ASR development efforts.

### IDSov

IDSov is essential when preparing for the evolving ASR and voice technology landscape. IDSov refers to the rights of indigenous peoples to own, control, access, and possess data collected about them, their communities, lands, and resources. Here are some critical points about IDSov and its relevance to ASR and voice technologies:

- IDSov provides a framework for ensuring indigenous peoples can benefit from and control how data about them is collected and used in ASR systems. As voice technologies increasingly capture and analyze speech data, indigenous communities must have sovereignty over whether and how their voices and languages are included.

- Preparing voice datasets for ASR training requires careful consideration of IDSov principles such as collective ownership, free prior and informed consent, and community-driven governance over indigenous data assets. Simply treating indigenous voice data as open data without restrictions violates IDSov.

- IDSov also extends to the governance of indigenous knowledge embedded in speech. ASR systems must have safeguards to protect culturally sensitive information and uphold indigenous intellectual property rights over traditional knowledge.

- To operationalize IDSov in the ASR development process, indigenous peoples must be engaged as decision-makers and co-developers, not just data subjects. Emerging approaches such as the *CARE Principles for Indigenous Data Governance* guide centering indigenous rights and interests.

Ultimately, IDSov is about indigenous peoples' inherent rights to self-determination and autonomy, which includes data self-determination. Preparing for the future of ASR responsibly means committing to IDSov and shifting power over indigenous data from colonial institutions back to Indigenous communities.

IDSov is essential as ASR and voice technologies rapidly advance. Embracing IDSov and its emphasis on indigenous rights, collective well-being, and community control over data is critical to developing ASR systems that are inclusive, ethical, and aligned with Indigenous peoples' self-determination. The ASR field must partner with indigenous communities to co-design data governance frameworks that enact IDSov principles.

## Implementing comprehensive data quality procedures

Robust data quality is a prerequisite for successful ASR development and deployment. Organizations should adopt process-driven and data-driven strategies to maintain high standards. Process-driven strategies involve redesigning and controlling data collection and annotation workflows to minimize errors and inconsistencies. On the other hand, a data-driven strategy focuses on analyzing and cleansing existing datasets to identify and address quality issues. Establishing well-defined data quality dimensions, such as completeness, timeliness, accuracy, and consistency, along with automated data profiling and monitoring capabilities, can help ensure the integrity and reliability of speech datasets.

## *Leveraging data augmentation techniques*

In addition to collecting diverse speech data, organizations can further enhance the robustness of their ASR models by leveraging data augmentation techniques. These techniques involve artificially expanding datasets by applying various transformations to existing audio samples. Methods such as adding background noise, applying audio effects, and using text-to-speech can synthetically increase the quantity and diversity of training data. Whisper's multilingual training approach also suggests incorporating speech data from multiple languages, even if not directly targeted, can boost overall performance. However, it is essential to carefully validate augmented data to ensure it aligns with the desired quality and diversity criteria.

Focusing on data quality and diversity is critical to building ASR systems that accurately transcribe speech across various speakers, accents, and environments. OpenAI Whisper's success highlights the importance of training on large, diverse datasets while also emphasizing the need for responsible and ethical data practices. By prioritizing data collection, implementing comprehensive quality procedures, leveraging data augmentation techniques, and adhering to ethical guidelines, organizations can develop robust and reliable ASR solutions that are well-equipped to handle the complexities of real-world speech.

Staying at the forefront of the rapidly advancing ASR field necessitates embracing emerging architectures and training techniques. OpenAI Whisper has demonstrated the potential of transformer-based models and self-supervised learning, setting a new standard for ASR performance.

# Embracing emerging architectures and training techniques

OpenAI Whisper has set a new standard for ASR performance, demonstrating the potential of emerging architectures and training techniques. Whisper utilizes transformer-based models, which have become the dominant architecture for ASR due to their ability to capture long-range dependencies and parallelize computations. To remain competitive, organizations must stay abreast of the emerging advancements as the field rapidly evolves. The Conformer model is a notable example of an emerging architecture that combines the strengths of transformers and CNNs. Explicitly developed for speech recognition tasks, the Conformer architecture employs convolutional layers to capture local features, self-attention layers to model long-range dependencies, and feedforward layers to process the learned representations further. By leveraging the complementary benefits of CNNs and transformers, the Conformer model has achieved state-of-the-art performance on various speech recognition benchmarks, surpassing previous approaches based on recurrent neural networks, RNNs, and standalone transformers or CNNs.

The Conformer model's success highlights the importance of exploring novel architectural designs that can effectively capture the unique characteristics of speech signals. Its ability to model local and global dependencies through convolutional and self-attention layers has proven particularly advantageous for speech recognition. As organizations look to embrace emerging architectures and stay at the forefront of ASR technology, considering models such as the Conformer can provide a competitive edge in terms of accuracy and efficiency.

Let's explore the three additional preparedness strategies for embracing these innovations in the context of Whisper and ASR systems.

### Leveraging self-supervised pre-training

Self-supervised learning, where models are pre-trained on large amounts of unlabeled data to learn general representations, has emerged as a powerful technique for improving ASR performance. Whisper's robustness can be attributed to its pre-training on diverse, multilingual web data. Organizations can leverage self-supervision by doing the following:

- Collecting large, diverse datasets spanning multiple languages, speakers, and domains
- Designing pre-training tasks that encourage learning of meaningful speech representations, such as contrastive predictive coding or masked language modeling
- Fine-tuning pre-trained models on target datasets using supervised objectives such as CTC or sequence-to-sequence loss

Wav2vec 2.0, which is used in Whisper, pre-trains on unlabeled speech by masking and reconstructing input segments, learning representations that transfer well to downstream ASR tasks.

### Exploring multitask and multilingual training

Training ASR models on multiple tasks and languages simultaneously can improve generalization and data efficiency. Whisper's multilingual training enables zero-shot transcription in 99 languages. Organizations can explore multitask and multilingual approaches by doing the following:

- Jointly training on speech recognition, speaker diarization, language identification, and other relevant tasks
- Combining data from multiple languages within the same language family or with shared linguistic properties
- Using language-agnostic input representations such as articulatory features or phonemes
- Applying meta-learning algorithms to adapt models to new languages with only a few examples

Microsoft's UniSpeech model achieves competitive performance across 51 languages by pre-training on unlabeled data from multiple languages and fine-tuning with a unified loss function.

### Applying efficient fine-tuning techniques

While Whisper has showcased impressive zero-shot capabilities, fine-tuning the model on high-quality, domain-specific datasets can further enhance its accuracy and adaptability. By exposing Whisper to targeted datasets from specific industries or domains, such as healthcare, legal, or technical fields, the model can learn and specialize in the unique vocabularies, jargon, and linguistic nuances prevalent in those areas. This approach enables the model to deliver more precise and contextually relevant transcriptions, making it a powerful tool for various industry-specific applications. However, fine-tuning large models can be computationally expensive. To address this challenge, organizations can apply efficient fine-tuning techniques such as the following:

- Adapter modules, which inject small, trainable layers between frozen pre-trained weights, reducing the memory footprint

- **Low-Rank Adaptation** (**LoRA**), which learns a low-rank update to the pre-trained weights, enabling efficient adaptation

- Prefix tuning, which prepends a small number of tunable tokens to the input sequence, keeping pre-trained parameters fixed

- Distillation approaches, which transfer knowledge from large pre-trained models to smaller, more efficient student models

Fine-tuning pre-trained models on target datasets is crucial for adapting to specific domains and optimizing performance. The `whisper-small` model performs comparably to the larger `whisper-medium` model by distilling knowledge during fine-tuning, reducing inference costs.

### The importance of embracing emerging architectures and training techniques

Transformer-based models have become the gold standard for ASR, consistently outperforming previous approaches like HMMs and **RNNs**. Self-supervised pre-training enables models to learn from vast amounts of unlabeled data, reducing the need for expensive, manually transcribed corpora. Multitasking and multilingual training allow for greater generalization and adaptability to new domains and languages.

Moreover, as models such as Whisper grow in size and complexity, efficient fine-tuning techniques become essential for practical deployment. Without these innovations, organizations risk falling behind in performance and scalability.

Embracing these emerging architectures and training techniques improves ASR quality. It opens up new possibilities for applications such as real-time transcription, low-resource language support, and speech-to-speech translation. As the demand for accurate, reliable ASR grows across industries, organizations that invest in these cutting-edge approaches will be well-positioned to meet the needs of their customers and stakeholders.

As ASR technology evolves, it is crucial to anticipate and prepare for the rise of multimodal interfaces and textless NLP. These emerging trends have the potential to revolutionize how we interact with and process spoken language, opening up new possibilities for more intuitive and efficient communication.

# Preparing for multimodal interfaces and textless NLP

Textless NLP refers to processing and understanding spoken language without relying on intermediate text representations. Instead of converting speech into text and applying traditional NLP techniques, textless NLP aims to directly learn and operate on speech signals. This approach can potentially capture the rich acoustic and prosodic information in speech, often lost in text-based representations. By eliminating the need for accurate speech-to-text conversion, textless NLP can enable more efficient and language-agnostic processing of spoken content. Let's explore the top four preparedness strategies to capitalize on the ongoing development and future potential breakthroughs in textless NLP and multimodal interfaces.

## Investing in scalable infrastructure

One key preparedness strategy is investing in scalable, cloud-based infrastructure that can handle the computational demands of large multimodal AI models such as Whisper. Cloud platforms such as Azure OpenAI Service now offer Whisper models that efficiently process time-sensitive workloads. Leveraging managed services allows developers to transcribe audio at scale without maintaining complex infrastructure. As multimodal models grow in size and capability, relying on scalable cloud infrastructure will be essential to deploying them in production applications.

## Collecting diverse training data

Another critical strategy is curating large, diverse datasets to train and fine-tune multimodal AI models. OpenAI Whisper's robustness stems from the breadth of its training data, which spans multiple languages, accents, and technical domains. Collecting representative data across different demographics, dialects, and subject areas is necessary to build inclusive models that perform well in real-world scenarios. Data diversity is especially critical when expanding ASR support to low-resource languages or niche verticals. Initiatives such as Common Voice are crowd-sourcing multilanguage audio datasets to make ASR technology more accessible. For domain-specific applications, gathering custom training data through audio data augmentation and weak supervision can help tailor models to individual use cases. Investing in scalable data pipelines to continuously enhance model performance is a crucial differentiator.

## Designing multimodal user experiences

Preparing for the era of multimodal interfaces requires a paradigm shift in **user experience** (**UX**) design. Rather than treating modalities such as speech and text as isolated input methods, designers must create cohesive experiences that blend multiple interaction modes. This involves studying the strengths and limitations of different modalities and understanding user contexts where each is most appropriate.

Multimodal UX design principles emphasize minimizing friction when switching between modalities and providing feedback across sensory channels. For example, a virtual assistant might visually highlight keywords in a transcription while the user is speaking, then seamlessly transition to a text chat interface for clarification or follow-up. Testing new interaction patterns with diverse user groups and measuring metrics such as task completion and cognitive load will help refine these experiences over time.

### Exploring representation learning techniques

Textless NLP, which aims to learn language representations directly from raw audio signals, is a rapidly advancing research area with significant implications for low-resource languages. OpenAI Whisper has shown promising results in this direction by demonstrating strong cross-lingual transfer performance and the ability to directly translate speech to English without relying on intermediate text representations.

More work is needed on unsupervised representation learning techniques that can uncover linguistic structures from unlabeled audio data to realize the full potential of textless NLP. Researchers are exploring approaches such as contrastive predictive coding, discrete unit discovery, and self-supervised pre-training to learn compressed speech representations that capture phonetic and semantic information. Combining these textless representations with other modalities, such as vision, could enable new applications for grounded language learning and visually-guided speech processing.

## Summary

In this final chapter, we explored the future of ASR and the exciting advancements shaping the landscape by focusing on OpenAI's groundbreaking Whisper model. We examined ongoing efforts to enhance Whisper's performance, including improving accuracy and robustness, expanding language support, achieving better punctuation and speaker diarization, and accelerating performance for real-time capabilities. Additionally, we delved into the critical ethical considerations and responsible AI practices crucial for developing and deploying ASR technologies, ensuring that these powerful tools benefit society while respecting individual rights and promoting fairness.

This chapter delved into cutting-edge techniques and strategies that can be used to enhance Whisper's performance, including increasing training data, fine-tuning for specific domains, and optimizing model architecture. We investigated approaches for making ASR more inclusive by collecting diverse datasets, applying transfer learning, and supporting low-resource languages. We also explored state-of-the-art methods for punctuation restoration, timestamp alignment, and speaker attribution to generate more readable and actionable transcripts.

Moreover, we highlighted the importance of scalable infrastructure, efficient fine-tuning techniques, and emerging architectures, such as multimodal interfaces and textless NLP, in preparing for the evolving ASR landscape. By adopting these preparedness strategies and prioritizing fairness, privacy, and responsible use, organizations can capitalize on the immense potential of voice technology while building robust, reliable, and inclusive ASR systems.

As we conclude this journey through the OpenAI Whisper and ASR world, I would like to express my sincere gratitude to you, the reader. Your dedication to learning and exploring this transformative technology has been truly inspiring. I hope the knowledge and insights you've gained throughout this book will empower you to shape the future of voice-enabled applications, unlocking new possibilities and positively impacting how we interact with spoken language in the digital world. Thank you for joining me on this exciting adventure, and I wish you all the best in your future endeavors with ASR and beyond.

Until next time, cheers!

# Index

## Symbols

## A

# ‹packt›

packtpub.com

Subscribe to our online digital library for full access to over 7,000 books and videos, as well as industry leading tools to help you plan your personal development and advance your career. For more information, please visit our website.

## Why subscribe?

- Spend less time learning and more time coding with practical eBooks and Videos from over 4,000 industry professionals
- Improve your learning with Skill Plans built especially for you
- Get a free eBook or video every month
- Fully searchable for easy access to vital information
- Copy and paste, print, and bookmark content

Did you know that Packt offers eBook versions of every book published, with PDF and ePub files available? You can upgrade to the eBook version at packtpub.com and as a print book customer, you are entitled to a discount on the eBook copy. Get in touch with us at customercare@packtpub.com for more details.

At www.packtpub.com, you can also read a collection of free technical articles, sign up for a range of free newsletters, and receive exclusive discounts and offers on Packt books and eBooks.

# Other Books You May Enjoy

If you enjoyed this book, you may be interested in these other books by Packt:

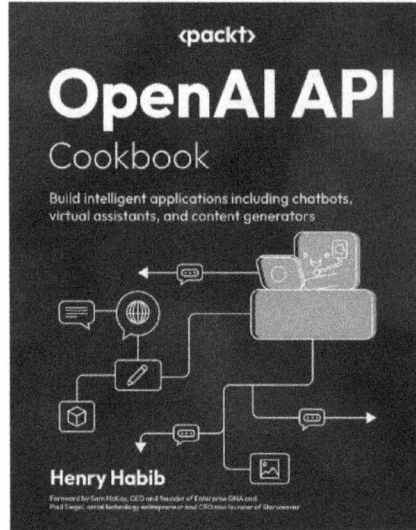

**OpenAI API Cookbook**

Henry Habib

ISBN: 978-1-80512-135-0

- Grasp the fundamentals of the OpenAI API

- Navigate the API's capabilities and limitations of the API

- Set up the OpenAI API with step-by-step instructions, from obtaining your API key to making your first call

- Explore advanced features such as system messages, fine-tuning, and the effects of different parameters

- Integrate the OpenAI API into existing applications and workflows to enhance their functionality with AI

- Design and build applications that fully harness the power of ChatGPT

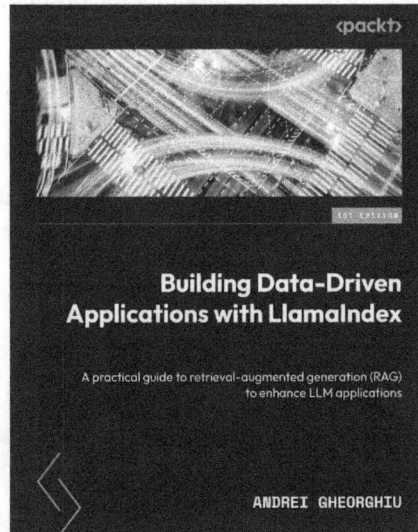

**Building Data-Driven Applications with LlamaIndex**

Andrei Gheorghiu

ISBN: 978-1-83508-950-7

- Integrate Whisper into voice assistants and chatbots
- Use Whisper for efficient, accurate transcription services
- Understand Whisper's transformer model structure and nuances
- Fine-tune Whisper for specific language requirements globally
- Implement Whisper in real-time translation scenarios
- Explore voice synthesis capabilities using Whisper's robust tech
- Execute voice diarization with Whisper and NVIDIA's NeMo
- Navigate ethical considerations in advanced voice technology

# Packt is searching for authors like you

If you're interested in becoming an author for Packt, please visit `authors.packtpub.com` and apply today. We have worked with thousands of developers and tech professionals, just like you, to help them share their insight with the global tech community. You can make a general application, apply for a specific hot topic that we are recruiting an author for, or submit your own idea.

# Share Your Thoughts

Now you've finished *Learn OpenAI Whisper*, we'd love to hear your thoughts! Scan the QR code below to go straight to the Amazon review page for this book and share your feedback or leave a review on the site that you purchased it from.

`https://packt.link/r/1-835-08592-X`

Your review is important to us and the tech community and will help us make sure we're delivering excellent quality content.

# Download a free PDF copy of this book

Thanks for purchasing this book!

Do you like to read on the go but are unable to carry your print books everywhere?

Is your eBook purchase not compatible with the device of your choice?

Don't worry, now with every Packt book you get a DRM-free PDF version of that book at no cost.

Read anywhere, any place, on any device. Search, copy, and paste code from your favorite technical books directly into your application.

The perks don't stop there, you can get exclusive access to discounts, newsletters, and great free content in your inbox daily

Follow these simple steps to get the benefits:

1. Scan the QR code or visit the link below

https://packt.link/free-ebook/9781835085929

2. Submit your proof of purchase
3. That's it! We'll send your free PDF and other benefits to your email directly